Novel Biomarkers in Non-Small Cell Lung Cancer (NSCLC)

Novel Biomarkers in Non-Small Cell Lung Cancer (NSCLC)

Guest Editor

Alfredo Tartarone

MDPI

Basel • Beijing • Wuhan • Barcelona • Belgrade • Novi Sad • Cluj • Manchester

Guest Editor
Alfredo Tartarone
Department of
Onco-Hematology
Centro di Riferimento
Oncologico della Basilicata,
IRCCS
Rionero in Vulture
Italy

Editorial Office
MDPI AG
Grosspeteranlage 5
4052 Basel, Switzerland

This is a reprint of the Special Issue, published open access by the journal *Cancers* (ISSN 2072-6694), freely accessible at: https://www.mdpi.com/journal/cancers/special_issues/3QFM802K04.

For citation purposes, cite each article independently as indicated on the article page online and as indicated below:

Lastname, A.A.; Lastname, B.B. Article Title. *Journal Name* **Year**, *Volume Number*, Page Range.

ISBN 978-3-7258-6670-0 (Hbk)
ISBN 978-3-7258-6671-7 (PDF)
https://doi.org/10.3390/books978-3-7258-6671-7

Contents

cancers

MDPI

Review

Variant Allele Frequency Analysis of Circulating Tumor DNA as a Promising Tool in Assessing the Effectiveness of Treatment in Non-Small Cell Lung Carcinoma Patients

Natalia Galant [1,*], Marcin Nicoś [1], Barbara Kuźnar-Kamińska [2] and Paweł Krawczyk [1]

1 Department of Pneumonology, Oncology and Allergology, Medical University of Lublin, 20-059 Lublin, Poland
2 Department of Pulmonology, Allergology and Respiratory Oncology, Poznan University of Medical Sciences, 61-710 Poznan, Poland; kaminska@ump.edu.pl
* Correspondence: nataliagalant@umlub.pl

Simple Summary: The non-invasive characteristic of liquid biopsy enables an increase in the potential of VAF analysis in monitoring tumor progression, remission, and recurrence during or after treatment. Moreover, the use of VAF analysis appears to be beneficial in making treatment decisions. Several studies have been performed on patients with NSCLC to evaluate the possibility of VAF usage. However, several issues require better understanding and standardization before VAF testing can be implemented in clinical practice. In this review, we discuss the difficulties in the application of ctDNA VAF analysis in clinical routine, discussing the diagnostic and methodological challenges in VAF measurement in liquid biopsy. We highlight the possible applications of VAF-based measurement in the monitoring of personalized treatment in patients with NSCLC who are under consideration in clinical trials.

Abstract: Despite the different possible paths of treatment, lung cancer remains one of the leading causes of death in oncological patients. New tools guiding the therapeutic process are under scientific investigation, and one of the promising indicators of the effectiveness of therapy in patients with NSCLC is variant allele frequency (VAF) analysis. VAF is a metric characterized as the measurement of the specific variant allele proportion within a genomic locus, and it can be determined using methods based on NGS or PCR. It can be assessed using not only tissue samples but also ctDNA (circulating tumor DNA) isolated from liquid biopsy. The non-invasive characteristic of liquid biopsy enables a more frequent collection of material and increases the potential of VAF analysis in monitoring therapy. Several studies have been performed on patients with NSCLC to evaluate the possibility of VAF usage. The research carried out so far demonstrates that the evaluation of VAF dynamics may be useful in monitoring tumor progression, remission, and recurrence during or after treatment. Moreover, the use of VAF analysis appears to be beneficial in making treatment decisions. However, several issues require better understanding and standardization before VAF testing can be implemented in clinical practice. In this review, we discuss the difficulties in the application of ctDNA VAF analysis in clinical routine, discussing the diagnostic and methodological challenges in VAF measurement in liquid biopsy. We highlight the possible applications of VAF-based measurements that are under consideration in clinical trials in the monitoring of personalized treatments for patients with NSCLC.

Keywords: NSCLC; liquid biopsy; circulating free DNA; circulating tumor DNA; variant allele frequency; personalized treatment

1. Introduction

Lung cancer is one of the leading causes of death in oncological patients worldwide. It is histologically classified as non-small cell lung carcinoma (NSCLC) or small cell lung carcinoma (SCLC). NSCLC occurs more frequently and represents 85% of all lung cancer cases. It is also divided into three main subtypes: adenocarcinoma (40%), squamous cell carcinoma (25–30%), and large cell carcinoma (5–10%). Despite the many possible therapeutic strategies that can be applied to patients with NSCLC, the responses to them are still very diverse [1–4].

To enable the selection of beneficial therapy to patients, new biomarkers with potential clinical applications are still being investigated. Circulating tumor DNA (ctDNA), among other biomarkers, is attracting much attention from researchers [5]. The significance of ctDNA analysis in patients with NSCLC has increased in recent years. In contrast to tumor tissue genotyping, ctDNA isolation from a liquid biopsy is non-invasive and may be performed at different time points; thus, patients can be monitored over the entire duration of therapy without the risk of biopsy-related side effects. Besides the possibility of more frequent examination, liquid biopsy enables the genetic testing of patients at a high risk of NSCLC when access to tissue material is limited [5,6]. Furthermore, ctDNA analysis also has the potential to become a marker for the detection and monitoring of minimal residual disease (MRD) [7–9].

The recent breakthrough in next-generation sequencing (NGS), and its implementations into clinical routines, has provided a large variety of measuring metrics that increase the sensitivity of liquid biopsy testing [10,11]. Variant allele frequency (VAF) in ctDNA is being considered one of the markers with prospective clinical utility [12]. The possibility of its usage to assess the effectiveness of therapy in patients with NSCLC is being evaluated [13]. Research results give hope that VAF monitoring may provide information about response to treatment and patient prognosis, and help in developing optimal therapy. Furthermore, VAF changes seem to occur quickly, even before radiological evidence of response is noticeable. However, it should be taken into account that the clinical usage of VAF requires more broad-based research [14].

Considering the increasing interest in searching for measurable parameters in liquid biopsy, in this review, we discuss the differentiation of cancer and non-cancer DNA fractions in liquid biopsy. Furthermore, we indicate the diagnostic and methodological challenges in VAF measurements in liquid biopsy. Finally, we highlight the possible applications of VAF-based measurement in the monitoring of personalized treatments for NSCLC that are already available in research or under consideration in clinical trials.

2. cfDNA

Circulating free DNA (cfDNA) fragments are double-stranded DNA fragments circulating in body fluids, such as peripheral blood, cerebrospinal fluid (CSF), and urine. cfDNA fragments are short (below 200–220 base pairs, usually 167 bp), and their half-life in circulation ranges from 5 to 150 min. In healthy individuals, cfDNA is normally at low concentrations and enters body fluids mostly via apoptosis, neutrophil extracellular traps (NETs), and erythroblast enucleation [15–18]. Moreover, it was indicated that smaller, average, and larger-than-average cfDNA fragments may be products of necrosis [19]. cfDNA concentrations may be influenced by many factors, such as age or smoking status, and, in general, are higher in older patients compared to younger ones and in smokers compared to non-smokers [20]. It has also been indicated that cfDNA levels may be increased physiologically, for example, after physical exercise. High cfDNA concentrations may also be affected by pathological processes, such as inflammation or cancer. A summary of the causes for increases in cfDNA concentrations is presented in Figure 1. The fraction of cfDNA released by cancer cells is called circulating tumor DNA (ctDNA) [18,21]. While in oncological patients, the terms cfDNA and ctDNA are commonly used interchangeably, it is important to understand the differences between them.

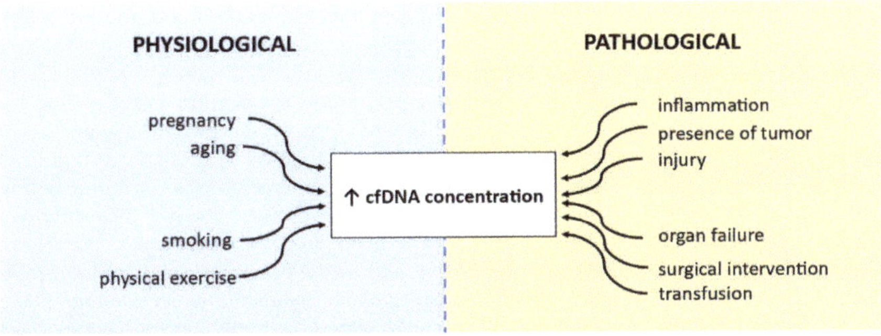

Figure 1. A summary of causes affecting cfDNA concentration increase. Based on [20,21].

2.1. ctDNA

ctDNA is a fraction of cfDNA released into the bloodstream by primary or metastatic tumor cells. Cancerous ctDNA and physiological cfDNA can be distinguished by a genetic background typical of tumors (mutations, rearrangements, or copy number alterations in selected genes) or by the abnormal methylation of gene promoters [22,23]. Furthermore, the important difference between ctDNA and cfDNA is their length, ctDNA can either be smaller or larger than cfDNA. Large ctDNA fragments (above 10,000 bp) may be released due to tumor cell necrosis [18,22]. However, of greatest interest to researchers are fragments with a length similar to cfDNA or shorter, as the selection of shorter fragments improves the ctDNA/cfDNA ratio. Moreover, as studies have proven, cfDNA fragments carrying mutant alleles are often shorter compared to those with wild-type alleles; thus, an analysis of shorter fragments may be beneficial during mutant allele frequency evaluation [24–26]. Additionally, the possibility of analyzing fragments with a length exceeding 10,000 bp is significantly limited due to methodological issues [27].

In oncological patients, the concentration of ctDNA may vary depending on the cancer type, stage of disease (especially tumor size), and presence of local or distant metastases [18,28,29]. A study performed on a group of 640 patients showed that ctDNA was detected more frequently (>75%) in patients with advanced pancreatic, colorectal, gastroesophageal, hepatocellular, bladder, ovarian, breast, and head and neck cancers or melanoma, while ctDNA was detected less often (<50%) in patients suffering from primary brain, prostate, thyroid, and renal cancers [30]. It was also found that ctDNA detection is positively related to cancer stage. Furthermore, the concentration of ctDNA is higher in patients with metastatic cancers compared to groups with localized disease [30,31]. Another study, including a group of 88 NSCLC patients, showed that ctDNA detection is more frequent in patients with more advanced cancer (ctDNA was detected before treatment in 24%, 77%, and 87% of patients with stages I, II, and III of the disease, respectively) [32].

2.2. cfDNA and ctDNA Differentiation

Nowadays, it is impossible to differentiate ctDNA from normal cfDNA during nucleic acid isolation. This is because the similar lengths of both ctDNA and cfDNA are extracted under the same conditions. The most common approach to discriminate those nucleic acid types is to analyze tumor-specific mutations or methylation patterns within the extracted material [17,18]. It is also possible to focus on quantitative changes. For instance, deleted regions in tumor cells' genomes should be underrepresented in ctDNA. Moreover, overrepresented sequences amplified in tumor cells may be observed at the copy number alteration (CNA) level in many patients with cancer [15]. However, to differentiate ctDNA from cfDNA at the molecular level, a proper assay must be selected; thus, knowledge about the type of cancer is indispensable [14].

Tumor-specific mutations and rearrangements in ctDNA may be analyzed during whole genome sequencing (WGS) or targeted sequencing. There are few assays targeted at specific patient cohorts that have been approved by the FDA (Food and Drug Administration) [18,33,34]. Currently, these tests apply for the assessment of prognosis and qualification for treatment but not for cancer screening [18,33]. During genotyping and distinguishing ctDNA from normal cfDNA, it is important to take into consideration the clonal hematopoiesis phenomenon, which stands behind the presence of somatic mutations in hematopoietic cells, as well as the fact that peripheral blood cells (PBCs) are a significant source of cfDNA in the bloodstream. Not considering clonal hematopoiesis in genotyping may lead to falsely categorizing white blood cell mutations as tumor-specific mutations. It is especially important in elderly patients in which clonal hematopoiesis occurs more often due to normal apoptotic processes [23,35]. One of the possibilities to eliminate false positive genotyping in liquid biopsies is a comparison of results between the sequencing of ctDNA and genomic DNA (gDNA) of tumor biopsy delivered from the same patient. This is viable due to approximately 60–80% mutation compliance in ctDNA and DNA from cancer cells of various cancers. The second option to distinguish ctDNA from normal cfDNA is a comparison of the mutation landscape between cfDNA and gDNA isolated from white blood cells, considered the normal match, which delivers a large repertoire of clonal hematopoietic alterations [18,36,37].

On the other hand, an analysis of methylation can be used to identify the kinds of cells that release cfDNA into the bloodstream and may become a powerful tool in distinguishing cfDNA from ctDNA. The discrimination of DNA released from "healthy" and tumor cells may be helpful in spotting tumor origin [38,39]. This kind of differentiation of normal cfDNA and ctDNA is possible due to the variety of methylation patterns in normal and cancer cells. In particular, the increased methylation of suppressor genes may be one of the first noticeable changes during tumorigenesis. Due to this fact, an analysis of CpG island (CGI) methylation seems to be a promising biomarker of neoplastic processes in various cancer types. Numerous DNA methylation biomarkers are already known and collected in The Cancer Genome Atlas (TCGA; http://cancergenome.nih.gov (accessed on 27 December 2023)), but it is necessary to increase their sensitivity and specificity before they can be widely clinically used. Nevertheless, several tests based on the analysis of suppressor gene methylation in liquid biopsies (e.g., peripheral blood and bronchoalveolar lavage) have been approved for in vitro diagnosis (IVD) of various cancers (e.g., colorectal cancer and NSCLC). Moreover, it has been considered that methylation haplotyping of ctDNA may be useful for early tumor detection, an assessment of progression, and the confirmation of the metastases' presence [40–42].

In the ctDNA of cancer patients, it is also possible to quantitatively detect large changes, such as chromosomal rearrangements and CNAs. Research conducted by Jiang et al. on hepatocellular carcinoma patients showed that the size of chromosomal arms (with and without deletions) in tumor tissue reflected the size in ctDNA. If the chromosomal arm was amplified in cancer cells, its contribution to plasma DNA was proportionally increased, and when it was deleted, its contribution was decreased [43]. Sivapalan et al. focused on, among others, plasma aneuploidy in small cell lung cancer (SCLC) patients as well. During their contributed research, they found plasma aneuploidy of specific chromosomes in ctDNA, and the aberrations detected in ctDNA were reflected in changes in tumor tissue. However, to differentiate tumor-derived aberrations from biological noise and changes related to clonal hematopoietic, a highly optimized NGS assay and computational error correction were applied [44].

Calculations relying on the dependencies described above are being used in some studies. The Tumor Fraction Estimator (TFE) is based on the measurement of tumor aneuploidy. In the TFE, the calculation of the ctDNA fraction is possible due to the comparison of tested sample sequencing with a set of samples with well-known tumor fractions [45]. The second algorithm, maximum somatic allele frequency (MSAF), is based on the measurement of the maximum somatic allele frequency of somatic and likely somatic mutations in ctDNA. MSAF considered that those frequencies' participation correspond to the abundance of ctDNA in cfDNA. Unfortunately, MSAF does not consider clonal hematopoiesis; thus, in many cases, TFE is more often recommended [45,46]. On the other hand, Schrock et al. observed in NSCLC patients that a lower MSAF level was associated with a higher risk of missing important genomic alterations in the plasma, such as exon 19 deletion and p.Thr790Met substitution in the *EGFR* gene [47]. Moreover, in a B-F1SRT study, an MSAF of <1% was associated with a higher response rate and better PFS during atezolizumab therapy in advanced NSCLC, but this effect was dependent on baseline tumor mutation burden (TMB) [48,49]. Gandara DR et al. suggested that a low MSAF could contribute to a poorer consistency between TMB in the bloodstream (bTMB) and tumor (tTMB) [50]. Incorporating MSAF with bTMB can partially improve the differentiation between patients with or without survival benefits from immune checkpoint inhibitors (ICIs). MSAF alone or in combination with bTMB can effectively distinguish NSCLC patients with or without OS (overall survival) and PFS (progression-free survival) benefit from atezolizumab compared with docetaxel. MSAF and the combined bTMB-MSAF classification may become practical, non-invasive biomarkers for atezolizumab efficacy in advanced NSCLC patients [49]. However, MSAF clinical value requires confirmation in a larger cohort within a standardized clinical trial.

3. Variant Allele Frequency (VAF)

The breakthrough in ultra-deep and sensitive metrics for mutation analysis in cancer patients has provided a modern approach to ctDNA analysis that focuses on variant allele frequency (VAF) [51]. However, in the literature, VAF is often substituted by mutant allele frequency (MAF), which carries the same meaning. However, the term MAF may also be used as minor allele frequency and refers to the frequency of germline alleles in large cohorts analyzed by genome-wide association studies (GWASs) [52]. For this reason, a good practice is to use VAF as a metric of cancerous variants. VAF is a metric characterized as the measurement of the specific variant allele proportion within a genomic locus determined by both NGS- and PCR-based methods (Figure 2) [17,53]. The prognostic and predictive roles of VAF have been evaluated across different studies [12,54–57]. However, any validated VAF thresholds may increase its clinical utility. This may be due to the limited standardization between diagnostic assays used in variant detection [53]. On the other hand, VAF provides insights into tumor clonality in somatic genomic testing, yielding a strong rationale for targeting dominant cancer cell populations that may pave the way for a new decision-making tool for targeted therapy selection [58].

$$\text{VAF} = \frac{\text{number of reads with a pathogenic variant}}{\text{number of reads with pathogenic variant} + \text{number of reads without variant}} \times 100\%$$

$$\text{Exemplary VAF} = \frac{100}{100 + 2000} \times 100\% = 4.76\%$$

Figure 2. Method of VAF calculation in cancer patients using NGS technology.

VAF refers to a fraction of alleles carrying a specific genomic alteration [13]. A high VAF value suggests that a high percentage of tumor cells carry a particular genomic alteration. In such situations, targeted therapy may be easier to select and more effective [17,53]. Conversely, low VAF values suggest the low clonality of the genomic alteration [59,60]; thus, targeted therapy may be less effective because of the presence of other subclones carrying distinct genomic profiles [53,61]. Moreover, VAF may allow distinguishing driver mutations from passenger mutations since driver mutations refer to genomic alterations that directly contribute to the development and progression of cancer [62,63]. In this situation, high VAF occurs in genes that confer a selective clonal selection of cancer cells [53]. By contrast, passenger mutations represent genetic alterations that occur randomly in cancer cells, resulting from genetic instability. For decades it was considered that passenger mutations do not directly contribute to the progression of cancer [53,59]; however, previous comprehensive pan-cancer studies have indicated that passenger mutations may have important functional roles in driving cancer, as well as playing roles in tumor progression [64,65]. Moreover, passenger mutation analysis may allow an accurate classification of human tumors [66]. On the other hand, passenger mutations are not so actively promoted or enriched in the cancer cell population as drivers [63]; however, an analysis of their VAF may allow the measurement of the fitness of cancer cells, as well as the anti-tumor effects of chemotherapies [67,68].

Both metrics obtained from PCR- and NGS-based methods enable VAF calculation. VAF analyses based on PCR, e.g., digital PCR-based method (ddPCR) and Beads, Emulsion, Amplification, and Magnetics (BEAMing), provide high sensitivity and are more cost-effective than VAF calculated by NGS results [69,70]. However, NGS-based methods (such as whole exome sequencing, whole genome sequencing, or targeted sequencing) enable better genome coverage and the selection of the appropriate sequencing depth, allowing for the distinction of sequencing artifacts, as well as germline and somatic mutations [53].

One of the most concerning issues of VAF usage in liquid biopsy is a low concentration of ctDNA carrying the driver mutations in plasma. This limitation may affect the sensitivity of the detection method and lead to false negative results. Potentially, this problem can be eliminated by using larger plasma volumes for ctDNA extraction [53,71]. However, the ctDNA yield depends on many factors, starting with the pre-analytic conditions and chosen isolation assay, through to the type and stage of the tumor, and ending with the patient's age. Moreover, cancer progression will also affect the ctDNA/cfDNA ratio [45,72]. At the step of sample preparation, it is important to centrifuge blood as soon as possible after collection to lower the risk of DNA contamination caused by cell lysis [73]. Various protocols suggest proceeding with centrifugation at different temperatures and speed conditions, depending on laboratory equipment; however, in general, centrifugation for 10 min at 1000–2000$\times$ g using a refrigerated centrifuge allows obtaining high-quality plasma, while centrifugation for 15 min at 2000$\times$ g removes platelets from the plasma [74]. Some protocols consider two cycles of centrifugation: the first cycle is performed at a lower speed to separate the plasma from morphotic elements, and the second one, at a higher speed, enables the reduction of cellular debris [75,76]. In the end, the prepared plasma can proceed immediately for further laboratory steps or be stored at −20 °C. Freezing at −80 °C is also acceptable, especially if long-term storage of the material is planned. However, it is recommended to avoid thawing and refreezing of plasma, which decreases its quality [76–78]. Although, in contrast to the isolation of DNA from FFPE (formalin-fixed paraffin-embedded) tissue, the usage of ctDNA does not involve formaldehyde contamination that may provide relatively fewer artifacts or unspecific results in analysis. Due to this, deamination does not affect ctDNA, and the risk of false positive results is lower in liquid biopsy [79]. Furthermore, single tissue collection by biopsy is related to a higher risk of tumor heterogeneity-related biases, while ctDNA is shed by all apoptotic or necrotic tumor cells, providing a representative material of the whole cancerous process for comprehensive genetic analysis [80–82].

Usage of ctDNA VAF Analysis

The potential clinical application of VAF is related to the possibility of prognosis assessment or treatment response monitoring. It is considered that a high VAF rate in blood correlates with a large tumor volume [83,84], and especially with TMB, it may correlate with a worse prognosis [53]. Research conducted on different groups of patients showed an inversely proportional relationship between VAF and overall survival. This correlation occurred regardless of tumor type, which suggests the relevance of pre-treatment VAF analysis in the prognosis assessment of various patient groups. Moreover, it has been indicated that a higher VAF may be associated with shorter PFS regardless of therapy type [55,85,86].

Assessing predictive value and utility in the therapy selection of VAF is a consideration of interest to researchers. Different studies have indicated that a decrease in VAF correlates with response to therapy. Analogously, an increase in VAF may result from disease progression or recurrence. Moreover, VAF change could be noticeable earlier when compared to the radiographic or clinical evidence of response. It is also considered that VAF analysis can be used during assessment if adjuvant therapy would be beneficial and help to determine optimal treatment management [14,87,88]. Also, the introduction of VAF into the analysis of copy number variants (CNVs) may provide the optimal strategy for calling somatic CNAs [89]. However, CNV analysis alone seems to have higher predictive value, while VAF alone may be valuable in monitoring response to treatment. This way, CNV and VAF may complement each other during therapy decision making in certain groups of patients [90].

4. Assessment of Therapy Effectiveness in NSCLC Patients Based on VAF

Research performed on NSCLC patients indicated the potential clinical utility of VAF as a marker of response to various types of treatment. Most studies have focused on monitoring the dynamics of changes in mean or maximum VAF detected in ctDNA [91–93]. At this time, information about specific variants that are useful in assessing the effectiveness of therapy is limited. This issue seems to be of great interest to researchers, and the results are promising [91,93]. Firstly, VAF dynamics changes in patients treated with adjuvant chemotherapy correlated with radiological recurrence. In addition, it was suggested that VAF analysis may correlate with the risk of MRD and could potentially be used in the real-time monitoring of recurrence in patients after resection [7]. Unfortunately, low VAF values in patients with MRD may not be sufficient to reach the detection limit of currently used methods [37,81,94].

In a study performed on a group of 22 advanced NSCLC patients treated with immunotherapy (the anti-PD-1 antibody camrelizumab) or immunotherapy combined with anti-angiogenic therapy (the VEGFR2 inhibitor apatinib), VAF seemed to have potential as a marker of tumor progression or relief [95]. Studies have shown that a VAF decrease during therapy corresponds with a reduction in tumor size [14,95]. In 1 patient, out of 22, the tumor maximal diameter did not change after the treatment with combined immunotherapy, while VAF decreased. After initial stabilization, this patient experienced partial remission (PR). This case suggests that the sensitivity of VAF changes may be higher when compared to the radiographic assessment [95]. Similarly, Raja et al. also observed a change in the frequency of variants before a radiological response to durvalumab (anti-PD-L1 antibody) therapy in 42% of patients with bladder or non-small cell lung cancer [14]. Chen et al. have suggested that monitoring only a few mutations in ctDNA may be sufficient in treatment response prediction [14,95].

In the third phase of the CameL-sq trial, which included advanced squamous lung cancer patients treated with camrelizumab in combination with platinum-based chemotherapy plus, VAF measurements after two cycles of therapy were useful in discriminating patients with PR from patients with stable disease (SD) or progressive disease (PD). Furthermore, VAF assessment, at this time, showed a stronger correlation with response to therapy compared to ctDNA concentration analysis at the beginning of treatment or after two cycles [96]. Thompson et al. analyzed NSCLC patients treated with pembrolizumab (anti-PD-1 antibody) in monotherapy or in combination with chemotherapy as a first or second line of treatment. The molecular response was assessed as the ratio of mean ctDNA VAF during treatment to the mean ctDNA VAF at the baseline for *TP53*, *KRAS*, and *STK11* genes. It was reported that patients with response to therapy had significantly lower VAF (<50%) in the 6th week of therapy, which was followed by longer PFS, OS, and duration of treatment with pembrolizumab. Mutations in *TP53* and *KRAS* genes occurred with the highest frequency, while the presence of mutations in the *STK11* gene, considered as associated with worse outcomes in patients treated with pembrolizumab, was also found in several patients [92].

A meta-analysis conducted by Vega et al. included five studies on NSCLC patients treated with ICIs in monotherapy or in combination with other agents. It was indicated that mean and maximum VAF values were strongly associated with outcomes, while median VAF gave inconsistent results. Moreover, the authors did not observe a correlation between baseline VAF and outcome, although they did not exclude the possibility of this value in clinical usage [91].

The IMPower 010 study assessed the presence of ctDNA in the blood serum of patients after surgery and after adjuvant chemotherapy before adjuvant atezolizumab (an anti-PD-L1 antibody) immunotherapy or best supportive care (BSC). The first analysis indicated that the use of adjuvant chemotherapy led to a decrease in ctDNA concentrations in blood sera in 62% of initially ctDNA-positive patients, which was then associated with a longer time to disease recurrence in the group of patients receiving BSC. It was shown that atezolizumab, when compared to BSC, reduced the risk of disease recurrence by 30% in the group of patients with decreased ctDNA concentration observed after adjuvant chemotherapy. However, no such significant differences occurred in the group of patients without a decrease in ctDNA concentration after chemotherapy. It also appears that in the group of patients with a negative ctDNA result after chemotherapy, atezolizumab delayed the conversion of patients to a positive ctDNA result [97].

Likewise, with regard to the research on ICIs mentioned previously, studies performed in NSCLC patients treated with ALK tyrosine kinase inhibitors (ALK-TKIs) and EGFR tyrosine kinase inhibitors (EGFR-TKIs) indicated the potential application of ctDNA and VAF analysis in therapy effectiveness prediction. Soo et al. investigated VAF for fusion or single nucleotide variants (SNVs) of the *ALK* gene in patients who received lorlatinib or crizotinib (CROWN clinical trial). Different somatic alternations, including somatic variants of the ALK gene, were analyzed. The VAF change (delta VAF (dVAF)) between the baseline value and the VAF measured between the 4th and 24th week of treatment was measured. A dVAF ≤ 0, both for *ALK* gene alternations and any other somatic alternations, in the lorlatinib-treated group, was associated with longer PFS and tumor size decrease. However, this correlation was observed for VAF measured at week 4, but it was not confirmed for VAF analyzed at week 24. This may suggest the usage of VAF of the *ALK* gene as a marker of early response [98].

In their case study, Begum et al. demonstrated that VAF analysis in ctDNA can be a tool for the detection and monitoring of response or resistance to treatment with crizotinib and lorlatinib in NSCLC patients with *ROS1* gene rearrangement. The acquisition of mutations, c.G2101A, in the *ROS1* gene and changes in VAF detected by NGS indicated a moment of progression during crizotinib treatment in this patient. The emphasis was mostly on acquired G2101A substitution in the *ROS1* gene, which seemed to be associated not only with resistance to crizotinib but also with sensitivity to lorlatinib; thus, it could guide therapy decision making. Regrettably, extended research and validation of the results on a larger cohort are still necessary [99].

The application of ctDNA VAF analysis was also examined on a group of NSCLC patients with *EGFR* gene mutations treated with osimertinib, the third generation of EGFR-TKI. It was observed that patients included in the group of non-responders had higher VAF values of specific, actionable *EGFR* mutations in comparison to responders (patients with PR or SD). The research suggested that tracking VAF dynamics in ctDNA may have utility in monitoring response to EGFR-TKI [100].

Vaclova et al. assessed the VAF of p.Thr790Met mutations in the *EGFR* gene in patients treated with EGFR-TKI and proposed 30% VAF as a cut-off value to differentiate patients with clonal (VAF $\geq$ 30%) and subclonal (VAF < 30%) p.Thr790Met mutations. However, regardless of the clonality of p.Thr790Met substitution, both groups benefited from the therapy with osimertinib. The follow-up indicated that after 3 weeks of treatment, the VAF value decreased, and after 6 weeks, it either decreased or the value remained at a very low level (VAF < 1%). These observations were independent of clonality status, but were more variable in the subclonal group [101]. Contrary to the aforementioned studies, Ai et al. did not observe a correlation between the ctDNA VAF of the *EGFR* gene and the effectiveness of EGFR-TKI treatment. However, their study indicated that clonal dominance of *EGFR* gene mutations is an independent factor associated with the efficiency of EGFR-TKI treatment in patients with advanced NSCLC [102]. Furthermore, many current clinical trials aim to evaluate the VAF clinical value in larger NSCLC patient groups (Table 1).

Table 1. A summary of clinical trials that are already recruiting NSCLC patients to evaluate VAF metrics for monitoring the effectiveness of personalized treatment. Data were collected from the ClinalTrials.gov database (http://clinicaltrials.gov/ (accessed on 4 January 2024)).

Clinical Trials ID (Duration)	Title	Main Location (Sponsor)	Number of Participants (Trial Type)	Primary Outcomes
NCT05708599 (02.2023–02.2028)	A Study to Compare Tissue and Liquid Biopsies in People With Different Types of Cancer	Germany (Boehringer Ingelheim)	180 (Interventional)	The mean VAF of mutations in ctDNA samples over the timescale of the patient's treatment course
NCT05429320 (06.2022–06.2025	A Study of Local Ablative Therapy (LAT) in People With Non-Small Cell Lung Cancer (NSCLC)	USA (Memorial Sloan Kettering Cancer Center)	117 (Interventional)	Measure the reduction in mean VAF by 6 months after local ablative therapy
NCT05921474 (04.2023–12.2023)	Detection of Circulating Tumor DNA After Stereotactic Ablative Radiotherapy in Patients With Unbiopsied Lung Tumors (SABR-DETECT)	Canada (Lawson Health Research Institute)	100 (Observational)	Increases in VAF or quantifiable ctDNA from baseline to post-treatment samples (in patients with detectable ctDNA at baseline)
NCT05221372 (02.2017–01.2031)	ProSpecTive sAmpling in dRiver muTation Pulmonary Oncology Patients on Tyrosine Kinase Inhibitors (START-TKI)	Netherlands (Erasmus Medical Center)	1300 (Observational)	The relative presence of primary mutations and resistance mutations in plasma levels under the treatment of a small molecule kinase inhibitor until the progression of disease measured in VAF
NCT04122833 (09.2019–12.2024)	Impact of Concomitant Genetic Alterations in *EGFR* Mutated Adenocarcinoma by NGS Analysis: A Multicenter Study	South Korea (Konkuk University Medical Center)	80 (Observational)	The correlation between the change in VAF and drug response in matched tumor tissues before and after TKI treatment
NCT05102110 (12.2021–12/2023)	Feasibility Study to Investigate Rectal Mucus in Aero-Digestive Tract Cancer (ORI-EGI-03)	United Kingdom (Origin Sciences)	300 (Observational)	The correlation of SNP allele frequency in genes associated with known aero-digestive cancers in paired samples of tumour type and rectal mucus
NCT05254795 (04.2022–12.2036)	Precision Medicine Randomized Clinical Trial Comparing Molecular Tumor Board Assisted Care to Usual Care (PRiMAL)	USA (Jill M Kolesar)	500 (Interventional)	The association of ctDNA VAF with 1-year overall survival
NCT05782361 (05.2023–02.2028)	POTENT-Tepotinib in Combination With Pembrolizumab in NSCLC	United Kingdom (Institute of Cancer Research)	38 (Interventional)	The determination of allele frequency of genomic aberrations including, but not limited to, the *MET*, *EGFR*, *BRAF*, and *KRAS* genes in plasma
NCT03778229 (01.2019–05.2025)	Osimertinib Plus Savolitinib in EGFRm+/ΔMET+ NSCLC Following Prior Osimertinib (SAVANNAH)	USA (AstraZeneca)	360 (Interventional)	Total clearance of *EGFR* mutations at 6 weeks after osimertinib and savolitinib therapy initiation (the percentage and absolute change from baseline in *EGFR* mutation allele frequencies)

5. Technical Aspects of VAF Evaluation in NSCLC Patients

5.1. Biological Factors Affecting VAF Measurement

During the analysis of VAF results, it is also important to take into account the possible influence of other factors on VAF value. For instance, it was shown that tumors with a high copy number of genes, especially amplification, may have significantly higher VAF values [103,104]. Nevertheless, it is being considered that the use of parameters such as ΔVAF can be valuable in the evaluation of ctDNA changes in patients with low or undetectable copy number variants and single nucleotide variants [105,106]. ctDNA concentration and VAF seem to correlate with tumor size; however, due to the detection limit of currently used platforms, VAF assessment appears to be the most optimal in patients with a tumor volume of at least 10 cm^3, while a smaller tumor volume might release a borderline amount of ctDNA for sensitive detection [107].

On the other hand, VAF and tumor size may also be dependent on occurring mutations. A stronger correlation between VAF and tumor size appears in *KRAS*- and *TP53*-mutated tumors compared to *EGFR*-mutated tumors. Contrarily, NSCLC tumors with *TP53* or *EGFR* mutations are most likely to shed ctDNA [103]. Moreover, the VAF value may vary in NSCLC patients depending on the presence and localization of metastatic lesions. Belloum et al. indicated that patients with oligo-brain metastases had lower VAF values compared to patients in whom metastases occurred not only in the brain but also apart from the central nervous system [108]. However, the highest VAF values were observed in NSCLC patients with metastases in locations other than the brain, especially metastases with high vascularity in the kidneys, adrenal glands, liver, or spleen [103]. This discrepancy may result from the presence of blood–brain barriers that hinder the release of tumor cells and their products, such as ctDNA, into the bloodstream [108].

5.2. Appropriate Methodology Selection

Choosing the optimal tool to calculate VAF seems to be an issue. Research by Cheng et al. indicated that tumor-informed assay (i.e., a personalized assay based on prior tissue genotyping) enabled the detection of VAF, rather than tumor-agnostic assay (i.e., an assay independent of tumor profiling), and frequently presented negative results. The results suggested the applicability of personalized ctDNA tumor-informed assay in the monitoring of patients treated with ICIs in monotherapy or combination with chemotherapy in cases when tissue is not available for examination [109–111]. Furthermore, in patients where low VAF values are expected, it may be beneficial to choose the ddPCR technique rather than NGS due to the higher sensitivity of ddPCR. However, it is important to consider that ddPCR enables the detection of fewer mutations than NGS approaches. Further validation of methods based on NGS could improve its sensitivity and expand the possibility of VAF analysis in clinical practice [100,112,113].

Finally, it is also important to remember that ctDNA may be extracted from different body fluids apart from peripheral blood. In NSCLC patients, bronchoalveolar lavage (BAL) or bronchial washing (BW) fluid could be analyzed. The concordance of driver mutations present in BAL or BW fluids with tumor tissue can reach 95% [114]. In the BAL, VAF values and ctDNA concentration are higher when compared to plasma; therefore, more tumor-driver mutations can be detected. As a result, a BAL assessment may be especially useful when blood samples have a low yield of ctDNA. Due to these factors, it is even considered that BAL or BW fluid analyses may be useful in lung cancer diagnosis, especially in patients with non-diagnostic biopsies. A crucial disadvantage of those materials' analyses is the necessity of performing bronchoscopy, which is an invasive procedure. However, bronchoscopy is often performed in patients with suspected NSCLC due to existing medical indications; therefore, material could be collected during standard procedures [114–116]. In patients with metastases in the central nervous system, cerebrospinal fluid (CSF) appears to be another useful source of ctDNA for VAF analysis. One of the most important possibilities of its usage seems to be facilitating the selection of therapy in patients with intracranial progression. However, lumbar puncture is an invasive procedure, and the possibility of regular monitoring of CSF ctDNA seems to be limited in metastatic NSCLC patients [108,117].

6. Conclusions

VAF seems to be a promising new tool in therapy effectiveness monitoring, but some key issues need further investigation before its potential implementation for clinical use. The first issue is the necessity of expanding and standardizing the method of VAF analysis. In particular, several or single genetic alterations should be identified for each treatment regimen as the measurable parameters for VAF tracking during clinical follow-up. In NSCLC patients, especially those treated with immunotherapy, the calculation of VAF and its monitoring in different studies has been based on various variants in several genes detected in a particular person. Even though it is possible to design a customized gene panel, it may be difficult to develop it optimally and then put it into widespread use due to the absence of fully consistent research results. Further investigation is also required to determine the optimal frequency of blood collection for VAF assessment. Another crucial matter is the development of the optimal calculation of VAF metrics—different studies are based on mean, median, or highest VAF, and values are calculated using different formulas. It should also be indicated whether it is preferable to calculate VAF once or multiple times, or concerning some baseline value, and how often VAF analysis should be performed.

The second limitation to the use of VAF testing is the methodological aspect. Many of the currently used methods are insufficient for assessing the effectiveness of therapy, especially MRD monitoring, due to the relatively low sensitivity or the limited number of mutations that can be detected. Further developments in sequencing and reductions in the costs of currently used methods could not only enable the expansion of research but also open the possibility of the clinical usage of ctDNA VAF analysis. Before implementation in clinical routine, methodologies need to be validated, including developing appropriate sensitivity and specificity. The presence of false negative and false positive results should be as low as possible.

In conclusion, despite the mentioned difficulties, VAF analysis seems to have great potential in the different aspects of cancer patients' diagnosis, especially due to the non-invasive procedure of ctDNA collection. Ongoing clinical trials and further sequencing methodology development, especially on single-cell resolution, may fill the gap in the current knowledge, paving the way for early cancer detection, cancer interception, and MRD monitoring, as well as measuring the treatment effect or tracking the metastatic spread at the molecular level by VAF analysis.

Author Contributions: Conceptualization, M.N. and P.K.; literature review, N.G.; formal analysis, M.N., B.K.-K. and P.K.; resources, N.G. and B.K.-K.; data curation, N.G. and M.N.; writing—original draft preparation, N.G.; writing—review and editing, M.N., B.K.-K. and P.K.; visualization, N.G.; supervision, M.N. and P.K.; project administration, M.N. and P.K.; funding acquisition, M.N. All authors have read and agreed to the published version of the manuscript.

Funding: This study was funded by the Polish National Centre for Research and Development (LIDER/46/0237/L-12/20/NCBR/2021) to M.N.

Conflicts of Interest: The authors declare no conflicts of interest.

Abbreviations

ACT	Adjuvant chemotherapy
ALK-TKI	ALK–tyrosine kinase inhibitor
BAL	Bronchoalveolar lavage
BSC	Best supportive care
BW	Bronchial washing
cfDNA	Circulating free DNA
ctDNA	Circulating tumor DNA

CNA	Copy number alteration
CNV	Copy number variation
CSF	Cerebrospinal fluid
EGFR-TKI	EGFR–tyrosine kinase inhibitor
FFPE	Formalin-fixed paraffin-embedded
ICI	Immune checkpoint inhibitor
LB	Liquid biopsy
MAF	Mutant allele frequency
MRD	Minimal residual disease
MSAF	Maximum somatic allele frequency
NET	Neutrophil extracellular trap
NSCLC	Non-small cell lung cancer
OS	Overall survival
PD	Progressive disease
PFS	Progression-free survival
PR	Partial response
SD	Stable disease
SNV	Single nucleotide variant
TCGA	The Cancer Genome Atlas
TFE	Tumor Fraction Estimator
VAF	Variant allele frequency
WES	Whole exome sequencing
WGS	Whole genome sequencing

References

1. Inamura, K. Lung Cancer: Understanding Its Molecular Pathology and the 2015 WHO Classification. *Front. Oncol.* **2017**, *7*, 193. [CrossRef] [PubMed]
2. Deshpand, R.; Chandra, M.; Rauthan, A. Evolving Trends in Lung Cancer: Epidemiology, Diagnosis, and Management. *Indian J. Cancer* **2022**, *59*, 90. [CrossRef]
3. Alduais, Y.; Zhang, H.; Fan, F.; Chen, J.; Chen, B. Non-Small Cell Lung Cancer (NSCLC): A Review of Risk Factors, Diagnosis, and Treatment. *Medicine* **2023**, *102*, e32899. [CrossRef] [PubMed]
4. Siegel, R.L.; Miller, K.D.; Jemal, A. Cancer Statistics, 2016. *CA Cancer J. Clin.* **2016**, *66*, 7–30. [CrossRef] [PubMed]
5. Rolfo, C.; Mack, P.; Scagliotti, G.V.; Aggarwal, C.; Arcila, M.E.; Barlesi, F.; Bivona, T.; Diehn, M.; Dive, C.; Dziadziuszko, R.; et al. Liquid Biopsy for Advanced NSCLC: A Consensus Statement From the International Association for the Study of Lung Cancer. *J. Thorac. Oncol.* **2021**, *16*, 1647–1662. [CrossRef] [PubMed]
6. Soliman, S.E.-S.; Alhanafy, A.M.; Habib, M.S.E.-D.; Hagag, M.; Ibrahem, R.A.L. Serum Circulating Cell Free DNA as Potential Diagnostic and Prognostic Biomarker in Non Small Cell Lung Cancer. *Biochem. Biophys. Rep.* **2018**, *15*, 45–51. [CrossRef]
7. Qiu, B.; Guo, W.; Zhang, F.; Lv, F.; Ji, Y.; Peng, Y.; Chen, X.; Bao, H.; Xu, Y.; Shao, Y.; et al. Dynamic Recurrence Risk and Adjuvant Chemotherapy Benefit Prediction by ctDNA in Resected NSCLC. *Nat. Commun.* **2021**, *12*, 6770. [CrossRef]
8. Filis, P.; Kyrochristos, I.; Korakaki, E.; Baltagiannis, E.G.; Thanos, D.; Roukos, D.H. Longitudinal ctDNA Profiling in Precision Oncology and Immuno-Oncology. *Drug Discov. Today* **2023**, *28*, 103540. [CrossRef]
9. Kemper, M.; Krekeler, C.; Menck, K.; Lenz, G.; Evers, G.; Schulze, A.B.; Bleckmann, A. Liquid Biopsies in Lung Cancer. *Cancers* **2023**, *15*, 1430. [CrossRef]
10. Chen, M.; Zhao, H. Next-Generation Sequencing in Liquid Biopsy: Cancer Screening and Early Detection. *Hum. Genom.* **2019**, *13*, 34. [CrossRef]
11. Garcia, J.; Kamps-Hughes, N.; Geiguer, F.; Couraud, S.; Sarver, B.; Payen, L.; Ionescu-Zanetti, C. Sensitivity, Specificity, and Accuracy of a Liquid Biopsy Approach Utilizing Molecular Amplification Pools. *Sci. Rep.* **2021**, *11*, 10761. [CrossRef] [PubMed]
12. Pairawan, S.; Hess, K.; Janku, F.; Sanchez, N.; Shaw, K.; Eng, C.; Damodaran, S.; Javle, M.; Kaseb, A.; Hong, D.; et al. Cell-Free Circulating Tumor DNA Variant Allele Frequency Associates with Survival in Metastatic Cancer. *Clin. Cancer Res.* **2020**, *26*, 1924–1931. [CrossRef]
13. Li, M.; Yang, L.; Hughes, J.; van den Hout, A.; Burns, C.; Woodhouse, R.; Dennis, L.; Hegde, P.; Oxnard, G.R.; Vietz, C. Driver Mutation Variant Allele Frequency in Circulating Tumor DNA and Association with Clinical Outcome in Patients with Non–Small Cell Lung Cancer and EGFR- and KRAS-Mutated Tumors. *J. Mol. Diagn.* **2022**, *24*, 543–553. [CrossRef] [PubMed]
14. Raja, R.; Kuziora, M.; Brohawn, P.Z.; Higgs, B.W.; Gupta, A.; Dennis, P.A.; Ranade, K. Early Reduction in ctDNA Predicts Survival in Patients with Lung and Bladder Cancer Treated with Durvalumab. *Clin. Cancer Res.* **2018**, *24*, 6212–6222. [CrossRef]
15. Snyder, M.W.; Kircher, M.; Hill, A.J.; Daza, R.M.; Shendure, J. Cell-Free DNA Comprises an In Vivo Nucleosome Footprint That Informs Its Tissues-Of-Origin. *Cell* **2016**, *164*, 57–68. [CrossRef]
16. Nikanjam, M.; Kato, S.; Kurzrock, R. Liquid Biopsy: Current Technology and Clinical Applications. *J. Hematol. Oncol.* **2022**, *15*, 131. [CrossRef]

17. Song, P.; Wu, L.R.; Yan, Y.H.; Zhang, J.X.; Chu, T.; Kwong, L.N.; Patel, A.A.; Zhang, D.Y. Limitations and Opportunities of Technologies for the Analysis of Cell-Free DNA in Cancer Diagnostics. *Nat. Biomed. Eng.* **2022**, *6*, 232–245. [CrossRef] [PubMed]
18. Dao, J.; Conway, P.J.; Subramani, B.; Meyyappan, D.; Russell, S.; Mahadevan, D. Using cfDNA and ctDNA as Oncologic Markers: A Path to Clinical Validation. *Int. J. Mol. Sci.* **2023**, *24*, 13219. [CrossRef]
19. Papadopoulos, N. Pathophysiology of ctDNA Release into the Circulation and Its Characteristics: What Is Important for Clinical Applications. In *Tumor Liquid Biopsies*; Schaffner, F., Merlin, J.-L., Von Bubnoff, N., Eds.; Recent Results in Cancer Research; Springer International Publishing: Cham, Switzerland, 2020; Volume 215, pp. 163–180. ISBN 978-3-030-26438-3.
20. Shen, H.; Jin, Y.; Zhao, H.; Wu, M.; Zhang, K.; Wei, Z.; Wang, X.; Wang, Z.; Li, Y.; Yang, F.; et al. Potential Clinical Utility of Liquid Biopsy in Early-Stage Non-Small Cell Lung Cancer. *BMC Med.* **2022**, *20*, 480. [CrossRef]
21. Aucamp, J.; Bronkhorst, A.J.; Badenhorst, C.P.S.; Pretorius, P.J. The Diverse Origins of Circulating Cell-free DNA in the Human Body: A Critical Re-evaluation of the Literature. *Biol. Rev.* **2018**, *93*, 1649–1683. [CrossRef]
22. Bardelli, A.; Pantel, K. Liquid Biopsies, What We Do Not Know (Yet). *Cancer Cell* **2017**, *31*, 172–179. [CrossRef] [PubMed]
23. Filipska, M.; Rosell, R. Mutated Circulating Tumor DNA as a Liquid Biopsy in Lung Cancer Detection and Treatment. *Mol. Oncol.* **2021**, *15*, 1667–1682. [CrossRef] [PubMed]
24. Nguyen, V.-C.; Nguyen, T.H.; Phan, T.H.; Tran, T.-H.T.; Pham, T.T.T.; Ho, T.D.; Nguyen, H.H.T.; Duong, M.-L.; Nguyen, C.M.; Nguyen, Q.-T.B.; et al. Fragment Length Profiles of Cancer Mutations Enhance Detection of Circulating Tumor DNA in Patients with Early-Stage Hepatocellular Carcinoma. *BMC Cancer* **2023**, *23*, 233. [CrossRef] [PubMed]
25. Underhill, H.R.; Kitzman, J.O.; Hellwig, S.; Welker, N.C.; Daza, R.; Baker, D.N.; Gligorich, K.M.; Rostomily, R.C.; Bronner, M.P.; Shendure, J. Fragment Length of Circulating Tumor DNA. *PLoS Genet.* **2016**, *12*, e1006162. [CrossRef]
26. Yamamoto, Y.; Uemura, M.; Fujita, M.; Maejima, K.; Koh, Y.; Matsushita, M.; Nakano, K.; Hayashi, Y.; Wang, C.; Ishizuya, Y.; et al. Clinical Significance of the Mutational Landscape and Fragmentation of Circulating Tumor DNA in Renal Cell Carcinoma. *Cancer Sci.* **2019**, *110*, 617–628. [CrossRef]
27. Jiang, P.; Lo, Y.M.D. The Long and Short of Circulating Cell-Free DNA and the Ins and Outs of Molecular Diagnostics. *Trends Genet.* **2016**, *32*, 360–371. [CrossRef] [PubMed]
28. Malapelle, U.; Pisapia, P.; Pepe, F.; Russo, G.; Buono, M.; Russo, A.; Gomez, J.; Khorshid, O.; Mack, P.C.; Rolfo, C.; et al. The Evolving Role of Liquid Biopsy in Lung Cancer. *Lung Cancer* **2022**, *172*, 53–64. [CrossRef] [PubMed]
29. Chaudhuri, A.A.; Chabon, J.J.; Lovejoy, A.F.; Newman, A.M.; Stehr, H.; Azad, T.D.; Khodadoust, M.S.; Esfahani, M.S.; Liu, C.L.; Zhou, L.; et al. Early Detection of Molecular Residual Disease in Localized Lung Cancer by Circulating Tumor DNA Profiling. *Cancer Discov.* **2017**, *7*, 1394–1403. [CrossRef]
30. Bettegowda, C.; Sausen, M.; Leary, R.J.; Kinde, I.; Wang, Y.; Agrawal, N.; Bartlett, B.R.; Wang, H.; Luber, B.; Alani, R.M.; et al. Detection of Circulating Tumor DNA in Early- and Late-Stage Human Malignancies. *Sci. Transl. Med.* **2014**, *6*, 224ra2. [CrossRef]
31. Sánchez-Herrero, E.; Serna-Blasco, R.; Robado de Lope, L.; González-Rumayor, V.; Romero, A.; Provencio, M. Circulating Tumor DNA as a Cancer Biomarker: An Overview of Biological Features and Factors That May Impact on ctDNA Analysis. *Front. Oncol.* **2022**, *12*, 943253. [CrossRef]
32. Gale, D.; Heider, K.; Ruiz-Valdepenas, A.; Hackinger, S.; Perry, M.; Marsico, G.; Rundell, V.; Wulff, J.; Sharma, G.; Knock, H.; et al. Residual ctDNA after Treatment Predicts Early Relapse in Patients with Early-Stage Non-Small Cell Lung Cancer. *Ann. Oncol.* **2022**, *33*, 500–510. [CrossRef]
33. Heidrich, I.; Ačkar, L.; Mossahebi Mohammadi, P.; Pantel, K. Liquid Biopsies: Potential and Challenges. *Int. J. Cancer* **2021**, *148*, 528–545. [CrossRef]
34. FDA. *FDA Approves Liquid Biopsy Next-Generation Sequencing Companion Diagnostic Test*; FDA: Silver Spring, MD, USA, 2020.
35. Hu, Y.; Ulrich, B.C.; Supplee, J.; Kuang, Y.; Lizotte, P.H.; Feeney, N.B.; Guibert, N.M.; Awad, M.M.; Wong, K.-K.; Jänne, P.A.; et al. False-Positive Plasma Genotyping Due to Clonal Hematopoiesis. *Clin. Cancer Res.* **2018**, *24*, 4437–4443. [CrossRef] [PubMed]
36. Razavi, P.; Li, B.T.; Brown, D.N.; Jung, B.; Hubbell, E.; Shen, R.; Abida, W.; Juluru, K.; De Bruijn, I.; Hou, C.; et al. High-Intensity Sequencing Reveals the Sources of Plasma Circulating Cell-Free DNA Variants. *Nat. Med.* **2019**, *25*, 1928–1937. [CrossRef] [PubMed]
37. Arisi, M.F.; Dotan, E.; Fernandez, S.V. Circulating Tumor DNA in Precision Oncology and Its Applications in Colorectal Cancer. *Int. J. Mol. Sci.* **2022**, *23*, 4441. [CrossRef] [PubMed]
38. Moss, J.; Magenheim, J.; Neiman, D.; Zemmour, H.; Loyfer, N.; Korach, A.; Samet, Y.; Maoz, M.; Druid, H.; Arner, P.; et al. Comprehensive Human Cell-Type Methylation Atlas Reveals Origins of Circulating Cell-Free DNA in Health and Disease. *Nat. Commun.* **2018**, *9*, 5068. [CrossRef] [PubMed]
39. Luo, H.; Wei, W.; Ye, Z.; Zheng, J.; Xu, R. Liquid Biopsy of Methylation Biomarkers in Cell-Free DNA. *Trends Mol. Med.* **2021**, *27*, 482–500. [CrossRef] [PubMed]
40. Guo, S.; Diep, D.; Plongthongkum, N.; Fung, H.-L.; Zhang, K.; Zhang, K. Identification of Methylation Haplotype Blocks Aids in Deconvolution of Heterogeneous Tissue Samples and Tumor Tissue-of-Origin Mapping from Plasma DNA. *Nat. Genet.* **2017**, *49*, 635–642. [CrossRef] [PubMed]
41. Luo, H.; Zhao, Q.; Wei, W.; Zheng, L.; Yi, S.; Li, G.; Wang, W.; Sheng, H.; Pu, H.; Mo, H.; et al. Circulating Tumor DNA Methylation Profiles Enable Early Diagnosis, Prognosis Prediction, and Screening for Colorectal Cancer. *Sci. Transl. Med.* **2020**, *12*, eaax7533. [CrossRef]

42. Lianidou, E. Detection and Relevance of Epigenetic Markers on ctDNA: Recent Advances and Future Outlook. *Mol. Oncol.* **2021**, *15*, 1683–1700. [CrossRef]
43. Jiang, P.; Chan, C.W.M.; Chan, K.C.A.; Cheng, S.H.; Wong, J.; Wong, V.W.-S.; Wong, G.L.H.; Chan, S.L.; Mok, T.S.K.; Chan, H.L.Y.; et al. Lengthening and Shortening of Plasma DNA in Hepatocellular Carcinoma Patients. *Proc. Natl. Acad. Sci. USA* **2015**, *112*, E1317–E1325. [CrossRef]
44. Sivapalan, L.; Iams, W.T.; Belcaid, Z.; Scott, S.C.; Niknafs, N.; Balan, A.; White, J.R.; Kopparapu, P.; Cann, C.; Landon, B.V.; et al. Dynamics of Sequence and Structural Cell-Free DNA Landscapes in Small-Cell Lung Cancer. *Clin. Cancer Res.* **2023**, *29*, 2310–2323. [CrossRef] [PubMed]
45. Huang, R.S.P.; Xiao, J.; Pavlick, D.C.; Guo, C.; Yang, L.; Jin, D.X.; Fendler, B.; Severson, E.; Killian, J.K.; Hiemenz, M.; et al. Circulating Cell-Free DNA Yield and Circulating-Tumor DNA Quantity from Liquid Biopsies of 12,139 Cancer Patients. *Clin. Chem.* **2021**, *67*, 1554–1566. [CrossRef] [PubMed]
46. Dong, Y.; Zhu, Y.; Zhuo, M.; Chen, X.; Xie, Y.; Duan, J.; Bai, H.; Hao, S.; Yu, Z.; Yi, Y.; et al. Maximum Somatic Allele Frequency-Adjusted Blood-Based Tumor Mutational Burden Predicts the Efficacy of Immune Checkpoint Inhibitors in Advanced Non-Small Cell Lung Cancer. *Cancers* **2022**, *14*, 5649. [CrossRef] [PubMed]
47. Schrock, A.B.; Welsh, A.; Chung, J.H.; Pavlick, D.; Bernicker, E.H.; Creelan, B.C.; Forcier, B.; Ross, J.S.; Stephens, P.J.; Ali, S.M.; et al. Hybrid Capture-Based Genomic Profiling of Circulating Tumor DNA from Patients with Advanced Non-Small Cell Lung Cancer. *J. Thorac. Oncol.* **2019**, *14*, 255–264. [CrossRef] [PubMed]
48. Socinski, M.A.; Paul, S.M.; Yun, C.; Hu, S.; Shen, V.; Velcheti, V.; Mok, T.S.; Gandara, D.R.; Chae, Y.K.; Schleifman, E.; et al. Abstract CT194: Exploratory Subgroup Analysis of Atezolizumab (Atezo) Clinical Characteristics in Patients (Pts) with Low Circulating Tumor DNA (ctDNA) in B-F1RST—A Phase II Trial Evaluating Blood-Based Tumor Mutational Burden (bTMB) in NSCLC. *Cancer Res.* **2019**, *79*, CT194. [CrossRef]
49. Chen, Y.; Seeruttun, S.R.; Wu, X.; Wang, Z. Maximum Somatic Allele Frequency in Combination With Blood-Based Tumor Mutational Burden to Predict the Efficacy of Atezolizumab in Advanced Non-Small Cell Lung Cancer: A Pooled Analysis of the Randomized POPLAR and OAK Studies. *Front. Oncol.* **2019**, *9*, 1432. [CrossRef] [PubMed]
50. Gandara, D.R.; Paul, S.M.; Kowanetz, M.; Schleifman, E.; Zou, W.; Li, Y.; Rittmeyer, A.; Fehrenbacher, L.; Otto, G.; Malboeuf, C.; et al. Blood-Based Tumor Mutational Burden as a Predictor of Clinical Benefit in Non-Small-Cell Lung Cancer Patients Treated with Atezolizumab. *Nat. Med.* **2018**, *24*, 1441–1448. [CrossRef] [PubMed]
51. Menon, V.; Brash, D.E. Next-Generation Sequencing Methodologies to Detect Low-Frequency Mutations: "Catch Me If You Can". *Mutat. Res./Rev. Mutat. Res.* **2023**, *792*, 108471. [CrossRef]
52. Uffelmann, E.; Huang, Q.Q.; Munung, N.S.; de Vries, J.; Okada, Y.; Martin, A.R.; Martin, H.C.; Lappalainen, T.; Posthuma, D. Genome-Wide Association Studies. *Nat. Rev. Methods Primers* **2021**, *1*, 59. [CrossRef]
53. Boscolo Bielo, L.; Trapani, D.; Repetto, M.; Crimini, E.; Valenza, C.; Belli, C.; Criscitiello, C.; Marra, A.; Subbiah, V.; Curigliano, G. Variant Allele Frequency: A Decision-Making Tool in Precision Oncology? *Trends Cancer* **2023**, *9*, 1058–1068. [CrossRef]
54. Manca, P.; Corallo, S.; Lonardi, S.; Fucà, G.; Busico, A.; Leone, A.G.; Corti, F.; Antoniotti, C.; Procaccio, L.; Smiroldo, V.; et al. Variant Allele Frequency in Baseline Circulating Tumour DNA to Measure Tumour Burden and to Stratify Outcomes in Patients with RAS Wild-Type Metastatic Colorectal Cancer: A Translational Objective of the Valentino Study. *Br. J. Cancer* **2022**, *126*, 449–455. [CrossRef]
55. Berchuck, J.E.; Facchinetti, F.; DiToro, D.F.; Baiev, I.; Majeed, U.; Reyes, S.; Chen, C.; Zhang, K.; Sharman, R.; Uson Junior, P.L.S.; et al. The Clinical Landscape of Cell-Free DNA Alterations in 1671 Patients with Advanced Biliary Tract Cancer. *Ann. Oncol.* **2022**, *33*, 1269–1283. [CrossRef] [PubMed]
56. Parkinson, C.A.; Gale, D.; Piskorz, A.M.; Biggs, H.; Hodgkin, C.; Addley, H.; Freeman, S.; Moyle, P.; Sala, E.; Sayal, K.; et al. Exploratory Analysis of TP53 Mutations in Circulating Tumour DNA as Biomarkers of Treatment Response for Patients with Relapsed High-Grade Serous Ovarian Carcinoma: A Retrospective Study. *PLoS Med.* **2016**, *13*, e1002198. [CrossRef]
57. Kujala, J.; Hartikainen, J.M.; Tengström, M.; Sironen, R.; Kosma, V.-M.; Mannermaa, A. High Mutation Burden of Circulating Cell-Free DNA in Early-Stage Breast Cancer Patients Is Associated with a Poor Relapse-Free Survival. *Cancer Med.* **2020**, *9*, 5922–5931. [CrossRef]
58. Dentro, S.C.; Leshchiner, I.; Haase, K.; Tarabichi, M.; Wintersinger, J.; Deshwar, A.G.; Yu, K.; Rubanova, Y.; Macintyre, G.; Demeulemeester, J.; et al. Characterizing Genetic Intra-Tumor Heterogeneity across 2658 Human Cancer Genomes. *Cell* **2021**, *184*, 2239–2254.e39. [CrossRef] [PubMed]
59. McGranahan, N.; Favero, F.; de Bruin, E.C.; Birkbak, N.J.; Szallasi, Z.; Swanton, C. Clonal Status of Actionable Driver Events and the Timing of Mutational Processes in Cancer Evolution. *Sci. Transl. Med.* **2015**, *7*, 283ra54. [CrossRef] [PubMed]
60. Fairchild, L.; Whalen, J.; D'Aco, K.; Wu, J.; Gustafson, C.B.; Solovieff, N.; Su, F.; Leary, R.J.; Campbell, C.D.; Balbin, O.A. Clonal Hematopoiesis Detection in Patients with Cancer Using Cell-Free DNA Sequencing. *Sci. Transl. Med.* **2023**, *15*, eabm8729. [CrossRef]
61. Shin, H.-T.; Choi, Y.-L.; Yun, J.W.; Kim, N.K.D.; Kim, S.-Y.; Jeon, H.J.; Nam, J.-Y.; Lee, C.; Ryu, D.; Kim, S.C.; et al. Prevalence and Detection of Low-Allele-Fraction Variants in Clinical Cancer Samples. *Nat. Commun.* **2017**, *8*, 1377. [CrossRef]
62. Vogelstein, B.; Papadopoulos, N.; Velculescu, V.E.; Zhou, S.; Diaz, L.A.; Kinzler, K.W. Cancer Genome Landscapes. *Science* **2013**, *339*, 1546–1558. [CrossRef]
63. Raphael, B.J.; Dobson, J.R.; Oesper, L.; Vandin, F. Identifying Driver Mutations in Sequenced Cancer Genomes: Computational Approaches to Enable Precision Medicine. *Genome Med.* **2014**, *6*, 5. [CrossRef]
64. Wodarz, D.; Newell, A.C.; Komarova, N.L. Passenger Mutations Can Accelerate Tumour Suppressor Gene Inactivation in Cancer Evolution. *J. R. Soc. Interface* **2018**, *15*, 20170967. [CrossRef]

65. Kumar, S.; Warrell, J.; Li, S.; McGillivray, P.D.; Meyerson, W.; Salichos, L.; Harmanci, A.; Martinez-Fundichely, A.; Chan, C.W.Y.; Nielsen, M.M.; et al. Passenger Mutations in 2500 Cancer Genomes: Overall Molecular Functional Impact and Consequences. *Cell* **2020**, *180*, 915–927.e16. [CrossRef]
66. Salvadores, M.; Mas-Ponte, D.; Supek, F. Passenger Mutations Accurately Classify Human Tumors. *PLoS Comput. Biol.* **2019**, *15*, e1006953. [CrossRef]
67. Pavel, A.B.; Korolev, K.S. Genetic Load Makes Cancer Cells More Sensitive to Common Drugs: Evidence from Cancer Cell Line Encyclopedia. *Sci. Rep.* **2017**, *7*, 1938. [CrossRef] [PubMed]
68. McFarland, C.D.; Yaglom, J.A.; Wojtkowiak, J.W.; Scott, J.G.; Morse, D.L.; Sherman, M.Y.; Mirny, L.A. The Damaging Effect of Passenger Mutations on Cancer Progression. *Cancer Res.* **2017**, *77*, 4763–4772. [CrossRef] [PubMed]
69. Diehl, F.; Li, M.; He, Y.; Kinzler, K.W.; Vogelstein, B.; Dressman, D. BEAMing: Single-Molecule PCR on Microparticles in Water-in-Oil Emulsions. *Nat. Methods* **2006**, *3*, 551–559. [CrossRef] [PubMed]
70. Denis, J.A.; Guillerm, E.; Coulet, F.; Larsen, A.K.; Lacorte, J.-M. The Role of BEAMing and Digital PCR for Multiplexed Analysis in Molecular Oncology in the Era of Next-Generation Sequencing. *Mol. Diagn. Ther.* **2017**, *21*, 587–600. [CrossRef]
71. Warburton, L.; Calapre, L.; Pereira, M.R.; Reid, A.; Robinson, C.; Amanuel, B.; Ziman, M.; Millward, M.; Gray, E. Circulating Tumour DNA in Advanced Melanoma Patients Ceasing PD1 Inhibition in the Absence of Disease Progression. *Cancers* **2020**, *12*, 3486. [CrossRef]
72. Casanova-Salas, I.; Athie, A.; Boutros, P.C.; Del Re, M.; Miyamoto, D.T.; Pienta, K.J.; Posadas, E.M.; Sowalsky, A.G.; Stenzl, A.; Wyatt, A.W.; et al. Quantitative and Qualitative Analysis of Blood-Based Liquid Biopsies to Inform Clinical Decision-Making in Prostate Cancer. *Eur. Urol.* **2021**, *79*, 762–771. [CrossRef]
73. Shields, M.D.; Chen, K.; Dutcher, G.; Patel, I.; Pellini, B. Making the Rounds: Exploring the Role of Circulating Tumor DNA (ctDNA) in Non-Small Cell Lung Cancer. *Int. J. Mol. Sci.* **2022**, *23*, 9006. [CrossRef]
74. Plasma and Serum Preparation. Available online: https://www.thermofisher.com/uk/en/home/references/protocols/cell-and-tissue-analysis/elisa-protocol/elisa-sample-preparation-protocols/plasma-and-serum-preparation.html (accessed on 5 February 2024).
75. Meddeb, R.; Pisareva, E.; Thierry, A.R. Guidelines for the Preanalytical Conditions for Analyzing Circulating Cell-Free DNA. *Clin. Chem.* **2019**, *65*, 623–633. [CrossRef]
76. Parpart-Li, S.; Bartlett, B.; Popoli, M.; Adleff, V.; Tucker, L.; Steinberg, R.; Georgiadis, A.; Phallen, J.; Brahmer, J.; Azad, N.; et al. The Effect of Preservative and Temperature on the Analysis of Circulating Tumor DNA. *Clin. Cancer Res.* **2017**, *23*, 2471–2477. [CrossRef]
77. Pittella-Silva, F.; Chin, Y.M.; Chan, H.T.; Nagayama, S.; Miyauchi, E.; Low, S.-K.; Nakamura, Y. Plasma or Serum: Which Is Preferable for Mutation Detection in Liquid Biopsy? *Clin. Chem.* **2020**, *66*, 946–957. [CrossRef]
78. Pös, Z.; Pös, O.; Styk, J.; Mocova, A.; Strieskova, L.; Budis, J.; Kadasi, L.; Radvanszky, J.; Szemes, T. Technical and Methodological Aspects of Cell-Free Nucleic Acids Analyzes. *Int. J. Mol. Sci.* **2020**, *21*, 8634. [CrossRef]
79. Thress, K.S.; Brant, R.; Carr, T.H.; Dearden, S.; Jenkins, S.; Brown, H.; Hammett, T.; Cantarini, M.; Barrett, J.C. EGFR Mutation Detection in ctDNA from NSCLC Patient Plasma: A Cross-Platform Comparison of Leading Technologies to Support the Clinical Development of AZD9291. *Lung Cancer* **2015**, *90*, 509–515. [CrossRef]
80. Lebofsky, R.; Decraene, C.; Bernard, V.; Kamal, M.; Blin, A.; Leroy, Q.; Rio Frio, T.; Pierron, G.; Callens, C.; Bieche, I.; et al. Circulating Tumor DNA as a Non-invasive Substitute to Metastasis Biopsy for Tumor Genotyping and Personalized Medicine in a Prospective Trial across All Tumor Types. *Mol. Oncol.* **2015**, *9*, 783–790. [CrossRef] [PubMed]
81. Scilla, K.A.; Rolfo, C. The Role of Circulating Tumor DNA in Lung Cancer: Mutational Analysis, Diagnosis, and Surveillance Now and into the Future. *Curr. Treat. Options Oncol.* **2019**, *20*, 61. [CrossRef] [PubMed]
82. Jee, J.; Lebow, E.S.; Yeh, R.; Das, J.P.; Namakydoust, A.; Paik, P.K.; Chaft, J.E.; Jayakumaran, G.; Rose Brannon, A.; Benayed, R.; et al. Overall Survival with Circulating Tumor DNA-Guided Therapy in Advanced Non-Small-Cell Lung Cancer. *Nat. Med.* **2022**, *28*, 2353–2363. [CrossRef] [PubMed]
83. An, Y.; Guan, Y.; Xu, Y.; Han, Y.; Wu, C.; Bao, C.; Zhou, B.; Wang, H.; Zhang, M.; Liu, W.; et al. The Diagnostic and Prognostic Usage of Circulating Tumor DNA in Operable Hepatocellular Carcinoma. *Am. J. Transl. Res.* **2019**, *11*, 6462–6474. [PubMed]
84. Van Velzen, M.J.M.; Creemers, A.; Van Den Ende, T.; Schokker, S.; Krausz, S.; Reinten, R.J.; Dijk, F.; Van Noesel, C.J.M.; Halfwerk, H.; Meijer, S.L.; et al. Circulating Tumor DNA Predicts Outcome in Metastatic Gastroesophageal Cancer. *Gastric Cancer* **2022**, *25*, 906–915. [CrossRef]
85. Zhang, Q.; Luo, J.; Wu, S.; Si, H.; Gao, C.; Xu, W.; Abdullah, S.E.; Higgs, B.W.; Dennis, P.A.; Van Der Heijden, M.S.; et al. Prognostic and Predictive Impact of Circulating Tumor DNA in Patients with Advanced Cancers Treated with Immune Checkpoint Blockade. *Cancer Discov.* **2020**, *10*, 1842–1853. [CrossRef]
86. Yang, Y.; Wang, J.; Wang, J.; Zhao, X.; Zhang, T.; Yang, Y.; Pang, J.; Ou, Q.; Wu, L.; Xu, X.; et al. Unrevealing the Therapeutic Benefits of Radiotherapy and Consolidation Immunotherapy Using ctDNA-Defined Tumor Clonality in Unresectable Locally Advanced Non-Small Cell Lung Cancer. *Cancer Lett.* **2023**, *582*, 216569. [CrossRef] [PubMed]
87. Goldberg, S.B.; Narayan, A.; Kole, A.J.; Decker, R.H.; Teysir, J.; Carriero, N.J.; Lee, A.; Nemati, R.; Nath, S.K.; Mane, S.M.; et al. Early Assessment of Lung Cancer Immunotherapy Response via Circulating Tumor DNA. *Clin. Cancer Res.* **2018**, *24*, 1872–1880. [CrossRef] [PubMed]
88. Wang, D.-S.; Yang, H.; Liu, X.-Y.; Chen, Z.-G.; Wang, Y.; Fong, W.P.; Hu, M.-T.; Zheng, Y.-C.; Zheng, Y.; Li, B.-K.; et al. Dynamic Monitoring of Circulating Tumor DNA to Predict Prognosis and Efficacy of Adjuvant Chemotherapy after Resection of Colorectal Liver Metastases. *Theranostics* **2021**, *11*, 7018–7028. [CrossRef] [PubMed]
89. Koboldt, D.C. Best Practices for Variant Calling in Clinical Sequencing. *Genome Med.* **2020**, *12*, 91. [CrossRef] [PubMed]

90. Jiang, T.; Jiang, L.; Dong, X.; Gu, K.; Pan, Y.; Shi, Q.; Zhang, G.; Wang, H.; Zhang, X.; Yang, N.; et al. Utilization of Circulating Cell-Free DNA Profiling to Guide First-Line Chemotherapy in Advanced Lung Squamous Cell Carcinoma. *Theranostics* **2021**, *11*, 257–267. [CrossRef] [PubMed]
91. Vega, D.M.; Nishimura, K.K.; Zariffa, N.; Thompson, J.C.; Hoering, A.; Cilento, V.; Rosenthal, A.; Anagnostou, V.; Baden, J.; Beaver, J.A.; et al. Changes in Circulating Tumor DNA Reflect Clinical Benefit Across Multiple Studies of Patients With Non–Small-Cell Lung Cancer Treated With Immune Checkpoint Inhibitors. *JCO Precis. Oncol.* **2022**, *6*, e2100372. [CrossRef] [PubMed]
92. Thompson, J.C.; Carpenter, E.L.; Silva, B.A.; Rosenstein, J.; Chien, A.L.; Quinn, K.; Espenschied, C.R.; Mak, A.; Kiedrowski, L.A.; Lefterova, M.; et al. Serial Monitoring of Circulating Tumor DNA by Next-Generation Gene Sequencing as a Biomarker of Response and Survival in Patients With Advanced NSCLC Receiving Pembrolizumab-Based Therapy. *JCO Precis. Oncol.* **2021**, *5*, 510–524. [CrossRef] [PubMed]
93. Zulato, E.; Del Bianco, P.; Nardo, G.; Attili, I.; Pavan, A.; Boscolo Bragadin, A.; Marra, L.; Pasello, G.; Fassan, M.; Calabrese, F.; et al. Longitudinal Liquid Biopsy Anticipates Hyperprogression and Early Death in Advanced Non-Small Cell Lung Cancer Patients Treated with Immune Checkpoint Inhibitors. *Br. J. Cancer* **2022**, *127*, 2034–2042. [CrossRef]
94. Bewicke-Copley, F.; Arjun Kumar, E.; Palladino, G.; Korfi, K.; Wang, J. Applications and Analysis of Targeted Genomic Sequencing in Cancer Studies. *Comput. Struct. Biotechnol. J.* **2019**, *17*, 1348–1359. [CrossRef]
95. Chen, Y.; Li, X.; Liu, G.; Chen, S.; Xu, M.; Song, L.; Wang, Y. ctDNA Concentration, MIKI67 Mutations and Hyper-Progressive Disease Related Gene Mutations Are Prognostic Markers for Camrelizumab and Apatinib Combined Multiline Treatment in Advanced NSCLC. *Front. Oncol.* **2020**, *10*, 1706. [CrossRef] [PubMed]
96. Ren, S.; Chen, J.; Xu, X.; Jiang, T.; Cheng, Y.; Chen, G.; Pan, Y.; Fang, Y.; Wang, Q.; Huang, Y.; et al. Camrelizumab Plus Carboplatin and Paclitaxel as First-Line Treatment for Advanced Squamous NSCLC (CameL-Sq): A Phase 3 Trial. *J. Thorac. Oncol.* **2022**, *17*, 544–557. [CrossRef] [PubMed]
97. Felip, E.; Srivastava, M.; Reck, M.; Wakelee, H.; Altorki, N.K.; Vallieres, E.; Liersch, R.; Harada, M.; Tanaka, H.; Hamm, J.T.; et al. 1O IMpower010: ctDNA Status in Patients (Pts) with Resected NSCLC Who Received Adjuvant Chemotherapy (Chemo) Followed by Atezolizumab (Atezo) or Best Supportive Care (BSC). *Immuno-Oncol. Technol.* **2022**, *16*, 100106. [CrossRef]
98. Soo, R.A.; Martini, J.-F.; Van Der Wekken, A.J.; Teraoka, S.; Ferrara, R.; Shaw, A.T.; Shepard, D.; Calella, A.M.; Polli, A.; Toffalorio, F.; et al. Early Circulating Tumor DNA Dynamics and Efficacy of Lorlatinib in Patients With Treatment-Naive, Advanced, ALK-Positive NSCLC. *J. Thorac. Oncol.* **2023**, *18*, 1568–1580. [CrossRef] [PubMed]
99. Begum, P.; Cui, W.; Popat, S. Crizotinib-Resistant ROS1 G2101A Mutation Associated With Sensitivity to Lorlatinib in ROS1-Rearranged NSCLC: Case Report. *JTO Clin. Res. Rep.* **2022**, *3*, 100376. [CrossRef]
100. Beagan, J.J.; Bach, S.; Van Boerdonk, R.A.; Van Dijk, E.; Thunnissen, E.; Van Den Broek, D.; Weiss, J.; Kazemier, G.; Pegtel, D.M.; Bahce, I.; et al. Circulating Tumor DNA Analysis of EGFR-Mutant Non-Small Cell Lung Cancer Patients Receiving Osimertinib Following Previous Tyrosine Kinase Inhibitor Treatment. *Lung Cancer* **2020**, *145*, 173–180. [CrossRef]
101. Vaclova, T.; Grazini, U.; Ward, L.; O'Neill, D.; Markovets, A.; Huang, X.; Chmielecki, J.; Hartmaier, R.; Thress, K.S.; Smith, P.D.; et al. Clinical Impact of Subclonal EGFR T790M Mutations in Advanced-Stage EGFR-Mutant Non-Small-Cell Lung Cancers. *Nat. Commun.* **2021**, *12*, 1780. [CrossRef]
102. Ai, X.; Cui, J.; Zhang, J.; Chen, R.; Lin, W.; Xie, C.; Liu, A.; Zhang, J.; Yang, W.; Hu, X.; et al. Clonal Architecture of EGFR Mutation Predicts the Efficacy of EGFR-Tyrosine Kinase Inhibitors in Advanced NSCLC: A Prospective Multicenter Study (NCT03059641). *Clin. Cancer Res.* **2021**, *27*, 704–712. [CrossRef]
103. Lam, V.K.; Zhang, J.; Wu, C.C.; Tran, H.T.; Li, L.; Diao, L.; Wang, J.; Rinsurongkawong, W.; Raymond, V.M.; Lanman, R.B.; et al. Genotype-Specific Differences in Circulating Tumor DNA Levels in Advanced NSCLC. *J. Thorac. Oncol.* **2021**, *16*, 601–609. [CrossRef]
104. Friedlaender, A.; Tsantoulis, P.; Chevallier, M.; De Vito, C.; Addeo, A. The Impact of Variant Allele Frequency in EGFR Mutated NSCLC Patients on Targeted Therapy. *Front. Oncol.* **2021**, *11*, 644472. [CrossRef]
105. Angeles, A.K.; Christopoulos, P.; Yuan, Z.; Bauer, S.; Janke, F.; Ogrodnik, S.J.; Reck, M.; Schlesner, M.; Meister, M.; Schneider, M.A.; et al. Early Identification of Disease Progression in ALK-Rearranged Lung Cancer Using Circulating Tumor DNA Analysis. *NPJ Precis. Oncol.* **2021**, *5*, 100. [CrossRef] [PubMed]
106. Dietz, S.; Christopoulos, P.; Yuan, Z.; Angeles, A.K.; Gu, L.; Volckmar, A.-L.; Ogrodnik, S.J.; Janke, F.; Fratte, C.D.; Zemojtel, T.; et al. Longitudinal Therapy Monitoring of ALK-Positive Lung Cancer by Combined Copy Number and Targeted Mutation Profiling of Cell-Free DNA. *EBioMedicine* **2020**, *62*, 103103. [CrossRef]
107. The TRACERx consortium; The PEACE consortium; Abbosh, C.; Birkbak, N.J.; Wilson, G.A.; Jamal-Hanjani, M.; Constantin, T.; Salari, R.; Le Quesne, J.; Moore, D.A.; et al. Phylogenetic ctDNA Analysis Depicts Early-Stage Lung Cancer Evolution. *Nature* **2017**, *545*, 446–451. [CrossRef] [PubMed]
108. Belloum, Y.; Janning, M.; Mohme, M.; Simon, R.; Kropidlowski, J.; Sartori, A.; Irwin, D.; Westphal, M.; Lamszus, K.; Loges, S.; et al. Discovery of Targetable Genetic Alterations in NSCLC Patients with Different Metastatic Patterns Using a MassARRAY-Based Circulating Tumor DNA Assay. *Cells* **2020**, *9*, 2337. [CrossRef] [PubMed]
109. Gong, J.; Hendifar, A.; Gangi, A.; Zaghiyan, K.; Atkins, K.; Nasseri, Y.; Murrell, Z.; Figueiredo, J.C.; Salvy, S.; Haile, R.; et al. Clinical Applications of Minimal Residual Disease Assessments by Tumor-Informed and Tumor-Uninformed Circulating Tumor DNA in Colorectal Cancer. *Cancers* **2021**, *13*, 4547. [CrossRef] [PubMed]

110. Chen, K.; Yang, F.; Shen, H.; Wang, C.; Li, X.; Chervova, O.; Wu, S.; Qiu, F.; Peng, D.; Zhu, X.; et al. Individualized Tumor-Informed Circulating Tumor DNA Analysis for Postoperative Monitoring of Non-Small Cell Lung Cancer. *Cancer Cell* **2023**, *41*, 1749–1762.e6. [CrossRef]
111. Cheng, L.; Gao, G.; Zhao, C.; Wang, H.; Yao, C.; Yu, H.; Yao, J.; Li, F.; Guo, L.; Jian, Q.; et al. Personalized Circulating Tumor DNA Detection to Monitor Immunotherapy Efficacy and Predict Outcome in Locally Advanced or Metastatic Non-small Cell Lung Cancer. *Cancer Med.* **2023**, *12*, 14317–14326. [CrossRef]
112. Plagnol, V.; Woodhouse, S.; Howarth, K.; Lensing, S.; Smith, M.; Epstein, M.; Madi, M.; Smalley, S.; Leroy, C.; Hinton, J.; et al. Analytical Validation of a next Generation Sequencing Liquid Biopsy Assay for High Sensitivity Broad Molecular Profiling. *PLoS ONE* **2018**, *13*, e0193802. [CrossRef]
113. González de Aledo-Castillo, J.M.; Serhir-Sgheiri, S.; Calbet-Llopart, N.; Arcocha, A.; Jares, P.; Reguart, N.; Puig-Butillé, J.A. Technical Evaluation of the COBAS EGFR Semiquantitative Index (SQI) for Plasma cfDNA Testing in NSCLC Patients with EGFR Exon 19 Deletions. *Diagnostics* **2021**, *11*, 1319. [CrossRef]
114. Ryu, J.-S.; Lim, J.H.; Lee, M.K.; Lee, S.J.; Kim, H.-J.; Kim, M.J.; Park, M.H.; Kim, J.S.; Nam, H.-S.; Park, N.; et al. Feasibility of Bronchial Washing Fluid-Based Approach to Early-Stage Lung Cancer Diagnosis. *Oncologist* **2019**, *24*, e603–e606. [CrossRef] [PubMed]
115. Otake, S.; Goto, T.; Higuchi, R.; Nakagomi, T.; Hirotsu, Y.; Amemiya, K.; Oyama, T.; Mochizuki, H.; Omata, M. The Diagnostic Utility of Cell-Free DNA from Ex Vivo Bronchoalveolar Lavage Fluid in Lung Cancer. *Cancers* **2022**, *14*, 1764. [CrossRef] [PubMed]
116. Nair, V.S.; Hui, A.B.-Y.; Chabon, J.J.; Esfahani, M.S.; Stehr, H.; Nabet, B.Y.; Zhou, L.; Chaudhuri, A.A.; Benson, J.; Ayers, K.; et al. Genomic Profiling of Bronchoalveolar Lavage Fluid in Lung Cancer. *Cancer Res.* **2022**, *82*, 2838–2847. [CrossRef] [PubMed]
117. Huang, R.; Xu, X.; Li, D.; Chen, K.; Zhan, Q.; Ge, M.; Zhou, X.; Liang, X.; Guan, M. Digital PCR-Based Detection of EGFR Mutations in Paired Plasma and CSF Samples of Lung Adenocarcinoma Patients with Central Nervous System Metastases. *Target. Oncol.* **2019**, *14*, 343–350. [CrossRef]

cancers

MDPI

Review

Gut Microbiota Are a Novel Source of Biomarkers for Immunotherapy in Non-Small-Cell Lung Cancer (NSCLC)

Teresa Del Giudice [1,*], Nicoletta Staropoli [2], Pierfrancesco Tassone [2], Pierosandro Tagliaferri [2] and Vito Barbieri [1]

1 Department of Hematology-Oncology, Azienda Ospedaliera Renato Dulbecco, 88100 Catanzaro, Italy; vitobarbieri@yahoo.it
2 Department of Experimental and Clinical Medicine, Magna Graecia University, 88100 Catanzaro, Italy; nicolettastaropoli@gmail.com (N.S.); tassone@unicz.it (P.T.); tagliaferri@unicz.it (P.T.)
* Correspondence: teresadelgiudice87@gmail.com

Simple Summary: Lung cancer is the most frequent cause of cancer-related death. Unfortunately, only 30% of patients treated with immunotherapy gain any benefit; it is, therefore, important to increase the number of patients who can benefit from immunotherapy. Biomarkers can help clinicians to reach this target and the gut microbiota is a potentially excellent source of predictive factors. All conditions that modify the gut microbiota may influence cancer onset and progression, its prognosis, and response to immunotherapy, with a relevant impact in clinical practice.

Abstract: Despite the recent availability of immune checkpoint inhibitors, not all patients affected by Non-Small-Cell Lung Cancer (NSCLC) benefit from immunotherapy. The reason for this variability relies on a variety of factors which may allow for the identification of novel biomarkers. Presently, a variety of biomarkers are under investigation, including the PD1/PDL1 axis, the tumor mutational burden, and the microbiota. The latter is made by all the bacteria and other microorganisms hosted in our body. The gut microbiota is the most represented and has been involved in different physiological and pathological events, including cancer. In this light, it appears that all conditions modifying the gut microbiota can influence cancer, its treatment, and its treatment-related toxicities. The aim of this review is to analyze all the conditions influencing the gut microbiota and, therefore, affecting the response to immunotherapy, iRAEs, and their management in NSCLC patients. The investigation of the landscape of these biological events can allow for novel insights into the optimal management of NSCLC immunotherapy.

Keywords: NSCLC; lung cancer; gut microbiota; immunotherapy; immune checkpoint inhibitors

1. Introduction

Lung cancer (LC) is one of the most common causes of cancer-related deaths worldwide. In the past years, new therapeutic approaches have been discovered and, presently, from 15 to 30% of Non-Small-Cell Lung Cancer (NSCLC) patients survive. These patients, in the absence of driver mutations, are treated with immune checkpoint inhibitors (ICIs), alone or in combination, or with chemotherapy. Among ICIs, monoclonal antibodies (mabs) against CTLA4 (cytotoxic T lymphocyte-associated protein 4) and against PD1 (programmed death type 1) or its ligand PDL1 (programmed death ligand type 1) can be used. Ipilimumab, Nivolumab, Pembrolizumab, Cemiplimab, Durvalumab, and Atezolizumab are at present the ICIs used in clinical practice [1]. Not all NSCLC patients benefit from immunotherapy at the beginning of the treatment, while others progress after the initial response to the treatment. At present, a suitable predictive marker for immunotherapy is not available. PDL1, as part of the PD1/PDL1 axis, seems to have a role in predicting the response to immunotherapy, but its expression is inducible and editable by different factors, and, for this reason, its role is challenging, even if it is still the unique predictive biomarker for patient selection by regulatory agencies [2].

Biomarker identification still represents an open challenge for the identification of immunotherapy predictors of response. Examples of different biomarkers include: the neutrophil-to-lymphocyte ratio, tumor-infiltrating lymphocytes, tumor mutation burden, and the gut microbiota [1–3]. This latter is composed by all commensal microorganisms present in our gastrointestinal tract, including bacteria, fungi, viruses, and protozoans, while the term microbiome refers to the total genetic material possessed by the microbiota. Alterations in the composition of this microbiota correlate with some diseases, like inflammatory intestinal or metabolic diseases. In the last decades, the gut microbiota has emerged as a crucial player in immunosurveillance and in both cancer onset and progression. In particular, it is possible to hypothesize a gut–lung axis; this theory could explain the correlation between the gut microbiota and an active immune response in LC. The advent of Next-Generation Sequencing (NGS) has allowed for the extensive investigation of the gut microbiome and has correlated it with cancer onset and response to immunotherapy [4].

The correlation between the gut microbiota and immune response in LC could be a potential biomarker in immunity against cancer and particularly in NSCLC patients. It is important to understand how basal microbiota diversity among different patients could affect prognosis and how the gut microbiota could be modified, and how this might change the immune response, and therefore impact the survival of NSCLC patients [5,6]. If this correlation is true, changes to the gut microbiota can potentially improve immunotherapy response, reduce immunotherapy-related adverse events (IRAEs), and prolong survival with immunotherapy treatment [6].

At this time, is important to discover the factors which might influence the gut microbiota, and particularly antibiotics and/or other agents used for different diseases (proton pump inhibitor, antidiabetics as insulin) in cancer patients and specifically in NSCLC patients [7].

The gut microbiota will be the candidate, given its role in cancer development and immunity, as a new cancer modulator; the objective of this review is to explain, from the current literature, the role of the gut microbiota in LC and how its modulation can improve cancer immunotherapy and IRAE management [6] (Figure 1).

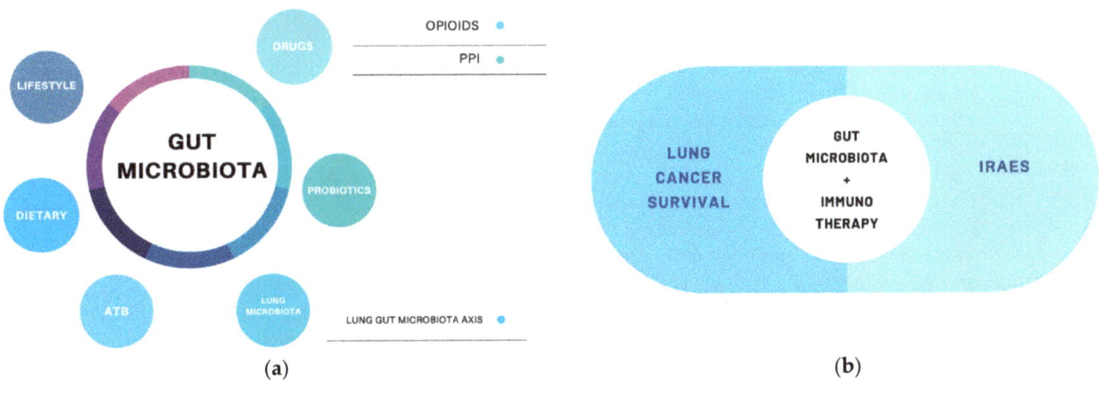

Figure 1. (**a**) Factors involved in microbiota modulation [1–3]. (**b**) The gut microbiota's relevance in LC [7–9].

The gut microbiota can be modified by the following different factors: lifestyle, dietary habits, some drugs such as opioids, antibiotics (ATB), PPis, and interactions with other human cells. All this conditions could also be involved, through the modulation of the microbiota, with lung cancer prognosis, survival, immunotherapeutic efficacy, and IRAE development.

2. Gut Microbiota and NSCLC

2.1. Gut–Lung–Microbiota Axis

The lung microbiota has not been as investigated as the gut microbiota, but its role in different respiratory diseases appears clear [10]. Compared to the gut microbiota, the lung microbiota is smaller, but not less important. The gut microbiota, lung microbiota, and other sites in the human host, in which there are Bacteroides and other elements, are defined as microbial communities [3]. The lung microbiota is composed of *Staphylococcus*, *Streptococcus*, and *Lactobacillus*, followed by *Proteobacteria* and *Actinobacteria*, a microbiota composition similar, in healthy patients, to the gut microbiota. The lung microbiota undergoes changes during inflammation and interacts with metabolites and other pathogens both external and internal to the host [2].

The correlation between the lung and gut microbiotas could depend on the similarity in the mucosa microenvironment, characterized by the same interactions between the microbiota and the immune system in the mucosal immune system (MIS). The MIS is the most important link between these two microbial systems and underlies the participation of the immune system and peptide and protein secretion, such as IgA and metabolite production [3].

Microbiota lung homeostasis is controlled not only within the lung but also by interactions with other organs, and particularly the gut microbiota. The gut microbiota and lung microbiota are linked in many ways: the lymphatic and blood circulation system through the gut microbiota could induce many respiratory diseases, such as asthma, respiratory infection, and others. The gut–lung–microbiota axis is a unique complex that maintains homeostasis; the alteration of this condition can lead to cancer development, tissue damage, and susceptibility to infections [11].

2.2. Gut Microbiota Composition, Anti-Tumor Activity, and Antibiotics

The gut microbiota composition seems to be more heterogeneous among different individuals due to different diets, genetic heritages, lifestyles, medical expositions, and other factors. It is clear that its composition correlates with many diseases, including autoimmune disease, inflammatory disease, and cancer [12]. The microbiota composition has an impact on disease pathogenesis, disease prognosis, and the response to therapy. All of these also depend on other factors; for example, it is described that Helicobacter infections are strongly related with gastric adenocarcinoma, but protective for Barret Esophagus development [3,13]. Some data demonstrate that a microbiota enriched with some bacteria, such as Akkermansia muciniphila and Ruminococcaceae, correlates with a more favorable outcome in melanoma and NSCLC patients than in head and neck patients, for which the same gut microbiota composition does not modify survival [13]. The presence of Phascolarbacterium is linked to a prolonged Progression-Free Survival (PFS) in NSCLC patients with treatment, while a microbiota enriched with Dialister bacteria in NSCLC patients has a worse prognosis [7]. Patients with a heterogenous gut microbiota composition at baseline have a better prognosis than those patients with a poor heterogeneous composition of microbiota [8]. It appears clear how all conditions modulating the gut microbiota, either with reduced variability or the elimination of good bacteria, can have a negative impact on the prognosis or treatment efficacy for LC patients, suggesting how relevant it is to learn how to modulate them [14].

Recently, some authors have demonstrated, through Mendelian Randomization, a potential correlation among gut microbiota phyla and lung carcinoma subtypes [15]. Three groups of protective microbiotas for the development of NSCLC and nine microbiota groups as risk factors have been identified. However, only one protective intestinal microbiota for the development of small-cell lung cancer (SCLC) and six groups of intestinal microbiotas potentially causing SCLC have been identified. The same authors have just identified some gut microbiota phyla predisposed to lung adenocarcinoma or squamous lung carcinoma. These findings, along with information from the retrospective trial, confirm the correlation between microbiota and lung cancer development, and are also linked to other conditions [15].

LC patients are sometimes treated with antibiotics. There is evidence that the exposure to antibiotics in the first days of life can modify microbiota characteristics, making children susceptible to future inflammatory and autoimmune diseases as compared to non-exposed children. This observation is due to the modification of the gut microbiota composition for months and sometimes for years [16]. It is not surprising, therefore, that antibiotic exposition correlates with cancer onset and progression or immunotherapeutic efficacy [17–19].

Potentially, antibiotics can, through the modulation and changing of the composition of the microbiota, reduce and alter the immunotherapeutic activity in LC patients [20]. Numerous data exist in the literature to support how antibiotics, through gut microbiota modulation, could have a negative impact on immune checkpoint activity and chemotherapy activity in cancer and NSCLC patients, and there is evidence that, despite microbiota alterations during antibiotic therapy, no changes in immunotherapeutic efficacy occurred thanks to the ability of the microbiota to return to baseline conditions. The use of antibiotics during immunotherapy in cancer patients could be correlated with primary or secondary immune resistance and, considering that 15–30% of NSCLC patients are treated with antibiotics in clinical practice, the problem is relevant [18].

Studies in mice models demonstrate that the anti-CTLA4 efficacy in cancer patients depends on the gut microbiota composition. A gut microbiota enriched with Bacteroides fragilis or Bacteroides Thetaiotaomicron through polysaccharide products and Th1 response-inducing dendritic cell maturation is related to an improvement in the effectiveness of anti-CTLA4 therapy; moreover, this effectiveness is restored by diet or the oral supplementation of this bacterium. At the same time, Bifidobacterioides improve anti-PD1 efficacy in melanoma-affected mice, while mice with a different gut microbiota, but undergoing fecal Bifidobacterioides-based microbiota transplantation, become anti-PD1 responders thanks to an increased T cell anti-tumor response [6].

Since 2017, several retrospective studies about the antibiotic effect on lung cancer patients treated with immunotherapy have been reported. The data from these trials are not completely consistent and the antibiotic role remains unclear. Particularly, some studies demonstrate a correlation between antibiotic exposure and worse prognosis and reduced immunotherapy efficacy, while other studies do not demonstrate such correlation, which could be associated to several factors, some associated to the host microbiota and host characteristics, and others to a selection study bias (Table 1).

The time to antibiotic exposure and different antibiotic types is reported in Figure 2 in order to describe the detrimental effect on the immunotherapeutic efficacy due to microbiome alteration. Figure 3 reports on the different antibiotics evaluated in retrospective studies with respect to correlations to immunotherapy efficacy.

The role of antibiotics in cancer treatment is controversial: antibiotics are essential for managing infections but their indiscriminate use can disrupt the gut microbiota, potentially impairing the effectiveness of immunotherapies in cancer patients, including those with NSCLC. However, we also acknowledge the conflicting data and the complex factors at play, including the ability of the microbiota to return to baseline conditions after antibiotic treatment and the varied impacts depending on the patient's unique microbial composition [19,20].

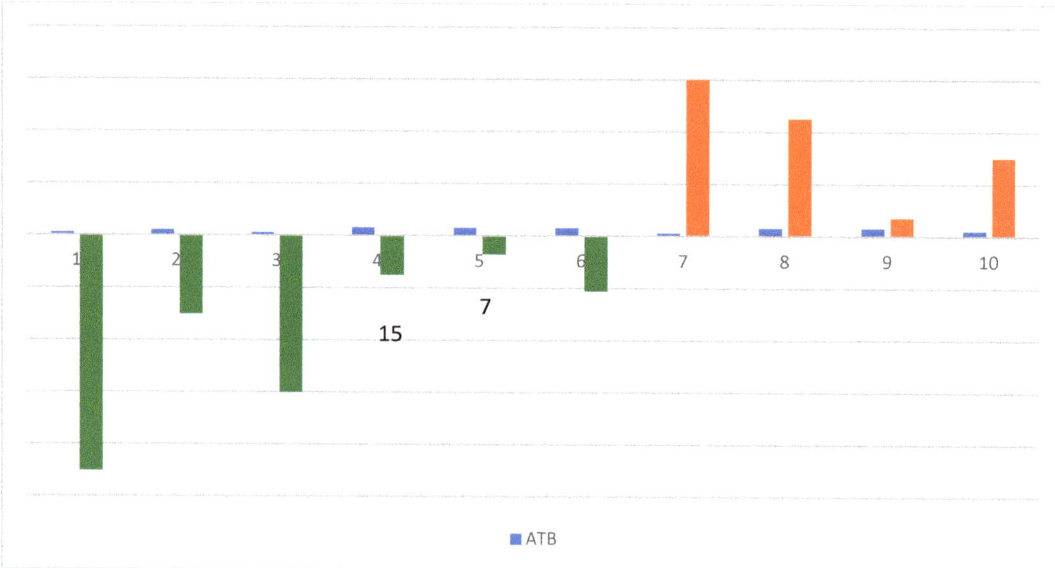

Figure 2. This figure explains the correlation between antibiotics (ATBs), shown with the blue bar, and its exposure time before (green bar) and after (orange bar) ICI (immune checkpoint inhibitor) treatment. The role of Fluoroquinolones (1) in ICI efficacy is evaluated at 90 and 60 days before the start of treatment. Fluoroquinolones, Vancomycin, Piperacillin–Tazobactam, Clindamycin, and other not-specified ATBs (2, 3) are evaluated at an exposure time of 30 days before and after the start of the ICIs. Finally, a group of not-specified ATBs (5, 6, 7, 9, 10) are evaluated in many exposure timings: 21, 15, and 7 days before and 45 and 7 days after.

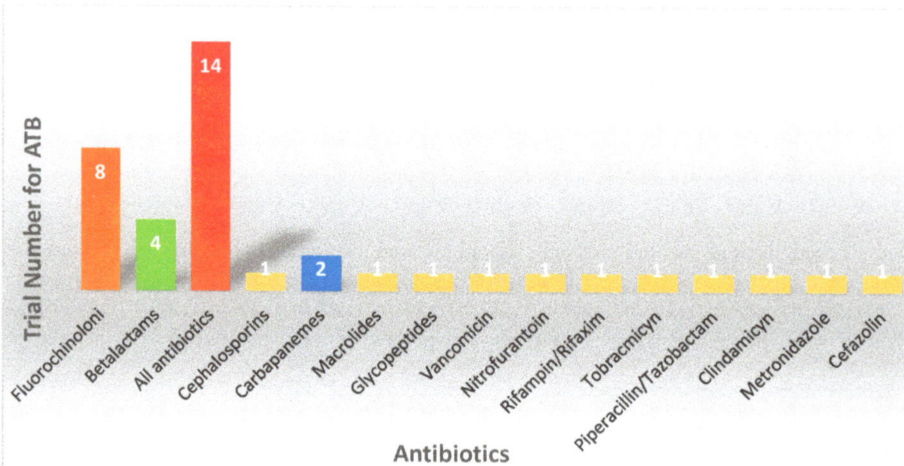

Figure 3. The different antibiotic (ATB) types evaluated in retrospective trials identified in the literature are reported. A total of 14 trials did not specify the ATB used (red bar), while Fluoroquinolones are the most represented (orange bar). The other ATBs are evaluated in poor retrospective trials (green bar, blue bar, and yellow bars). Several ATBs are evaluated, but the resulting data are controversial.

Table 1. Retrospective studies from 2017 to 2023, including lung cancer patients treated with immunotherapy and those exposed to antibiotics.

Year	Author	Patients	Treatment	Antibiotic Typologies	Antibiotic Exposition Timing	Reference
2017	Kaderbhai et al.	74	Anti-PD1, Nivolumab	Fluoroquinolones	3 months before starting ICIs	[21]
2018	Derosa et al.	239	Anti-PD1, Anti-CTLA4 Monotherapy or combination	Fluoroquinolones, Betalactams	30 days before starting immunotherapy	[22]
2018	Hakozaki et al.	90	Anti-PD1, Nivolumab	Not specified	30 days before starting immunotherapy	[18]
2018	Huemer et al.	30	Anti-PD1, Nivolumab, Pembrolizumab	Not specified	30 days before and after starting immunotherapy	[23]
2019	Zhao S et al.	109	Anti-PD1, Anti-PDL1	Not specified	Not specified	[24]
2019	Kim H et al.	131	Anti-PD1, Anti-PDL1, Anti-CTLA4 Monotherapy or combination	Fluoroquinolones, Betalactams, Cephalosporins	60 days before starting immunotherapy	[19]
2019	Galli et al.	157	Anti-PD1, Anti-PDL1, Anti-CTLA4 Monotherapy or combination	Not specified	Before and during immunotherapy	[25]
2020	PH Lu et al.	340	Anti-PD1, Anti-PD1, Anti-CTLA4 Monotherapy or combination	Fluoroquinolones	30 days before starting immunotherapy	[26]
2020	E Pérez-Ruiz et al.	120	Anti-PD1, Anti-CTLA4 Monotherapy or combination	Not specified	Not specified	[27]
2020	Svaton M et al.	224	Anti-PD1, Nivolumab	Not specified	Not specified	[28]
2020	Chalabi M et al.	757	Anti-PDL1, Atezolizumab	Fluoroquinolones, Carbapanems, Macrolides, Glycopeptides	30 days before and after starting immunotherapy	[29]
2020	Tinsley et al.	64	Anti-PD1	Not specified	15 days before and 45 days after starting immunotherapy	[30]
2020	Kulkarni et al.	140	Anti-PD1	Vancomicyn, Nitrofurantoin, Rifampin, Rifaximin, Tobramicyn	30 days before and after starting immunotherapy	[31]
2021	Geum et al.	140	Anti-PD1, Nivolumab	Not specified	Not specified	[32]

Table 1. *Cont.*

Year	Author	Patients	Treatment	Antibiotic Typologies	Antibiotic Exposistion Timing	Reference
2021	Cortellini et al.	302	Chemotherapy, Immunotherapy	Not specified	7 days before and after starting immunotherapy	[33]
2021	Giordan et al.	65	Anti-PD1, Anti-CTLA4, Monotherapy or combination	Not specified	60 days before starting immunotherapy	[34]
2021	Cortellini et al.	950	Anti-PD1, Pembrolizumab	Piperacillin-Tazobactam, Clindamycin, Metronidazole, Meropenem	30 days before starting immunotherapy	[35]
2021	Hamada et al.	69	Anti-PD1	Not specified	21 days before starting immunotherapy	[36]
2022	Hopkins et al.	2723	Anti-PDL1, Atezolizumab	Not specified	30 days before starting immunotherapy	[37]
2022	Barbarosa et al.	140	Anti-PD1, Anti-PD1, Anti-CTLA4, Monotherapy or combination	Fluoroquinolones, Betalactams	2 months before and after starting immunotherapy	[17]
2022	Nyein et al.	256	Anti-PD1, Anti-PDL1, Anti-CTLA4, Monotherapy or combination	Fluoroquinolones, Cefazolin, Azithromicin	60 days before and after starting immunotherapy	[38]
2022	Qiu H et al.	148	Anti-PD1, Anti-PDL1, Chemotherapy	Fluoroquinolones, Betalactams	60 days before and after starting immunotherapy	[39]
2023	Manning-Bennett et al.	2724	Anti-PDL1, Atezolizumab, Alone or in combination with chemotherapy	Not specified	Not specified	[40]
2023	Vihinen et al.	199	Anti-PD1, Anti-PDL1	Not specified	3 months before and 1 months after starting immunotherapy	[41]

A retrospective analysis of 70 NSCLC patients treated with ICIs investigated the gut microbiota diversity in patients with an OS (Overall Survival) >12 months and <12 months. The gut microbiota of long survivors was enriched with *Lachnospiraceae*, a member of *Clostridiale*, with increased circulating CD4 and CD8 T cell and CD8 T cell-infiltrating tumors. The study confirmed that a diversified microbiota correlates with a better prognosis and that the use of antibiotics also reduces the diversity of the gut microbiota [42].

The antibiotic exposure from 60 days before starting immunotherapy and 30 days after the last immunotherapy correlated with a poor prognosis and immunotherapy resistance [14]. Scarce information on the antibiotic type, antibiotic route, and the duration of therapy are available. Greater information would help us to use microbiota modulation to improve immunotherapy in cancer. As just mentioned, the microbiota has the ability to return to homeostasis after antibiotic damage: different time frames of the reconstitution of the baseline status might explain the unclear data on the prognosis and exposure to antibiotics in different patients with lung cancer [11,13,16].

There is evidence about the relevance of the gut microbiota on the efficacy and toxicity of chemotherapy and the intestinal microbiota; the maximum effectiveness of chemotherapy in treated cancer is mediated by a good balance between the intestinal microbiota and the immune system [3]. The relevance of the gut microbiota in chemotherapy management and efficacy represents today an important issue considering NSCLC treatment based not only on ICI monotherapy but also on the chemo–immunotherapy association [43,44].

It is clear that the microbiota have a role in cancer from onset to therapy response, but the knowledge of the conditions implicated in the change in the intestinal microbiota that should be avoided unless necessary, such as the use of antibiotics, remains to be clarified [8,13,20].

2.3. Gut Microbiota and Probiotic Use

Since the modulation of the gut microbiota can modify the effectiveness of immunotherapy in cancer patients, finding a way to remodulate the microbiota and restore it so to improve the effectiveness of immunotherapy could be an option for our patients. Oral probiotic supplements have been associated with the improved efficacy of immunotherapy for cancer patients [45,46]. Probiotics are a bacterial strain that do not alter antibiotic resistance; they reach the colon and the entire intestine, where they carry out their metabolism. Probiotics may be safe in animals, resistant to acids, and able to colonize the intestine [3]. Probiotics can modulate the gut microbiota by (a) modifying humoral, cellular, and innate immunity; (b) improving NK (Natural Killer) immune activity; (c) macrophage and neutrophil activation; (d) IgA secretion; (e) inflammatory cytokine inhibition. Moreover, probiotics can modulate chemotherapy toxicity and iRAEs development. To date, the more used probiotics are composed of *Bifidobacterium* spp. and *Lactobacillus* spp.; their role appears marginal in NSCLC patients treated with immunotherapy because they are not specifically chosen for this reason. The discovery of probiotics able to modulate the immune response with immunotherapy and able to prevent and improve iRAEs and chemotherapy toxicities [3], or limit the unfavorable effect of antibiotics [47], could be very interesting.

With the recent advent of Next-Generation Sequencing (NGS), new species of probiotics have been identified and called next-generation probiotics (NGPs), which are presently under evaluation in the context of specific diseases. NGPs are able to modulate the gut microbiota to improve the immunotherapy and control iRAEs. *Eubacterium limosum*, *E. hirae*, *Enterococcus faecium*, *Collinsella aerofaciens*, and *Burkholderia cepacia* appear to have promising efficacy in this setting [3].

A recent metanalysis underlined the role of probiotics and their effect on the survival of NSCLC patients treated with immunotherapy. This study demonstrated a positive correlation between probiotic exposure and OS and PFS. There was no correlation with ORR, but it can be demonstrated by the types of studies and the sample size [46].

Probiotic use can improve immunotherapeutic efficacy through the modulation of inflammation. The data about this correlation are limited by study design, cancer types, sample size, and the duration of oral probiotic implementation [46]. Certainly, the use of probiotics increases the heterogeneity of the intestinal microbiota, which is the basis of a better prognosis, a better response, and fewer iRAEs in cancer patients [9]. It is important to understand whether a single bacterial species can modulate the entire microbiota or if the presence of different bacterial species is necessary at the same time. It is also important to understand the amount of single species that are needed for a beneficial effect to a well-balanced gut microflora [8].

2.4. Gut Microbiota and iRAEs

In the era of immunotherapy, in which cancer patients undergo ICI treatment for long periods, toxicity becomes the most relevant issue. Few data have evidenced correlations between the gut microbiota composition and iRAEs: some bacterial strains seem to be protective against iRAEs, while other strains could increase the risk of iRAEs [48].

In a retrospective analysis, a link between the gut microbiota, antibiotic exposure, and iRAEs was investigated: a correlation between antibiotic use and iRAEs was not observed, but a gut microbiota enriched with Akkermansia Mucinipihila correlated with fewer iRAEs [42].

Microbiota diversity correlates with the development of iRAEs during immunotherapy. As demonstrated for the survival and prognosis, it appears that patients with a low diversity of gut microbiota exhibit skin iRAEs more often than patients with a gut microbiota enriched with many bacterial types [1,9].

In the gut microbiota composition of mouse models and cancer patients, the presence of Bacterioides and other microbes implicated in vitamin B production seem to be protective against colitis development during immunotherapy; the detection of some types of Bacterioides in the gut microbiota can help us to predict colitis presentation during immunotherapy, while bacterial supplementation with *Bacteroidales* and *Burkholderiales* can improve colitis, particularly in cancer patients treated with antibiotics during immunotherapy [48].

The gut microbiota composition could be implicated in skin iRAEs during immunotherapy in cancer. This theory is supported by modulatory effects on the skin mediated by the gut microbiota through immunity regulation and metabolite products. There is evidence of the improvement of dermatitis with the use of oral Lactobacillales and Bifidobacteriales. The oral use of Bifidobacteriodes in humans reduces inflammatory markers such as peptide C and TNF-alpha, with an improvement in psoriasis [9]. It is necessary to gain more information and data about the gut microbiota and iRAEs to improve immunotherapy management [9].

2.5. Other Conditions Modifying Gut Microbiota

In addition to the role of antibiotics and probiotics in the modulation of the intestinal microbiota for the improvement in the efficacy of ICIs in LC patients, there are other molecules that are potentially implicated and that have been recently studied, with more hypotheses and a few relevant points that need to be confirmed by other studies.

One of the retrospective studies in the literature about gut microbiota modulation mediated by drugs, describing 132 lung patients treated with immunotherapy, presents a shorter PFS and OS when exposed to opioids through the impairment of T cell function, upregulating Treg cells modulating the gut microbiota; the opioid exposition did not correlate with different iRAE incidence. This exploratory data could be important in considering the high percentage of cancer and LC patients exposed to opioid drugs for pain management, and this research area needs to be focused on. It could be very important to acquire data to confirm these results, considering that pain is one of the most important causes of quality-of-life reduction, and that this could be correlated to a worse response to immunotherapy. This could mean, as suggested in a small study, that worse PFS and OS are not related to opioid exposure but to a poor Performance Status (PS) [49].

In 2021, one of the first retrospective papers about the correlation between PPi (Proton Pump Inhibitor) exposition and LC patient survival treated with immunotherapy was presented. Lung cancer patients treated with ICIs and exposed to PPis have a 28% increased risk of death and a shorter survival compared to unexposed patients; this relationship is not observed in subgroup patients exposed to PPis but treated with chemotherapy only. The PPi/survival association is consistent considering the sample size, and the data are confirmed when the exposition window changes. The possible cause of this negative relationship can be linked to PPi-mediated acid reduction, which alters the intestinal microbiota [50]. Recently, another author published a metanalysis regarding PPi exposition and survival in cancer patients treated with immunotherapy that confirmed the negative relationship; also, in this case, the relationship could be modified by the poor quality of life in cancer patients exposed to PPis [51]. In both works presented, there are many limitations;

one of the limitations is the presence of retrospective trials, as well as the PPi exposition (dose, time, and PPi use). However, other retrospective studies and metanalyses did not confirm the relationship between PPis and immunotherapy efficacy [50,51].

3. Future Directions

Despite the recent huge impact of cancer immunotherapy, only 30% of lung cancer patients gain any benefit. This condition depends on different factors that are intrinsic and extrinsic to LC patients [11]. For these reasons, it is relevant to identify biomarkers to select responder patients, improve the immunotherapeutic efficacy, and manage iRAEs. Among the well-known biomarkers, such as PDL1 and the PD1–PDL1 axis, TMB (tumor mutational burden), and others, the microbiota potentially appears to be the most relevant predictor of immunotherapeutic efficacy. There are many factors that can influence and modulate the microbiota, and all of them could play an important role in cancer development as well as in the efficacy and toxicity of immunotherapy [48,52].

Interestingly, among the numerous factors implicated, several authors have demonstrated that different bacteria defined as "intratumoral microorganisms" appear to be related to the progression of different types of cancer. Bacteria were detected both in immune cells and tumor cells, determining a strong tumor microenvironment involvement. Recent data confirmed a potential prognostic role of "intratumoral microorganisms" and the potential therapeutic implications [53].

The evaluation of the "intratumor microbiome" was conducted with several methodologies based on the deep sequencing of the amplicons of prokaryote and eukaryote rRNA genes or metagenome-based shotgun sequencing (WMS). Moreover, immunohistochemistry (IHC), fluorescent in situ hybridization (FISH), and D-alanine-based methods have allowed authors to evaluate the presence and characteristics of specific tumor bacteria. Finally, the QIIME 2 method was described as a reproducible "intratumoral" microbiome data analysis system [54].

The intestinal microbiota plays a role in anti-tumor immunotherapy and, therefore, all conditions that modify the intestinal microbiota, reducing its diversity, might potentially modify the efficacy and also the occurrence of iRAEs. We have relevant information about the microbiota–immunotherapy relationship, but we have no clear-cut information that presently allows us to improve our approach to cancer therapy management [10,48]. It is not only important to know the composition of the microbiota but also the interplay between all players of the intestinal microbiota and the microenvironment that surrounds it, taking into account that all elements are regulated by different intrinsic and extrinsic factors, including the genetic host features, diet, lifestyle, age, and concomitant medication [10]. This is why the microbiota cannot be considered a weapon to be used, but rather a biological entity whose relevance is still underestimated.

More controversial findings focus attention on the potential role of microbiota transplantation in order to influence the tumor microenvironment and the "intratumoral microbiome". However, no evidence supports this practice to improve immunotherapeutic efficacy [55,56].

At present, what we know about the microbiota and its relationship with cancer immunotherapy is derived from retrospective analyses or meta-analyses based on retrospective studies. All derived information could be defined as hypothesis-generating and not clear-cut findings, which can be derived only by prospective and randomized trials. For example, we know that the exposure to antibiotic exposition can adversely modulate the gut microbiota, thus reducing the efficacy of immunotherapy [25]. However, we do not know which class of antibiotics are implicated, the timing of the exposure, or the differences in the routes of administration; at the same time, we know that not all immunotherapy-treated patients have a worse prognosis related to exposure, and many sources of biases can be identified. The same difficulty occurs in understanding the gut microbiota–immunotherapeutic efficacy relationship with opioids and PPis, whose role has always been investigated retrospectively.

Therefore, the more relevant question is the following: what is the next step for translating microbiota knowledge into LC management?

Again, a major role is played by the retrospective study design versus prospective trials, which are pivotal trials for the adequate stratification of the administered drugs.

Considering the huge world represented by the microbiota, it is clear that the current methodologies are not completely applicable and new methodologies might be considered in the future. The machine learning (ML) recently used in cancer immunotherapy for predicting the development of iRAEs in cancer patients during immunotherapy [57] could help us to identify the role of the exposure to many drugs through microbiota modulation. Instead of predicting iRAEs, modifying the microbiota conditions depends on various elements which are not simple to manage in clinical practice: for this reason, AI (artificial intelligence) and ML could help us in our objective. There are several reports about using ML and microbiota. From the data on the microbiota produced by omics-based methods (metagenomics, meta transcriptomics, and metabolomics), ML can predict and find new non-theoretically inferred information to help us to increase the efficacy and reduce iRAEs [58,59].

Alongside AI and ML, which could be of help in the future on this topic, it needs to be highlighted that microbiota and microbiomes not are the same in all populations, in all persons, and in all cancers. We know today about LC microbiota and that its correlation with immunotherapy efficacy is different from head and neck microbiota, SCLC microbiota, or bladder microbiota. It is important to make clear that the relevance of microbiota in LC does not necessarily translate in other cancers which grow in a different tumor microenvironment.

4. Conclusions

To date, the intestinal microbiota appears to be an important biomarker of immunotherapy in LC. However, taking into account the complexity of the whole scenario, it is necessary to make a great effort to gain more functional information. It is also necessary that the data available are confirmed by more robust trials or in an alternative way through new methodologies, such as AI. While waiting for all of this information, what we can tell from the microbiota, cancer prevention, and immunotherapy is that it is important to limit the use of antibiotics, PPis, and opioids, which can be used if necessary, and that probiotic use could improve some conditions such as colitis management and other iRAEs. More adequate probiotics might be identified with NGS and NPS. A healthy diet and a better lifestyle might be, in any case, the mainstream proposal to our patients.

Author Contributions: T.D.G. and V.B. designed the paper. T.D.G. wrote the paper and constructed the images and tables. P.T. (Pierfrancesco Tassone), P.T. (Pierosandro Tagliaferri), and N.S. revised the final version of the manuscript. All authors have read and agreed to the published version of the manuscript.

Funding: This research received no external funding.

Conflicts of Interest: The authors declare no conflicts of interest.

References

1. Rocco, D.; Della Gravara, L.; Ragone, A.; Sapio, L.; Naviglio, S.; Gridelli, C. Prognostic Factors in Advanced Non-Small Cell Lung Cancer Patients Treated with Immunotherapy. *Cancers* **2023**, *15*, 4684. [CrossRef]
2. Shah, H.; Ng, T.L. A Narrative Review from Gut to Lungs: Non-Small Cell Lung Cancer and the Gastrointestinal Microbiome. *Transl. Lung Cancer Res.* **2023**, *12*, 909–926. [CrossRef]
3. Pizzo, F.; Maroccia, Z.; Hammarberg Ferri, I.; Fiorentini, C. Role of the Microbiota in Lung Cancer: Insights on Prevention and Treatment. *Int. J. Mol. Sci.* **2022**, *23*, 6138. [CrossRef] [PubMed]
4. Botticelli, A.; Vernocchi, P.; Marini, F.; Quagliariello, A.; Cerbelli, B.; Reddel, S.; Del Chierico, F.; Di Pietro, F.; Giusti, R.; Tomassini, A.; et al. Gut Metabolomics Profiling of Non-Small Cell Lung Cancer (NSCLC) Patients under Immunotherapy Treatment. *J. Transl. Med.* **2020**, *18*, 49. [CrossRef]
5. Zhou, Y.; Liu, Z.; Chen, T. Gut Microbiota: A Promising Milestone in Enhancing the Efficacy of PD1/PD-L1 Blockade Therapy. *Front. Oncol.* **2022**, *12*, 847350. [CrossRef]

6. Cheng, W.Y.; Wu, C.-Y.; Yu, J. The Role of Gut Microbiota in Cancer Treatment: Friend or Foe? *Gut* **2020**, *69*, 1867–1876. [CrossRef]
7. Rahman, M.M.; Islam, M.R.; Shohag, S.; Ahasan, M.T.; Sarkar, N.; Khan, H.; Hasan, A.M.; Cavalu, S.; Rauf, A. Microbiome in Cancer: Role in Carcinogenesis and Impact in Therapeutic Strategies. *Biomed. Pharmacother.* **2022**, *149*, 112898. [CrossRef]
8. Jin, Y.; Dong, H.; Xia, L.; Yang, Y.; Zhu, Y.; Shen, Y.; Zheng, H.; Yao, C.; Wang, Y.; Lu, S. The Diversity of Gut Microbiome Is Associated with Favorable Responses to Anti–Programmed Death 1 Immunotherapy in Chinese Patients With NSCLC. *J. Thorac. Oncol.* **2019**, *14*, 1378–1389. [CrossRef]
9. Bredin, P.; Naidoo, J. Correction to: The Gut Microbiome, Immune Check Point Inhibition and Immune-related Adverse Events in Non-small Cell Lung Cancer. *Cancer Metastasis Rev.* **2022**, *41*, 347–366. [CrossRef]
10. Zhang, H.; Xu, Z. Gut-Lung Axis: Role of the Gut Microbiota in Non-Small Cell Lung Cancer Immunotherapy. *Front. Oncol.* **2023**, *13*, 1257515. [CrossRef]
11. Georgiou, K.; Marinov, B.; Farooqi, A.A.; Gazouli, M. Gut Microbiota in Lung Cancer: Where Do We Stand? *Int. J. Mol. Sci.* **2021**, *22*, 10429. [CrossRef] [PubMed]
12. Souza, V.G.P.; Forder, A.; Pewarchuk, M.E.; Telkar, N.; de Araujo, R.P.; Stewart, G.L.; Vieira, J.; Reis, P.P.; Lam, W.L. The Complex Role of the Microbiome in Non-Small Cell Lung Cancer Development and Progression. *Cells* **2023**, *12*, 2801. [CrossRef]
13. Vernocchi, P.; Gili, T.; Conte, F.; Del Chierico, F.; Conta, G.; Miccheli, A.; Botticelli, A.; Paci, P.; Caldarelli, G.; Nuti, M.; et al. Network Analysis of Gut Microbiome and Metabolome to Discover Microbiota-Linked Biomarkers in Patients Affected by Non-Small Cell Lung Cancer. *Int. J. Mol. Sci.* **2020**, *21*, 8730. [CrossRef]
14. Xin, Y.; Liu, C.-G.; Zang, D.; Chen, J. Gut Microbiota and Dietary Intervention: Affecting Immunotherapy Efficacy in Non–Small Cell Lung Cancer. *Front. Immunol.* **2024**, *15*, 1343450. [CrossRef]
15. Ma, Y.; Deng, Y.; Shao, T.; Cui, Y.; Shen, Y. Causal Effects of Gut Microbiota in the Development of Lung Cancer and Its Histological Subtypes: A Mendelian Randomization Study. *Thorac. Cancer* **2024**, *15*, 486–495. [CrossRef]
16. Francino, M.P. Antibiotics and the Human Gut Microbiome: Dysbioses and Accumulation of Resistances. *Front. Microbiol.* **2016**, *6*, 1543. [CrossRef]
17. Barbosa, C.M.-M.; Lletí, A.C.C.; Sánchez, R.P.; Román, C.D.; Alonso, P.T.; González, B.F. Impact of the Use of Antibiotics on the Clinical Response to Immune Checkpoint Inhibitors in Patients with Non-Small Cell Lung Cancer. *Rev. Esp. Quimioter.* **2022**, *35*, 551–558. [CrossRef]
18. Hakozaki, T.; Okuma, Y.; Omori, M.; Hosomi, Y. Impact of Prior Antibiotic Use on the Efficacy of Nivolumab for Non-Small Cell Lung Cancer. *Oncol. Lett.* **2019**, *17*, 2946–2952. [CrossRef]
19. Kim, H.; Lee, J.E.; Hong, S.H.; Lee, M.A.; Kang, J.H.; Kim, I.-H. The Effect of Antibiotics on the Clinical Outcomes of Patients with Solid Cancers Undergoing Immune Checkpoint Inhibitor Treatment: A Retrospective Study. *BMC Cancer* **2019**, *19*, 1100. [CrossRef]
20. Jiang, S.; Geng, S.; Chen, Q.; Zhang, C.; Cheng, M.; Yu, Y.; Zhang, S.; Shi, N.; Dong, M. Effects of Concomitant Antibiotics Use on Immune Checkpoint Inhibitor Efficacy in Cancer Patients. *Front. Oncol.* **2022**, *12*, 823705. [CrossRef]
21. Kaderbhai, C.; Richard, C.; Fumet, J.D.; Aarnink, A.; Foucher, P.; Coudert, B.; Favier, L.; Lagrange, A.; Limagne, E.; Boidot, R.; et al. Antibiotic Use Does Not Appear to Influence Response to Nivolumab. *Anticancer Res.* **2017**, *37*, 3195–3200.
22. Derosa, L.; Hellmann, M.D.; Spaziano, M.; Halpenny, D.; Fidelle, M.; Rizvi, H.; Long, N.; Plodkowski, A.J.; Arbour, K.C.; Chaft, J.E.; et al. Negative Association of Antibiotics on Clinical Activity of Immune Checkpoint Inhibitors in Patients with Advanced Renal Cell and Non-Small-Cell Lung Cancer. *Ann. Oncol.* **2018**, *29*, 1437–1444. [CrossRef]
23. Huemer, F.; Rinnerthaler, G.; Westphal, T.; Hackl, H.; Hutarew, G.; Gampenrieder, S.P.; Weiss, L.; Greil, R. Impact of Antibiotic Treatment on Immune-Checkpoint Blockade Efficacy in Advanced Non-Squamous Non-Small Cell Lung Cancer. *Oncotarget* **2018**, *9*, 16512–16520. [CrossRef]
24. Zhao, S.; Gao, G.; Li, W.; Li, X.; Zhao, C.; Jiang, T.; Jia, Y.; He, Y.; Li, A.; Su, C.; et al. Antibiotics Are Associated with Attenuated Efficacy of Anti-PD-1/PD-L1 Therapies in Chinese Patients with Advanced Non-Small Cell Lung Cancer. *Lung Cancer* **2019**, *130*, 10–17. [CrossRef]
25. Galli, G.; Triulzi, T.; Proto, C.; Signorelli, D.; Imbimbo, M.; Poggi, M.; Fucà, G.; Ganzinelli, M.; Vitali, M.; Palmieri, D.; et al. Association between Antibiotic-Immunotherapy Exposure Ratio and Outcome in Metastatic Non Small Cell Lung Cancer. *Lung Cancer* **2019**, *132*, 72–78. [CrossRef]
26. Lu, P.-H.; Tsai, T.-C.; Chang, J.W.-C.; Deng, S.-T.; Cheng, C.-Y. Association of Prior Fluoroquinolone Treatment with Survival Outcomes of Immune Checkpoint Inhibitors in Asia. *J. Clin. Pharm. Ther.* **2021**, *46*, 408–414. [CrossRef]
27. Pérez-Ruiz, E.; Jiménez-Castro, J.; Berciano-Guerrero, M.-A.; Valdivia, J.; Estalella-Mendoza, S.; Toscano, F.; Rodriguez de la Borbolla Artacho, M.; Garrido-Siles, M.; Martínez-Bautista, M.J.; Villatoro Roldan, R.; et al. Impact of Intestinal Dysbiosis-Related Drugs on the Efficacy of Immune Checkpoint Inhibitors in Clinical Practice. *Clin. Transl. Oncol.* **2020**, *22*, 1778–1785. [CrossRef]
28. Svaton, M.; Zemanova, M.; Zemanova, P.; Kultan, J.; Fischer, O.; Skrickova, J.; Jakubikova, L.; Cernovska, M.; Hrnciarik, M.; Jirousek, M.; et al. Impact of Concomitant Medication Administered at the Time of Initiation of Nivolumab Therapy on Outcome in Non-Small Cell Lung Cancer. *Anticancer Res.* **2020**, *40*, 2209–2217. [CrossRef]
29. Chalabi, M.; Cardona, A.; Nagarkar, D.R.; Dhawahir Scala, A.; Gandara, D.R.; Rittmeyer, A.; Albert, M.L.; Powles, T.; Kok, M.; Herrera, F.G.; et al. Efficacy of Chemotherapy and Atezolizumab in Patients with Non-Small-Cell Lung Cancer Receiving Antibiotics and Proton Pump Inhibitors: Pooled Post Hoc Analyses of the OAK and POPLAR Trials. *Ann. Oncol.* **2020**, *31*, 525–531. [CrossRef]

30. Tinsley, N.; Zhou, C.; Tan, G.; Rack, S.; Lorigan, P.; Blackhall, F.; Krebs, M.; Carter, L.; Thistlethwaite, F.; Graham, D.; et al. Cumulative Antibiotic Use Significantly Decreases Efficacy of Checkpoint Inhibitors in Patients with Advanced Cancer. *Oncologist* **2020**, *25*, 55–63. [CrossRef]
31. Kulkarni, A.A.; Ebadi, M.; Zhang, S.; Meybodi, M.A.; Ali, A.M.; DeFor, T.; Shanley, R.; Weisdorf, D.; Ryan, C.; Vasu, S.; et al. Comparative Analysis of Antibiotic Exposure Association with Clinical Outcomes of Chemotherapy versus Immunotherapy across Three Tumour Types. *ESMO Open* **2020**, *5*, e000803. [CrossRef] [PubMed]
32. Geum, M.J.; Kim, C.; Kang, J.E.; Choi, J.H.; Kim, J.S.; Son, E.S.; Lim, S.M.; Rhie, S.J. Broad-Spectrum Antibiotic Regimen Affects Survival in Patients Receiving Nivolumab for Non-Small Cell Lung Cancer. *Pharmaceuticals* **2021**, *14*, 445. [CrossRef]
33. Cortellini, A.; Ricciuti, B.; Facchinetti, F.; Alessi, J.V.M.; Venkatraman, D.; Dall'Olio, F.G.; Cravero, P.; Vaz, V.R.; Ottaviani, D.; Majem, M.; et al. Antibiotic-Exposed Patients with Non-Small-Cell Lung Cancer Preserve Efficacy Outcomes Following First-Line Chemo-Immunotherapy. *Ann. Oncol.* **2021**, *32*, 1391–1399. [CrossRef] [PubMed]
34. Impact of Antibiotics and Proton Pump Inhibitors on Efficacy and Tolerance of Anti-PD-1 Immune Checkpoint Inhibitors-PubMed. Available online: https://pubmed.ncbi.nlm.nih.gov/34777340/ (accessed on 7 March 2024).
35. Cortellini, A.; Di Maio, M.; Nigro, O.; Leonetti, A.; Cortinovis, D.L.; Aerts, J.G.; Guaitoli, G.; Barbieri, F.; Giusti, R.; Ferrara, M.G.; et al. Differential Influence of Antibiotic Therapy and Other Medications on Oncological Outcomes of Patients with Non-Small Cell Lung Cancer Treated with First-Line Pembrolizumab versus Cytotoxic Chemotherapy. *J. Immunother. Cancer* **2021**, *9*, e002421. [CrossRef]
36. Hamada, K.; Yoshimura, K.; Hirasawa, Y.; Hosonuma, M.; Murayama, M.; Narikawa, Y.; Ariizumi, H.; Ohkuma, R.; Shida, M.; Kubota, Y.; et al. Antibiotic Usage Reduced Overall Survival by over 70% in Non-Small Cell Lung Cancer Patients on Anti-PD-1 Immunotherapy. *Anticancer Res.* **2021**, *41*, 4985–4993. [CrossRef]
37. Hopkins, A.M.; Badaoui, S.; Kichenadasse, G.; Karapetis, C.S.; McKinnon, R.A.; Rowland, A.; Sorich, M.J. Efficacy of Atezolizumab in Patients With Advanced NSCLC Receiving Concomitant Antibiotic or Proton Pump Inhibitor Treatment: Pooled Analysis of Five Randomized Control Trials. *J. Thorac. Oncol.* **2022**, *17*, 758–767. [CrossRef] [PubMed]
38. Nyein, A.F.; Bari, S.; Hogue, S.; Zhao, Y.; Maller, B.; Sha, S.; Gomez, M.F.; Rollison, D.E.; Robinson, L.A. Effect of Prior Antibiotic or Chemotherapy Treatment on Immunotherapy Response in Non-Small Cell Lung Cancer. *BMC Cancer* **2022**, *22*, 101. [CrossRef] [PubMed]
39. Qiu, H.; Ma, Q.-G.; Chen, X.-T.; Wen, X.; Zhang, N.; Liu, W.-M.; Wang, T.-T.; Zhang, L.-Z. Different Classes of Antibiotics Exhibit Disparate Negative Impacts on the Therapeutic Efficacy of Immune Checkpoint Inhibitors in Advanced Non-Small Cell Lung Cancer Patients. *Am. J. Cancer Res.* **2022**, *12*, 3175–3184. [PubMed]
40. Manning-Bennett, A.T.; Cervesi, J.; Bandinelli, P.-A.; Sorich, M.J.; Hopkins, A.M. Prognostic Associations of Concomitant Antibiotic Use in Patients with Advanced NSCLC Treated with Atezolizumab: Sensitivity Analysis of a Pooled Investigation of Five Randomised Control Trials. *Biomedicines* **2023**, *11*, 528. [CrossRef]
41. Vihinen, H.; Jokinen, A.; Laajala, T.D.; Wahid, N.; Peltola, L.; Kettunen, T.; Rönkä, A.; Tiainen, L.; Skyttä, T.; Kohtamäki, L.; et al. Antibiotic Treatment Is an Independent Poor Risk Factor in NSCLC But Not in Melanoma Patients Who Had Received Anti-PD-1/L1 Monotherapy. *Clin. Lung Cancer* **2023**, *24*, 295–304. [CrossRef] [PubMed]
42. Hakozaki, T.; Richard, C.; Elkrief, A.; Hosomi, Y.; Benlaïfaoui, M.; Mimpen, I.; Terrisse, S.; Derosa, L.; Zitvogel, L.; Routy, B.; et al. The Gut Microbiome Associates with Immune Checkpoint Inhibition Outcomes in Patients with Advanced Non–Small Cell Lung Cancer. *Cancer Immunol. Res.* **2020**, *8*, 1243–1250. [CrossRef]
43. Grenda, A.; Iwan, E.; Krawczyk, P.; Frąk, M.; Chmielewska, I.; Bomba, A.; Giza, A.; Rolska-Kopińska, A.; Szczyrek, M.; Kieszko, R.; et al. Attempting to Identify Bacterial Allies in Immunotherapy of NSCLC Patients. *Cancers* **2022**, *14*, 6250. [CrossRef]
44. Song, P.; Yang, D.; Wang, H.; Cui, X.; Si, X.; Zhang, X.; Zhang, L. Relationship between Intestinal Flora Structure and Metabolite Analysis and Immunotherapy Efficacy in Chinese NSCLC Patients. *Thorac. Cancer* **2020**, *11*, 1621–1632. [CrossRef]
45. Takada, K.; Shimokawa, M.; Takamori, S.; Shimamatsu, S.; Hirai, F.; Tagawa, T.; Okamoto, T.; Hamatake, M.; Tsuchiya-Kawano, Y.; Otsubo, K.; et al. Clinical Impact of Probiotics on the Efficacy of Anti-PD-1 Monotherapy in Patients with Nonsmall Cell Lung Cancer: A Multicenter Retrospective Survival Analysis Study with Inverse Probability of Treatment Weighting. *Int. J. Cancer* **2021**, *149*, 473–482. [CrossRef]
46. Wan, L.; Wu, C.; Wu, Q.; Luo, S.; Liu, J.; Xie, X. Impact of Probiotics Use on Clinical Outcomes of Immune Checkpoint Inhibitors Therapy in Cancer Patients. *Cancer Med.* **2022**, *12*, 1841–1849. [CrossRef]
47. Tomita, Y.; Sakata, S.; Imamura, K.; Iyama, S.; Jodai, T.; Saruwatari, K.; Hamada, S.; Akaike, K.; Anai, M.; Fukusima, K.; et al. Association of Clostridium Butyricum Therapy Using the Live Bacterial Product CBM588 with the Survival of Patients with Lung Cancer Receiving Chemoimmunotherapy Combinations. *Cancers* **2023**, *16*, 47. [CrossRef] [PubMed]
48. Inamura, K. Roles of Microbiota in Response to Cancer Immunotherapy. *Semin. Cancer Biol.* **2020**, *65*, 164–175. [CrossRef] [PubMed]
49. Yu, X.; Zhao, L.; Song, B. Impact of Opioid Analgesics on the Efficacy of Immune Checkpoint Inhibitors in a Lung Cancer Population. *BMC Pulm. Med.* **2022**, *22*, 431. [CrossRef] [PubMed]
50. Baek, Y.-H.; Kang, E.J.; Hong, S.; Park, S.; Kim, J.H.; Shin, J.-Y. Survival Outcomes of Patients with Nonsmall Cell Lung Cancer Concomitantly Receiving Proton Pump Inhibitors and Immune Checkpoint Inhibitors. *Int. J. Cancer* **2022**, *150*, 1291–1300. [CrossRef]

51. Lopes, S.; Pabst, L.; Dory, A.; Klotz, M.; Gourieux, B.; Michel, B.; Mascaux, C. Do Proton Pump Inhibitors Alter the Response to Immune Checkpoint Inhibitors in Cancer Patients? A Meta-Analysis. *Front. Immunol.* **2023**, *14*, 1070076. [CrossRef]
52. Duttagupta, S.; Hakozaki, T.; Routy, B.; Messaoudene, M. The Gut Microbiome from a Biomarker to a Novel Therapeutic Strategy for Immunotherapy Response in Patients with Lung Cancer. *Curr. Oncol.* **2023**, *30*, 9406–9427. [CrossRef] [PubMed]
53. Xue, C.; Chu, Q.; Zheng, Q.; Yuan, X.; Su, Y.; Bao, Z.; Lu, J.; Li, L. Current Understanding of the Intratumoral Microbiome in Various Tumors. *Cell Rep. Med.* **2023**, *4*, 100884. [CrossRef] [PubMed]
54. Bolyen, E.; Rideout, J.R.; Dillon, M.R.; Bokulich, N.A.; Abnet, C.C.; Al-Ghalith, G.A.; Alexander, H.; Alm, E.J.; Arumugam, M.; Asnicar, F.; et al. Reproducible, Interactive, Scalable and Extensible Microbiome Data Science Using QIIME 2. *Nat. Biotechnol.* **2019**, *37*, 852–857. [CrossRef]
55. Kang, X.; Lau, H.C.-H.; Yu, J. Modulating Gut Microbiome in Cancer Immunotherapy: Harnessing Microbes to Enhance Treatment Efficacy. *Cell Rep. Med.* **2024**, *5*, 101478. [CrossRef]
56. Ren, S.; Feng, L.; Liu, H.; Mao, Y.; Yu, Z. Gut Microbiome Affects the Response to Immunotherapy in Non-Small Cell Lung Cancer. *Thorac. Cancer* **2024**. [CrossRef] [PubMed]
57. Lippenszky, L.; Mittendorf, K.F.; Kiss, Z.; LeNoue-Newton, M.L.; Napan-Molina, P.; Rahman, P.; Ye, C.; Laczi, B.; Csernai, E.; Jain, N.M.; et al. Prediction of Effectiveness and Toxicities of Immune Checkpoint Inhibitors Using Real-World Patient Data. *JCO Clin. Cancer Inform.* **2024**, *8*, e2300207. [CrossRef] [PubMed]
58. Giuffrè, M.; Moretti, R.; Tiribelli, C. Gut Microbes Meet Machine Learning: The Next Step towards Advancing Our Understanding of the Gut Microbiome in Health and Disease. *Int. J. Mol. Sci.* **2023**, *24*, 5229. [CrossRef]
59. Liang, H.; Jo, J.-H.; Zhang, Z.; MacGibeny, M.A.; Han, J.; Proctor, D.M.; Taylor, M.E.; Che, Y.; Juneau, P.; Apolo, A.B.; et al. Predicting Cancer Immunotherapy Response from Gut Microbiomes Using Machine Learning Models. *Oncotarget* **2022**, *13*, 876–889. [CrossRef]

cancers

MDPI

Review

HER2-Altered Non-Small Cell Lung Cancer: A Journey from Current Approaches to Emerging Strategies

Giorgia Ferrari †, Benedetta Del Rio †, Silvia Novello and Francesco Passiglia *

Department of Oncology, University of Turin, San Luigi Hospital, 10124 Orbassano, Italy; giorgia.ferrari@unito.it (G.F.); benedetta.delrio@unito.it (B.D.R.); silvia.novello@unito.it (S.N.)
* Correspondence: francesco.passiglia@unito.it
† These authors contributed equally to this work.

Simple Summary: The introduction of trastuzumab deruxtecan is significantly changing the therapeutic landscape of advanced HER2-mutated non-small cell lung cancer (NSCLC). The results of the DESTINY-Lung04 trial are highly anticipated for their potential to redefine the first-line therapeutic standard for HER2-mutant disease. Furthermore, several studies evaluating combination therapy regimes are currently ongoing. This review outlines the current state of the art in the clinical management of HER2-altered NSCLC and explores potential future perspectives in the field of HER2 targeted strategies.

Abstract: For patients diagnosed with advanced HER2-altered non-small cell lung cancer (NSCLC), the current standard of care is represented by a platinum-pemetrexed-based chemotherapy, eventually in combination with immunotherapy. Different pan-HER tyrosine kinase inhibitors have been evaluated in limited phase II trials, yielding generally unsatisfactory outcomes, although certain genotypes demonstrated some clinical benefit. Conversely, antibody-drug conjugates (ADCs) targeting HER2, particularly trastuzumab-deruxtecan, have shown promising results against HER2-mutant disease, including a great intracranial activity in patients with brain metastasis. Based on the results obtained from DESTINY-Lung01 and DESTINY-Lung02 trials, trastuzumab deruxtecan received regulatory approval as the first targeted therapy for pre-treated, HER2-mutant, advanced NSCLC patients. More recently, the Food and Drug Administration (FDA) granted the accelerated approval of trastuzumab deruxtecan for advanced, pre-treated HER2-positive solid tumours with no other treatment options. In this scenario, emerging evidence is increasingly pointing towards the exploration of combination regimens with synergistic effects in the advanced disease. In this review, we provide a detailed summary of current approaches and emerging strategies in the management of HER2-altered NSCLC, also focusing on unmet needs, including the treatment of patients with brain metastases.

Keywords: non-small cell lung cancer; HER2; antibody-drug conjugates; targeted therapy

1. Introduction

In recent years, remarkable progress has been made in non-small cell lung cancer (NSCLC) treatment by identifying oncogenic drivers and developing targeted therapies. Among these oncogenic drivers, epidermal growth factor 2 receptor (HER2) has recently emerged as a promising but challenging oncogenic driver and therapeutic target in NSCLC patients.

HER2, along with HER1 (EGFR), HER3, and HER4, is part of the HER/ERBB receptor tyrosine kinase family, which encodes receptors consisting of three domains: an extracellular domain, a transmembrane domain, and an intracellular tyrosine kinase domain. When inactive, HER receptors are in the form of monomers, and only when the ligand binding occurs is the dimerisation domain exposed, and the dimerisation allowed. The hetero- or homodimerisation between the receptors of HER family results in the activation of the downstream PI3KAKT and MEK-ERK signalling pathways to promote cell proliferation, differentiation, and migration. HER2 is the only member of the HER/erbB family that lacks a specific ligand. Consequently, it remains in a perpetually active conformational state, always making it readily available for dimerisation [1]. HER2 heterodimerises with HER1 (also known as EGFR), HER3 and HER 4 and is only able to homodimerise when overexpressed. Alterations of HER2 causing an overexpression of the receptor result in an excess of ERBB2-mediated signalling, promoting cell survival and proliferation [2,3].

HER2 alterations are present in different cancer types such as: bladder cancer, colorectal, lung, breast, and uterine cervix cancers. In recent years, anti-HER2 therapies have been developed for many of these tumours, showing great efficacy and tolerability profiles [2,4]. In this review we provide a summary of the current state of the art and future directions in the treatment of HER2-altered NSCLC.

2. HER2 Alterations in NSCLC

Different types of HER2 alterations have been identified in NSCLC: gene mutation, gene amplification, and protein overexpression (Figure 1). The overlap between these alterations is rare. Each alteration identifies a subgroup with different biological behaviour, affecting different patient subsets and showing different responses to anti-HER2 drugs. For this reason, it is imperative to investigate them as distinct clinicopathological as well as molecular entities [5,6].

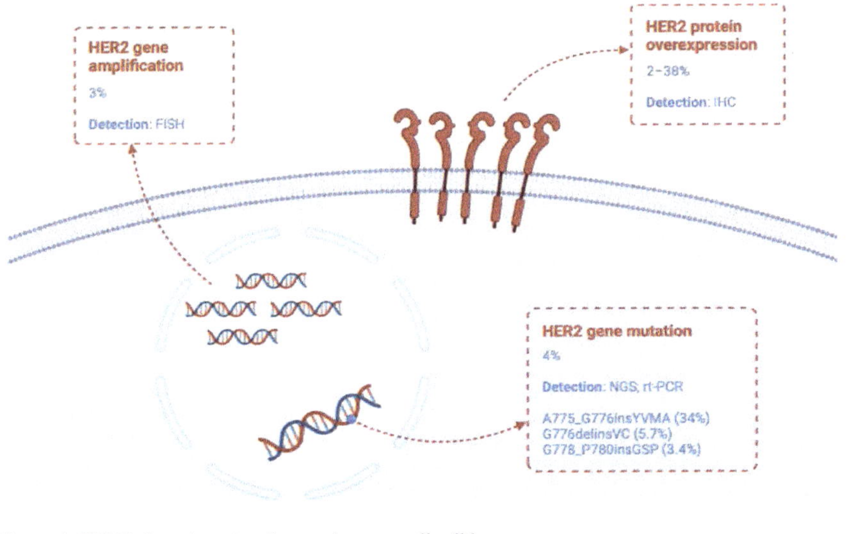

Figure 1. HER2 alterations in advanced non-small cell lung cancer.

2.1. HER2 Mutation

HER2 mutations, occurring in approximately 4% of NSCLC and detectable by NGS or rt-PCR, are more prevalent in females, non-smokers, and adenocarcinoma histology, similar to EGFR mutations [7]. HER2 mutations are predominantly mutually exclusive with other oncogenic drivers; only a minority of patients display concurrent EGFR mutations, ALK translocations, or ROS-1 translocations [8].

The most common mutation occurs in exon 20 within the tyrosine kinase domain. It involves a 12 base-pair insertion coding for the amino acids YVMA (A775_G776insYVMA) and accounts for 34% of all HER2 mutations. Other common exon 20 mutations affecting the tyrosine kinase domain include G776delinsVC and G778_P780insGSP, which account for 5.7% and 3.4% of all HER2 mutations, respectively [9]. On the other hand, mutations such as I655V (4.5%), *P122L* (2.3%), and *G222C* (1.1%) affect the extracellular domain and

S310F (5.1%) the transmembrane domain [6]. Different mutations give the disease different characteristics; the YMVA mutation is associated with a higher risk of developing brain metastases than the non-YMVA subgroup [10]. Additionally, the diverse conformational landscape of HER2 ex20ins, resulting from different mutations, contributes to the high heterogeneity in terms of clinical response to anti-HER2 agents [11].

2.2. HER2 Amplification

The frequency of HER2 amplification in NSCLC is approximately 3%. The most widely accepted definition of HER2 amplification is the HER2/CEP17 ratio $\geq$ 2.0 by fluorescence in situ hybridisation (FISH) testing, which is used in breast cancer as well as in most clinical trials. De novo amplifications of HER2 have been found to be more common in males and in smokers [12,13].

In addition, HER2 amplification has been identified as an acquired resistance mechanism in EGFR-mutant NSCLC cancers treated with EGFR inhibitors, occurring in approximately 15% of patients resistant to EGFR TKIs. HER2 amplification has been observed as a resistance mechanism to first-generation EGFR TKIs and is mutually exclusive with the more common resistance mechanism, T790M. In addition, HER2 amplification has been identified as a resistance mechanism to the third-generation EGFR inhibitor osimertinib, whether given sequentially after a first-generation inhibitor or as first-line therapy [14].

HER2-amplified disease correlates with certain particularly aggressive features, notably, larger tumour size and a higher propensity to develop pleural metastases and lymphovascular invasion. However, despite this correlation, the prognostic and predictive role of HER2 amplification in NSCLC remains unclear [15].

2.3. HER2 Overexpression

HER2 overexpression occurs in 2–38% of NSCLC and is generally assessed by immunohistochemistry. This alteration is predominantly observed in males and in smokers, as HER2 amplification, and in adenocarcinoma with papillary histology [12,15].

In NSCLC, HER2 overexpression is typically not driven by HER2 amplification but rather by polysomy, in contrast to breast cancer where HER2 amplification is strongly correlated with HER2 expression. Polysomy is considered to be the presence of a HER2 gene copy number greater than 5 or 6, but HER2/CEP17 < 2 [16]. However, there are conflicting findings regarding the actual overlap between overexpression and HER2 amplification [17]. While it remains unclear regarding HER2 mutations and amplifications, overexpression is associated with a worse prognosis. The reason for this correlation is not yet clear, but it may be related to increased chemoresistance resulting from this molecular alteration [13].

3. Current Therapeutic Approaches

3.1. Chemotherapy

Platinum-pemetrexed-based chemotherapy, optionally in combination with immunotherapy, currently represents the standard of care for first-line treatment of patients with HER2-altered advanced NSCLC [18]. Similar to patients harbouring KRAS and EGFR mutations, pemetrexed-based chemotherapy showed an overall response rate (ORR) of 36% and a progression-free survival (PFS) of 5.1 months [10,19].

Several clinical trials have evaluated the efficacy of chemotherapy in combination with HER2-targeted agents; these studies will be analysed in the subsequent paragraphs.

3.2. Immune Checkpoint Inhibitors (ICIs)

Immune checkpoint inhibitors (ICIs) have become a standard treatment option for advanced non-oncogene-addicted NSCLC. However, ICI efficacy in patients with oncogenic drivers is limited, which is attributed to a "cold" tumour microenvironment characterised by reduced PD-L1 expression and a lower tumour mutation burden (TMB) [20].

A retrospective analysis evaluating the efficacy of ICI monotherapy in patients with advanced HER2-mutant NSCLC showed an ORR of 12% with median PFS and OS of

1.9 and 10.4 months, respectively [21]. Similar outcomes were reported by the French Group on Lung Cancer (GFPC) [22] in patients who had received one previous treatment, and in the IMMUNOTARGET registry, where smoker patients with HER2-mutant NSCLC exhibited a significantly longer median PFS compared with nonsmokers (3.4 months vs. 2.0 months, $p = 0.04$) [23].

Regarding the use of ICIs in combination with chemotherapy in HER2-mutated patients in a first-line setting, the observed outcomes are similar to those of the non-selective NSCLC cohort in the KEYNOTE-189 trial with an ORR, median PFS and a one-year OS of 52%, 6 months, and 88%, respectively [24,25]. A subsequent retrospective analysis demonstrated an ORR of 28.9% and a median PFS of 5.2 months. Despite the high ORR, the increase in median PFS was not statistically significant compared to chemotherapy alone (5.2 vs. 4.03 months, $p = 0.20$) [26].

Presently, the available data do not encourage the use of ICI monotherapy for the treatment of HER2-altered advanced NSCLC. On the other hand, chemoimmunotherapy combinations, despite limited evidence, continue to be a feasible first-line treatment alternative.

3.3. *Tyrosine Kinase Inhibitors (TKIs)*

Initial efforts to target HER2 in NSCLC involved the employment of second-generation irreversible tyrosine kinase inhibitors (TKIs) designed for EGFR mutation treatment.

Afatinib was evaluated in a compassionate use program in heavily pretreated patients with HER2-mutant NSCLC and showed a median time-to-treatment failure (TTF) of 2.9 months, with interesting variations among subtypes; notably, the HER2 exon 20 YVMA insertion subtype showed a median TTF of 9.6 months [27]. Conversely, other studies identified a clinical association between HER2 exon 20 YVMA insertion and a reduced efficacy to afatinib, whereas mutations such as G778_P780dup and G776delinsVC demonstrated improved outcomes [28,29]. In the EUHER2 study, afatinib showed an ORR of 18.2% and a median PFS of 3.9 months in 11 patients with HER2-mutant NSCLC [8]. Additionally, the NICHE phase II trial confirmed the modest clinical activity of afatinib in patients with HER2-mutant NSCLC, demonstrating an ORR of 7.7% and a median PFS of 15.9 week [30].

Dacomitinib, an irreversible pan-HER TKI, was evaluated in a phase II trial in pretreated HER2-altered NSCLC patients and showed a 12% partial response rate in HER2-mutant patients, compared with 0% in those with HER2 amplification. Notably, responses were limited to patients with specific mutations (P780_Y781insGSP and M774delinsWLV) and absent in those with the A775_G776insYVMA insertion [31].

Neratinib, a pan-HER TKI, was evaluated in the phase II SUMMIT basket trial, in patients with refractory HER2-mutant NSCLC, achieving a partial response in only one patient (ORR 3.8%) [32]. In a randomised phase II trial comparing neratinib alone or in combination with temsirolimus in HER2-mutant advanced NSCLC, combination therapy achieved an ORR of 19% vs. 0% with neratinib alone [33].

Non-selective HER2-TKIs demonstrated limited efficacy in treating patients with HER2-altered NSCLC, prompting the development of next-generation TKIs to improve clinical outcomes.

Poziotinib, a novel covalent and irreversible EGFR/HER2 inhibitor, was evaluated in the ZENITH20-2 trial in NSCLC patients with HER2 exon 20 insertions. The treatment was administered at a dose of 16 mg QD and showed an ORR of 27.8% and a disease control rate (DCR) of 70.0%, achieving a median PFS of 5.5 months. However, it exhibited a suboptimal safety profile, with 97.8% of patients experiencing treatment-related adverse events (TRAEs), including 78.9% with grade 3 TRAEs [34]. With the aim of improving the drug's tolerability profile, the ZENITH20-4 trial compared poziotinib 16 mg QD to 8 mg BID in treatment-naïve patients with HER2-mutant NSCLC. The study revealed an ORR of 41% and a median PFS of 5.6 months, and it found that the twice-daily regimen led to fewer dose reductions and interruptions compared to the once-daily regimen, while maintaining a comparable incidence of grade ≥ 3 TRAEs [35]. In consideration of the complex evaluation of the overall risk-benefit analysis, the FDA has determined that additional data from a randomised controlled trial are needed before poziotinib can be approved in pre-treated patients with NSCLC harboring HER2 exon 20 insertion mutations.

Pyrotinib, an irreversible pan-HER TKI, was evaluated in a phase II trial including platinum-pretreated advanced NSCLC patients, achieving an ORR of 30% and a median PFS of 6.9 months. Subgroup analyses revealed that patients with different HER2 mutations, including those with BMs, benefited from the treatment [36]. In a different phase II trial, pyrotinib was administered as either first or subsequent line in patients with advanced HER2-mutant NSCLC, yielding an ORR of 19.2% and a median PFS of 5.6 months. Notably, treatment-naive patients experienced a higher median PFS (8.9 months) [37]. Additionally, the PATHER2 phase II trial evaluated pyrotinib in combination with apatinib in pretreated advanced HER2-altered NSCLC patients, showing an ORR of 51.5%, a median PFS of 6.9 months, and a median OS of 14.8 months [38]. New efficacy and safety data will be provided by the ongoing phase III PYRAMID-1 trial comparing pyrotinib with docetaxel in patients with advanced NSCLC harbouring a HER2 exon 20 mutation who have been previously treated with platinum-based chemotherapy (NCT04447118).

Mobocertinib (TAK-788), an oral EGFR/HER2 inhibitor targeting exon 20 insertions, demonstrated antitumor activity in a phase I/II trial involving advanced NSCLC patients with EGFR exon 20 insertion, while results from an expansion cohort focusing on patients harbouring HER2 exon 20 alterations are still pending [39,40].

Currently, phase I and II clinical studies are evaluating other non-selective HER2 TKIs, including furmonertinib (NCT05364073), BAY2927088 (NCT05099172), and tucatinib in combination with trastuzumab (NCT04579380).

Ongoing efforts are also focused on developing novel and highly-HER2 selective TKIs, with the aim of improving treatment efficacy. BI 1810361 (zongertinib), a potent covalent HER2 inhibitor, demonstrated an ORR of 46% in an ongoing phase I trial (NCT04886804) involving HER2-mutant NSCLC patients refractory to platinum-based chemotherapy [41]. Furthermore, ELVN-002 has demonstrated efficacy against common HER2 mutations along with the ability to penetrate the CNS [42] and is currently under evaluation in a phase I clinical trial (NCT05650879). Table 1 lists the major ongoing clinical trials of TKIs in HER2-altered NSCLC.

Table 1. Ongoing clinical trial of TKIs in HER2-altered NSCLC.

Trial (ClinicalTrials.gov Identifier)	Description	Phase	Drug	HER2 Alteration
NCT05378763	A Study of Poziotinib in Previously Treated Participants With Locally Advanced or Metastatic NSCLC Harboring HER2 Exon 20 Mutations (PINNACLE)	III	Poziotinib 8 mg BID; Docetaxel 75 mg/mq	Exon 20 mutations
NCT04447118	Phase 3 Study of Pyrotinib Versus Docetaxel in Patients With Advanced Non-squamous NSCLC Harboring a HER2 Exon 20 Mutation Who Failed Platinum Based Chemotherapy (PYRAMID-1)	III	Pyrotinib 400 mg QD; Docetaxel 75 mg/mq	Exon 20 mutations
NCT05364073	Study of Furmonertinib in Patients With Advanced or Metastatic Non-Small Cell Lung Cancer (NSCLC) With Activating, Including Uncommon, Epidermal Growth Factor Receptor (EGFR) or Human Epidermal Growth Factor Receptor 2 (HER2) Mutations	Ib	Furmonertinib	Exon 20 mutation; EGFR exon 20 mutations and uncommon mutations
NCT05099172	First in Human Study of BAY2927088 in Participants Who Have Advanced Non-small Cell Lung Cancer (NSCLC) With Mutations in the Genes of Epidermal Growth Factor Receptor (EGFR) and/or Human Epidermal Growth Factor Receptor 2 (HER2)	I	BAY2927088	HER2 mutations; EGFR mutations

Table 1. *Cont.*

Trial (ClinicalTrials.gov Identifier)	Description	Phase	Drug	HER2 Alteration
NCT04579380	Basket Study of Tucatinib and Trastuzumab in Solid Tumors With HER2 Alterations	II	Tucatinib 300 mg BID + Trastuzumab 6 mg/kg	HER2 mutations, overexpression, or amplification
NCT04886804	Beamion LUNG-1: An Open Label, Phase I Dose Escalation Trial, With Dose Confirmation and Expansion, of Zongertinib (BI 1810631) as Monotherapy in Patients With Advanced or Metastatic Solid Tumors With HER2	I	Zongertinib	HER2 mutations, overexpression, or amplification
NCT05650879	A Phase 1a/1b Study of ELVN-002 for the Treatment of Patients With HER2 Mutant Non-Small Cell Lung Cancer	I	ELVN-002	HER2 mutations
NCT04144569	PD-1 Combined With Pyrotinib for Chemotherapy Failure HER2 Insertion Mutation Advanced NSCLC	II	PD-1 + Pyrotinib 400 mg QD	HER2 insertion mutations

Abbreviations: TKI, tyrosine kinase inhibitors; NSCLC, non-small cell lung cancer; QD, once daily; BID, twice daily.

3.4. Monoclonal Antibodies

Trastuzumab, a humanised IgG monoclonal antibody targeting HER2 directly, has been evaluated for its efficacy in advanced NSCLC patients with HER-2/neu positivity across various chemotherapy regimens. Its addition to carboplatin and paclitaxel in a Phase II clinical trial demonstrated an ORR of 24.5%, with a median PFS and OS of 3.3 and 10.1 months, respectively [43]. Subgroup analysis revealed improved survival outcomes in patients with HER-2/neu 3+ expression, consistent with findings from other chemotherapy regimens such as cisplatin-gemcitabine [44]. However, the limited number of HER2 3+ or FISH-positive patients precludes a definitive confirmation of these results.

Trastuzumab monotherapy was evaluated in the phase II HOT1303-B trial, which enrolled pre-treated HER2-altered NSCLC patients and resulted in no therapeutic response (ORR 0%), although it achieved a disease control ratio (DCR) of 70.0% and a median PFS of 5.2 months [45].

The combination of trastuzumab and pertuzumab showed a modest ORR of 11% in heavily pre-treated patients with HER2-mutant or amplified NSCLC, particularly in those harbouring HER2 exon 20 mutations [46].

Given the favourable results observed with the combination of trastuzumab, pertuzumab, and docetaxel in breast cancer, its efficacy was assessed in the single-arm phase II IFCT 1703-R2D2 trial in patients with HER2-mutant NSCLC who had previously been treated with platinum-based chemotherapy. This regimen demonstrated an ORR of 29% and a median PFS of 6.8 months, with equally promising results [47]. Notably, this combination demonstrated a median duration of response (mDoR) of 11.0 months, significantly exceeding the duration of response observed with other HER2-targeted monoclonal antibody regimens.

3.5. Antibody-Drug Conjugates (ADCs)

Trastuzumab Emtansine (T-DM1) is an antibody-drug conjugate (ADC) combining the HER2-targeted monoclonal antibody trastuzumab with the cytotoxic microtubule inhibitor emtansine (DM1). In a phase II basket trial, T-DM1 was administered at a dose of 3.6 mg/kg to 18 heavily pre-treated HER2-mutant patients with advanced NSCLC, achieving an ORR of 44% and a PFS of 5 months. Biomarker analysis identified variations in response rates by HER2 mutation subtype, with the highest observed in exon 20 insertion mutations, and no significant correlation between HER2 IHC and clinical outcomes [48]. Updated analyses involving 28 HER2-mutant and 11 HER2-amplified pre-treated NSCLC patients revealed an ORR of 50% and 55%, respectively [49]. A study conducted on patients with HER2-overexpressed NSCLC treated with T-DM1 reported no therapeutic response in the

IHC 2+ group and an ORR of 20% in the IHC 3+ group, despite a similar PFS and OS. Further analysis indicated that most responders had HER2 amplification, with a subset also showing HER2 mutations, highlighting the insufficiency of exclusively relying on IHC to forecast the efficacy of T-DM1, differently from what is observed in breast and gastric cancer [50]. In conclusion, a cohort of 22 NSCLC patients with HER2 exon 20 insertion mutations was evaluated in a phase II study revealing an ORR of 38.1% and a median PFS of 2.8 months, highlighting the limited response duration associated with T-DM1 treatment [51].

Trastuzumab deruxtecan (T-DXd) is a novel anti-HER2 antibody-drug conjugate (ADC) consisting of deruxtecan, a topoisomerase I inhibitor, conjugated to trastuzumab via a cleavable linker. In the phase III DESTINY-Breast03 trial, T-DXd achieved an ORR of 79% compared to 35% with standard treatment in patients with unresectable or metastatic HER2-positive breast cancer previously treated with anti-HER2 regimens, leading to its FDA and EMA approval [52]. Based on these encouraging results, its efficacy has also been studied in patients with HER2-altered NSCLC. In a phase I study, T-DXd showed promising activity in patients with HER-altered NSCLC, with an ORR of 55.8% and a median PFS of 11.3 months. Of particular note was the improved efficacy seen in the HER2-mutant subgroup, with an ORR of 72.7% [53]. Thereafter, the phase II DESTINY-Lung01 trial evaluated the efficacy and safety of T-DXd at the dose of 6.4 mg/kg in HER2-mutant or overexpressed, recurrent, or refractory NSCLC. Notable outcomes were seen in the HER2-mutant subgroup, where T-DXd monotherapy showed an ORR of 55%, median PFS of 8.2 months, and median OS of 17.8 months [54]. These findings, supported by preclinical research showing enhanced receptor internalisation and trastuzumab deruxtecan intracellular uptake in activating HER2 mutations, confirm the heightened activity of trastuzumab deruxtecan in HER2-mutant versus HER2-overexpressing NSCLC [49]. The toxicity profile of T-DXd was characterised by a 97% incidence of TRAEs, 46% of which were grade ≥ 3, leading to dose interruption, reduction, or treatment discontinuation in 53.1%, 34.7%, and 22.4% of cases, respectively. Notably, interstitial lung disease (ILD) emerged as the main adverse effect, affecting 26% of the patient population and leading to two deaths [54]. The greater frequency of ILD observed in lung cancer versus breast or gastric cancers may reflect underlying pulmonary conditions in lung cancer patients, including smoking-related damage or reduced lung capacity from prior treatments, but further investigation for definitive conclusions is needed [55]. The subsequent phase II DESTINY-Lung02 trial revealed that for HER2-mutant, pre-treated NSCLC patients, a 5.4 mg/kg Q3W dosage of T-DXd outperforms the 6.4 mg/kg Q3W regimen, achieving higher ORR (53.8% vs. 42.9%), with reduced severe TRAEs (31.7% vs. 58%) and ILD incidence (5.9% vs. 14%) [56]. These results led to the FDA and EMA approval of trastuzumab deruxtecan as the first and only approved targeted therapy for the treatment of metastatic NSCLC with HER2 mutations previously treated with platinum-based chemotherapy. Moreover, on 5 April 2024, the FDA granted accelerated approval to T-DXd for patients with advanced, pre-treated HER2-positive (IHC3+) solid tumours with no other treatment options.

T-DXd is currently under evaluation also in first-line setting. The ongoing phase III DESTINY-Lung04 trial (NCT05048797) will assess the efficacy and safety of T-DXd as single-agent therapy compared to chemotherapy plus pembrolizumab for first-line treatment in advanced NSCLC patients with HER2 exon 19 or 20 mutations.

4. Brain Metastases in HER2-Altered NSCLC: Incidence and Treatment Strategies

The occurrence of brain metastases in patients with HER2-altered NSCLC ranges from 6% to 29% [8,57,58], with exon 20 YVMA insertion associated with a significantly higher baseline and lifetime incidence [10]. It is, therefore, undeniable that the development of HER2-targeted therapies capable of crossing the blood-brain barrier (BBB) is necessary to achieve more effective treatments and longer lasting responses over time.

As previously mentioned, although non-selective HER2 TKIs like afatinib, dacomitinib, and pyrotinib showed systemic efficacy in HER2-mutant NSCLC, data on their impact on the central nervous system (CNS) are lacking. On the other hand, poziotinib revealed an ORR of 28.6% and a median PFS of 7.4 months in patients with BMs, but the reliability of these results is limited by a small cohort size, the omission of a baseline brain MRI, and the prevalent use of prior brain radiation [59].

T-DXd demonstrated significant intracranial efficacy in HER2-positive breast cancer. In the phase III DESTINY-Breast03 trial, T-DXd achieved an intracranial response rate of 63.8%, compared to 33.3% for T-DM1 among patients with stable BMs at baseline [60]. Furthermore, a subgroup analysis from DESTINY-Breast01 recently underscored the efficacy of T-DXd on stable BMs in patients previously treated with T-DM1 [61]. T-DXd demonstrated promising efficacy also in the treatment of HER2-positive metastatic breast cancer patients with active BMs. In this scenario, the phase II DEBBRAH trial recorded an intracranial response rate of 44.4% among individuals treated with T-DXd with either HER2-positive or HER2-low breast cancer experiencing progression of BMs after local therapy [62]. Moreover, in the phase II TUXEDO-1 trial, T-DXd achieved an intracranial response rate of 100% in patients with de novo and 66.7% in patients with progressive BMs [63].

Consistent with what observed in breast cancer, T-DXd exhibited encouraging intracranial activity also in patients with HER2-mutant NSCLC. Specifically, a pooled analysis of the DESTINY-Lung01 and DESTINY-Lung02 trials demonstrated that 86% of patients with measurable BMs receiving a 5.4 mg/kg dose and 78% receiving a 6.4 mg/kg dose of T-DXd showed a reduction in brain lesions size [54,56]. Moreover, the ongoing DESTINY-Lung04 trial (NCT05048797) will evaluate CNS PFS as secondary endpoint.

5. Future Directions in HER2-Positive NSCLC Treatment

The treatment of HER2-driven NSCLC is expected to evolve from monotherapy to combinations of agents with synergistic effects, significantly changing the current therapeutic landscape.

Preclinical studies have revealed that T-DXd increases the expression of PD-L1 by the major histocompatibility complex class I and potentiates the infiltration of CD8 T cells into breast cancer cells, outlining the therapeutic potential of the combination of T-DXd and ICI [64]. In light of these findings, several clinical trials are underway. The phase II HUDSON basket trial evaluated the combination of T-DXd plus durvalumab in patients previously treated with anti-PD1/PD-L1 therapy, including those with HER2-mutant NSCLC and HER2-overexpressed NSCLC, showing greater efficacy in the former subgroup [65]. The phase Ib DESTINY-Lung03 trial (NCT04686305) is evaluating the safety, tolerability and efficacy of T-DXd in combination with durvalumab and chemotherapy as first-line treatment, but only in patients with advanced HER2-overexpressed NSCLC [66]. In addition, an ongoing phase Ib trial (NCT04042701) is investigating T-DXd in association with pembrolizumab in HER2-positive and mutant NSCLC patients without previous exposure to anti-PD-1/PD-L1 or HER2 therapy. However, the concurrent administration of Durvalumab and T-DXd, both associated with pulmonary toxicity, presents significant risks. In the HUDSON trial, 55% of patients experienced grade ≥ 3 TEAEs, including pneumonitis, pulmonary embolism, and anaemia, with pneumonitis being the most prevalent [65].

ICIs are also under evaluation in combination with TKIs in HER2-mutant NSCLC after failure of first-line chemotherapy, in a phase II study involving pyrotinib and PD-1 inhibitors (NCT04144569). The combination of HER2 TKIs and ICIs is supported by their synergistic effects: TKIs induce immunogenic cell death and cytokine release, enhancing tumour antigen presentation and immune cell activation; on the other hand, ICIs amplify the immune response by preventing T-cell inactivation, resulting in a more effective antitumour response [67].

Improving the architectural configuration of ADCs to amplify their efficacy and reduce associated toxicities is a promising direction. Within this framework, novel HER2-targeted ADCs such as A166, ARX788, SHRA1811, and MRG002 are currently under evaluation in phase I/II trials in patients with HER2-altered NSCLC (NCT03602079, NCT03255070, NCT04818333, and NCT05141786). Additionally, an ongoing phase IB/II clinical trial will evaluate SHR-A1811 in combination with pyrotinib or SHR-1316 in patients with HER-2 altered advanced NSCLC (NCT05482568). In conclusion, the NCT04235101 trial is exploring the safety profile of the combination therapy involving SYD985 (trastuzumab duocarmazine) and niraparib in individuals diagnosed with solid tumours. Table 2 lists the main ongoing clinical trials of ADCs in HER2-altered NSCLC.

Table 2. Ongoing clinical trial of ADCs in HER2-altered NSCLC.

Trial (ClinicalTrials.gov Identifier)	Description	Phase	Drug	HER2 Alteration
NCT05048797	An Open-label, Randomized, Multicenter, Phase 3 Study to Assess the Efficacy and Safety of Trastuzumab Deruxtecan as First-line Treatment of Unresectable, Locally Advanced, or Metastatic NSCLC Harboring HER2 Exon 19 or 20 Mutations (DESTINY-Lung04)	III	T-DXd	Exon 19 or 20 mutations
NCT04686305	A Phase Ib Multicenter, Open-label Study to Evaluate the Safety and Tolerability of Trastuzumab Deruxtecan (T-DXd) and Immunotherapy Agents With and Without Chemotherapy Agents in First-line Treatment of Patients With Advanced or Metastatic Non-squamous Non-small Cell Lung Cancer (NSCLC) and Human Epidermal Growth Factor Receptor 2 (HER2) Overexpression (OE) (DESTINY-Lung03)	Ib	T-DXd; T-DXd + Durvalumab + Cisplatin, Carboplatin or Pemetrexed; T-DXd + MEDI5752; T-DXd + MEDI5752 + Carboplatin	HER2 overexpression
NCT04042701	A Phase 1b, Multicenter, Two-Part, Open-Label Study of Trastuzumab Deruxtecan (DS-8201a), An Anti-Human Epidermal Growth Factor Receptor-2 (HER2)-Antibody Drug Conjugate (ADC), In Combination With Pembrolizumab, An Anti-PD-1 Antibody, For Subjects With Locally Advanced/Metastatic Breast Or Non-Small Cell Lung Cancer (NSCLC)	Ib	T-DXd + Pembrolizumab	HER2-expressing, HER2 mutations
NCT03602079	A Phase I-II, FIH Study of A166 in Locally Advanced/Metastatic Solid Tumors Expressing Human Epidermal Growth Factor Receptor 2 (HER2) or Are HER2 Amplified That Did Not Respond or Stopped Responding to Approved Therapies	I-II	A166	HER2-expressing, HER2 amplification
NCT03255070	A Phase 1, Multicenter, Open-label, Multiple Dose-escalation and Expansion Study of ARX788, as Monotherapy in Advanced Solid Tumors With HER2 Expression	I	ARX788	HER2-expressing
NCT04818333	Phase I/II Clinical Study of the Safety, Tolerability, Pharmacokinetics, and Efficacy of SHR-A1811 for Injection in Subjects With Advanced Non-small Cell Lung Cancer Who Have HER2 Expression, Amplification, or Mutation	I	SHR-A1811	HER2-expresssing, HER2 amplification, HER2 mutations
NCT05141786	An Open-label, Multi-center, Non-randomized Phase II Clinical Study to Evaluate the Efficacy and Safety of MRG002 in Patients With HER2-mutated Unresectable/ Metastatic Non-small Cell Lung Cancer (NSCLC)	II	MRG002	HER2 mutations

Table 2. *Cont.*

Trial (ClinicalTrials.gov Identifier)	Description	Phase	Drug	HER2 Alteration
NCT05482568	Phase IB/II Clinical Study of the Safety, Tolerability, Pharmacokinetics, and Efficacy of Injectable SHR-A1811 in Combination With Pyrotinib or SHR-1316 in Subjects With Advanced Non-small Cell Lung Cancer With HER2	Ib/II	SHR-A1811 + Pyrotinib; SHR-A1811 + SHR-1316	HER2-expresssing, HER2 amplification, HER2 mutations
NCT04235101	A Two-part Phase I Study With the Antibody-drug Conjugate SYD985 in Combination With Niraparib to Evaluate Safety, Pharmacokinetics, and Efficacy in Patients With HER2-expressing Locally Advanced or Metastatic Solid Tumours	I	SYD985 + Niraparib	HER2-expressing

Abbreviations: ADC, antibody; NSCLC, non-small cell lung cancer; T-DXd, trastuzumab deruxtecan.

Emerging evidence highlights the importance of liquid biopsy in monitoring treatment efficacy, with recent studies showing that fluctuations in circulating tumour DNA (ctDNA) levels correlate with treatment outcomes and survival rates in solid tumours undergoing targeted therapy [68].

Consistent with previous findings, a recent analysis of advanced HER2-mutant NSCLC patients receiving pyrotinib from two phase II clinical trials reported improved treatment outcomes in those with ctDNA clearance after 40 days of therapy. However, the limited number of patients enrolled in the analysis makes further investigation necessary [69]. Finally, an interesting case report showed how the kinetics of HER2 mutation allele frequency assessed by plasma ctDNA analysis at different timepoints during treatment with T-DXd exactly matched the clinical course of the disease [70].

6. Conclusions

This review highlights the dynamic and evolving landscape of HER2-targeted therapies in NSCLC, marking significant advances from the initial use of monotherapies to more recent strategies of combination regimen. While platinum-pemetrexed-based chemotherapy still represents the standard of care for first-line treatment, emerging therapies, including T-DXd and novel TKIs, showed promising results also in treating patients with BMs. Notably, the outcomes of the DESTINY-Lung04 trial are keenly awaited as they hold the potential to significantly alter the treatment paradigm for patients with advanced HER2-mutant NSCLC. Moreover, emerging evidence is pointing to the potential role of liquid biopsy in monitoring disease progression, but further studies are needed.

Author Contributions: Conceptualization, F.P., G.F. and B.D.R.; methodology, F.P.; investigation, F.P., G.F. and B.D.R.; resources, F.P., G.F. and B.D.R.; data curation, F.P., G.F. and B.D.R.; writing—original draft preparation, F.P., G.F. and B.D.R.; writing—review and editing, F.P., G.F. and B.D.R.; visualization, G.F. and B.D.R.; supervision, F.P. and S.N.; project administration, F.P. All authors have read and agreed to the published version of the manuscript.

Funding: This research received no external funding.

Conflicts of Interest: F.P. declared consultant/advisory fees from Astra Zeneca, Janssen, Sanofi, Amgen, Roche, Bristol Myer Squibb, Beigene, and Thermofisher Scientific. S.N. declared speaker bureau/advisor's fees from Boehringer Ingelheim, Roche, Merck Sharp, Dohme, Amgen, Thermo Fisher Scientific, Eli Lilly, GlaxoSmithKline, Merck, AstraZeneca, Janssen, Novartis, Takeda, Bayer, Pfizer. The other authors have no conflicts of interest to declare.

References

1. Baselga, J.; Swain, S.M. Novel Anticancer Targets: Revisiting ERBB2 and Discovering ERBB3. *Nat. Rev. Cancer* **2009**, *9*, 463–475. [CrossRef] [PubMed]
2. Pahuja, K.B.; Nguyen, T.T.; Jaiswal, B.S.; Prabhash, K.; Thaker, T.M.; Senger, K.; Chaudhuri, S.; Kljavin, N.M.; Antony, A.; Phalke, S.; et al. Actionable Activating Oncogenic ERBB2/HER2 Transmembrane and Juxtamembrane Domain Mutations. *Cancer Cell* **2018**, *34*, 792–806.e5. [CrossRef] [PubMed]
3. Riudavets, M.; Sullivan, I.; Abdayem, P.; Planchard, D. Targeting HER2 in Non-Small-Cell Lung Cancer (NSCLC): A Glimpse of Hope? An Updated Review on Therapeutic Strategies in NSCLC Harbouring HER2 Alterations. *ESMO Open* **2021**, *6*, 100260. [CrossRef] [PubMed]
4. Cocco, E.; Lopez, S.; Santin, A.D.; Scaltriti, M. Prevalence and Role of HER2 Mutations in Cancer. *Pharmacol. Ther.* **2019**, *199*, 188–196. [CrossRef] [PubMed]
5. Li, B.T.; Ross, D.S.; Aisner, D.L.; Chaft, J.E.; Hsu, M.; Kako, S.L.; Kris, M.G.; Varella-Garcia, M.; Arcila, M.E. HER2 Amplification and Her2 Mutation Are Distinct Molecular Targets in Lung Cancers. *J. Thorac. Oncol.* **2016**, *11*, 414–419. [CrossRef] [PubMed]
6. Ren, S.; Wang, J.; Ying, J.; Mitsudomi, T.; Lee, D.H.; Wang, Z.; Chu, Q.; Mack, P.C.; Cheng, Y.; Duan, J.; et al. Consensus for HER2 Alterations Testing in Non-Small-Cell Lung Cancer. *ESMO Open* **2022**, *7*, 100395. [CrossRef] [PubMed]
7. Yu, Y.; Yang, Y.; Li, H.; Fan, Y. Targeting HER2 Alterations in Non-Small Cell Lung Cancer: Therapeutic Breakthrough and Challenges. *Cancer Treat. Rev.* **2023**, *114*, 102520. [CrossRef] [PubMed]
8. Mazières, J.; Barlesi, F.; Filleron, T.; Besse, B.; Monnet, I.; Beau-Faller, M.; Peters, S.; Dansin, E.; Früh, M.; Pless, M.; et al. Lung Cancer Patients with HER2 Mutations Treated with Chemotherapy and HER2-Targeted Drugs: Results from the European EUHER2 Cohort. *Ann. Oncol.* **2016**, *27*, 281–286. [CrossRef] [PubMed]
9. Pillai, R.N.; Behera, M.; Berry, L.D.; Rossi, M.R.; Kris, M.G.; Johnson, B.E.; Bunn, P.A.; Ramalingam, S.S.; Khuri, F.R. HER2 Mutations in Lung Adenocarcinomas: A Report from the Lung Cancer Mutation Consortium. *Cancer* **2017**, *123*, 4099–4105. [CrossRef]
10. Yang, S.; Wang, Y.; Zhao, C.; Li, X.; Liu, Q.; Mao, S.; Liu, Y.; Yu, X.; Wang, W.; Tian, Q.; et al. Exon 20 YVMA Insertion Is Associated with High Incidence of Brain Metastasis and Inferior Outcome of Chemotherapy in Advanced Non-Small Cell Lung Cancer Patients with HER2 Kinase Domain Mutations. *Transl. Lung Cancer Res.* **2021**, *10*, 753–765. [CrossRef]
11. Zhao, S.; Fang, W.; Pan, H.; Yang, Y.; Liang, Y.; Yang, L.; Dong, X.; Zhan, J.; Wang, K.; Zhang, L. Conformational Landscapes of HER2 Exon 20 Insertions Explain Their Sensitivity to Kinase Inhibitors in Lung Adenocarcinoma. *J. Thorac. Oncol.* **2020**, *15*, 962–972. [CrossRef] [PubMed]
12. Yu, X.; Ji, X.; Su, C. HER2-Altered Non-Small Cell Lung Cancer: Biology, Clinicopathologic Features, and Emerging Therapies. *Front. Oncol.* **2022**, *12*, 860313. [CrossRef] [PubMed]
13. Nützinger, J.; Bum Lee, J.; Li Low, J.; Ling Chia, P.; Talisa Wijaya, S.; Chul Cho, B.; Min Lim, S.; Soo, R.A. Management of HER2 Alterations in Non-Small Cell Lung Cancer—The Past, Present, and Future. *Lung Cancer* **2023**, *186*, 107385. [CrossRef] [PubMed]
14. Takezawa, K.; Pirazzoli, V.; Arcila, M.E.; Nebhan, C.A.; Song, X.; de Stanchina, E.; Ohashi, K.; Janjigian, Y.Y.; Spitzler, P.J.; Melnick, M.A.; et al. HER2 Amplification: A Potential Mechanism of Acquired Resistance to EGFR Inhibition in EGFR-Mutant Lung Cancers That Lack the Second-Site EGFRT790M Mutation. *Cancer Discov.* **2012**, *2*, 922–933. [CrossRef] [PubMed]
15. Kim, E.K.; Kim, K.A.; Lee, C.Y.; Shim, H.S. The Frequency and Clinical Impact of HER2 Alterations in Lung Adenocarcinoma. *PLoS ONE* **2017**, *12*, e0171280. [CrossRef] [PubMed]
16. Bunn, P.A.; Helfrich, B.; Soriano, A.F.; Franklin, W.A.; Varella-Garcia, M.; Hirsch, F.R.; Baron, A.; Zeng, C.; Chan, D.C. Expression of Her-2/Neu in Human Lung Cancer Cell Lines by Immunohistochemistry and Fluorescence in Situ Hybridization and Its Relationship to in Vitro Cytotoxicity by Trastuzumab and Chemotherapeutic Agents. *Clin. Cancer Res.* **2001**, *7*, 3239–3250. [PubMed]
17. Hirsch, F.R.; Varella-Garcia, M.; Franklin, W.A.; Veve, R.; Chen, L.; Helfrich, B.; Zeng, C.; Baron, A.; Bunn, P.A. Evaluation of HER-2/Neu Gene Amplification and Protein Expression in Non-Small Cell Lung Carcinomas. *Br. J. Cancer* **2002**, *86*, 1449–1456. [CrossRef] [PubMed]
18. Hendriks, L.E.; Kerr, K.M.; Menis, J.; Mok, T.S.; Nestle, U.; Passaro, A.; Peters, S.; Planchard, D.; Smit, E.F.; Solomon, B.J.; et al. Oncogene-Addicted Metastatic Non-Small-Cell Lung Cancer: ESMO Clinical Practice Guideline for Diagnosis, Treatment and Follow-Up. *Ann. Oncol.* **2023**, *34*, 339–357. [CrossRef] [PubMed]
19. Wang, Y.; Zhang, S.; Wu, F.; Zhao, J.; Li, X.; Zhao, C.; Ren, S.; Zhou, C. Outcomes of Pemetrexed-Based Chemotherapies in HER2-Mutant Lung Cancers. *BMC Cancer* **2018**, *18*, 326. [CrossRef]
20. Negrao, M.V.; Skoulidis, F.; Montesion, M.; Schulze, K.; Bara, I.; Shen, V.; Xu, H.; Hu, S.; Sui, D.; Elamin, Y.Y.; et al. Oncogene-Specific Differences in Tumor Mutational Burden, PD-L1 Expression, and Outcomes from Immunotherapy in Non-Small Cell Lung Cancer. *J. Immunother. Cancer* **2021**, *9*, e002891. [CrossRef]
21. Lai, W.-C.V.; Feldman, D.L.; Buonocore, D.J.; Brzostowski, E.B.; Rizvi, H.; Plodkowski, A.J.; Ni, A.; Sabari, J.K.; Offin, M.D.; Kris, M.G.; et al. PD-L1 Expression, Tumor Mutation Burden and Response to Immune Checkpoint Blockade in Patients with *HER2*-Mutant Lung Cancers. *J. Clin. Oncol.* **2018**, *36*, 9060. [CrossRef]
22. Guisier, F.; Dubos-Arvis, C.; Viñas, F.; Doubre, H.; Ricordel, C.; Ropert, S.; Janicot, H.; Bernardi, M.; Fournel, P.; Lamy, R.; et al. Efficacy and Safety of Anti–PD-1 Immunotherapy in Patients with Advanced NSCLC with BRAF, HER2, or MET Mutations or RET Translocation: GFPC 01-2018. *J. Thorac. Oncol.* **2020**, *15*, 628–636. [CrossRef] [PubMed]

23. Mazieres, J.; Drilon, A.; Lusque, A.; Mhanna, L.; Cortot, A.B.; Mezquita, L.; Thai, A.A.; Mascaux, C.; Couraud, S.; Veillon, R.; et al. Immune Checkpoint Inhibitors for Patients with Advanced Lung Cancer and Oncogenic Driver Alterations: Results from the IMMUNOTARGET Registry. *Ann. Oncol.* **2019**, *30*, 1321–1328. [CrossRef] [PubMed]
24. Saalfeld, F.C.; Wenzel, C.; Christopoulos, P.; Merkelbach-Bruse, S.; Reissig, T.M.; Laßmann, S.; Thiel, S.; Stratmann, J.A.; Marienfeld, R.; Berger, J.; et al. Efficacy of Immune Checkpoint Inhibitors Alone or in Combination With Chemotherapy in NSCLC Harboring ERBB2 Mutations. *J. Thorac. Oncol.* **2021**, *16*, 1952–1958. [CrossRef] [PubMed]
25. Gandhi, L.; Rodríguez-Abreu, D.; Gadgeel, S.; Esteban, E.; Felip, E.; De Angelis, F.; Domine, M.; Clingan, P.; Hochmair, M.J.; Powell, S.F.; et al. Pembrolizumab plus Chemotherapy in Metastatic Non–Small-Cell Lung Cancer. *N. Engl. J. Med.* **2018**, *378*, 2078–2092. [CrossRef]
26. Yang, G.; Yang, Y.; Liu, R.; Li, W.; Xu, H.; Hao, X.; Li, J.; Xing, P.; Zhang, S.; Ai, X.; et al. First-Line Immunotherapy or Angiogenesis Inhibitor plus Chemotherapy for *HER2*-Altered NSCLC: A Retrospective Real-World POLISH Study. *Ther. Adv. Med. Oncol.* **2022**, *14*, 175883592210823. [CrossRef] [PubMed]
27. Peters, S.; Curioni-Fontecedro, A.; Nechushtan, H.; Shih, J.-Y.; Liao, W.-Y.; Gautschi, O.; Spataro, V.; Unk, M.; Yang, J.C.-H.; Lorence, R.M.; et al. Activity of Afatinib in Heavily Pretreated Patients With ERBB2 Mutation–Positive Advanced NSCLC: Findings From a Global Named Patient Use Program. *J. Thorac. Oncol.* **2018**, *13*, 1897–1905. [CrossRef]
28. Fang, W.; Zhao, S.; Liang, Y.; Yang, Y.; Yang, L.; Dong, X.; Zhang, L.; Tang, Y.; Wang, S.; Yang, Y.; et al. Mutation Variants and Co-Mutations as Genomic Modifiers of Response to Afatinib in *HER2*-Mutant Lung Adenocarcinoma. *Oncologist* **2020**, *25*, e545–e554. [CrossRef] [PubMed]
29. Yuan, B.; Zhao, J.; Zhou, C.; Wang, X.; Zhu, B.; Zhuo, M.; Dong, X.; Feng, J.; Yi, C.; Yang, Y.; et al. Co-Occurring Alterations of ERBB2 Exon 20 Insertion in Non-Small Cell Lung Cancer (NSCLC) and the Potential Indicator of Response to Afatinib. *Front. Oncol.* **2020**, *10*, 729. [CrossRef]
30. Dziadziuszko, R.; Smit, E.F.; Dafni, U.; Wolf, J.; Wasąg, B.; Biernat, W.; Finn, S.P.; Kammler, R.; Tsourti, Z.; Rabaglio, M.; et al. Afatinib in NSCLC With HER2 Mutations: Results of the Prospective, Open-Label Phase II NICHE Trial of European Thoracic Oncology Platform (ETOP). *J. Thorac. Oncol.* **2019**, *14*, 1086–1094. [CrossRef]
31. Kris, M.G.; Camidge, D.R.; Giaccone, G.; Hida, T.; Li, B.T.; O'Connell, J.; Taylor, I.; Zhang, H.; Arcila, M.E.; Goldberg, Z.; et al. Targeting HER2 Aberrations as Actionable Drivers in Lung Cancers: Phase II Trial of the Pan-HER Tyrosine Kinase Inhibitor Dacomitinib in Patients with HER2-Mutant or Amplified Tumors. *Ann. Oncol.* **2015**, *26*, 1421–1427. [CrossRef] [PubMed]
32. Hyman, D.M.; Piha-Paul, S.A.; Won, H.; Rodon, J.; Saura, C.; Shapiro, G.I.; Juric, D.; Quinn, D.I.; Moreno, V.; Doger, B.; et al. HER Kinase Inhibition in Patients with HER2- and HER3-Mutant Cancers. *Nature* **2018**, *554*, 189–194. [CrossRef] [PubMed]
33. Gandhi, L.; Besse, B.; Mazieres, J.; Waqar, S.; Cortot, A.; Barlesi, F.; Quoix, E.; Otterson, G.; Ettinger, D.; Horn, L.; et al. MA04.02 Neratinib ± Temsirolimus in HER2-Mutant Lung Cancers: An International, Randomized Phase II Study. *J. Thorac. Oncol.* **2017**, *12*, S358–S359. [CrossRef]
34. Le, X.; Cornelissen, R.; Garassino, M.; Clarke, J.M.; Tchekmedyian, N.; Goldman, J.W.; Leu, S.-Y.; Bhat, G.; Lebel, F.; Heymach, J.V.; et al. Poziotinib in Non–Small-Cell Lung Cancer Harboring *HER2* Exon 20 Insertion Mutations After Prior Therapies: ZENITH20-2 Trial. *J. Clin. Oncol.* **2022**, *40*, 710–718. [CrossRef] [PubMed]
35. Sun, S.; Prelaj, A.; Baik, C.; Le, X.; Garassino, M.; Wollner, M.; Haura, E.; Piotrowska, Z.; Socinski, M.; Dreiling, L.; et al. 26MO Efficacy and Safety of Poziotinib in Treatment-Naïve HER2 Exon 20 Insertion (Ex20ins) Mutated Non-Small Cell Lung Cancer (NSCLC): ZENITH20-4. *Ann. Oncol.* **2022**, *33*, S13. [CrossRef]
36. Zhou, C.; Li, X.; Wang, Q.; Gao, G.; Zhang, Y.; Chen, J.; Shu, Y.; Hu, Y.; Fan, Y.; Fang, J.; et al. Pyrotinib in *HER2*-Mutant Advanced Lung Adenocarcinoma After Platinum-Based Chemotherapy: A Multicenter, Open-Label, Single-Arm, Phase II Study. *J. Clin. Oncol.* **2020**, *38*, 2753–2761. [CrossRef] [PubMed]
37. Song, Z.; Li, Y.; Chen, S.; Ying, S.; Xu, S.; Huang, J.; Wu, D.; Lv, D.; Bei, T.; Liu, S.; et al. Efficacy and Safety of Pyrotinib in Advanced Lung Adenocarcinoma with HER2 Mutations: A Multicenter, Single-Arm, Phase II Trial. *BMC Med.* **2022**, *20*, 42. [CrossRef]
38. Yang, G.; Xu, H.; Yang, Y.; Zhang, S.; Xu, F.; Hao, X.; Li, J.; Xing, P.; Hu, X.; Liu, Y.; et al. Pyrotinib Combined with Apatinib for Targeting Metastatic Non-Small Cell Lung Cancer with HER2 Alterations: A Prospective, Open-Label, Single-Arm Phase 2 Study (PATHER2). *BMC Med.* **2022**, *20*, 277. [CrossRef] [PubMed]
39. Neal, J.; Doebele, R.; Riely, G.; Spira, A.; Horn, L.; Piotrowska, Z.; Costa, D.; Zhang, S.; Bottino, D.; Zhu, J.; et al. P1.13-44 Safety, PK, and Preliminary Antitumor Activity of the Oral EGFR/HER2 Exon 20 Inhibitor TAK-788 in NSCLC. *J. Thorac. Oncol.* **2018**, *13*, S599. [CrossRef]
40. Riely, G.J.; Neal, J.W.; Camidge, D.R.; Spira, A.; Piotrowska, Z.; Horn, L.; Costa, D.B.; Tsao, A.; Patel, J.; Gadgeel, S.; et al. 1261MO Updated Results from a Phase I/II Study of Mobocertinib (TAK-788) in NSCLC with EGFR Exon 20 Insertions (Exon20ins). *Ann. Oncol.* **2020**, *31*, S815–S816. [CrossRef]
41. Heymach, J.; Opdam, F.; Barve, M.A.; Tu, H.-Y.; Wu, Y.-L.; Berz, D.; Gibson, N.; Sadrolhefazi, B.; Serra, J.; Yoh, K.; et al. Phase I Beamion Lung 1 Trial of BI 1810631, a HER2 Tyrosine Kinase Inhibitor (TKI), as Monotherapy in Patients (Pts) with Advanced/Metastatic Solid Tumors with HER2 Aberrations: Updated Data. *J. Clin. Oncol.* **2023**, *41*, 8545. [CrossRef]
42. Aujay, M.; Broad, A.J.; Gross, S.D.; Ren, L.; Lyssikatos, J.P.; Kintz, S.; Wang, Q.; Collins, H. Abstract 4019: Preclinical Activity of ELVN-002: A Potent, Selective, and Irreversible HER2 and Pan-HER2 Mutant Small Molecule Inhibitor for the Treatment of HER2 Driven Malignancies. *Cancer Res.* **2023**, *83*, 4019. [CrossRef]

43. Langer, C.J.; Stephenson, P.; Thor, A.; Vangel, M.; Johnson, D.H. Trastuzumab in the Treatment of Advanced Non-Small-Cell Lung Cancer: Is There a Role? Focus on Eastern Cooperative Oncology Group Study 2598. *J. Clin. Oncol.* **2004**, *22*, 1180–1187. [CrossRef] [PubMed]
44. Gatzemeier, U.; Groth, G.; Butts, C.; Van Zandwijk, N.; Shepherd, F.; Ardizzoni, A.; Barton, C.; Ghahramani, P.; Hirsh, V. Randomized Phase II Trial of Gemcitabine–Cisplatin with or without Trastuzumab in HER2-Positive Non-Small-Cell Lung Cancer. *Ann. Oncol.* **2004**, *15*, 19–27. [CrossRef] [PubMed]
45. Kinoshita, I.; Goda, T.; Watanabe, K.; Maemondo, M.; Oizumi, S.; Amano, T.; Hatanaka, Y.; Matsuno, Y.; Nishihara, H.; Asahina, H.; et al. A Phase II Study of Trastuzumab Monotherapy in Pretreated Patients with Non-Small Cell Lung Cancers (NSCLCs) Harboring HER2 Alterations: HOT1303-B Trial. *Ann. Oncol.* **2018**, *29*, viii540. [CrossRef]
46. Ganti, A.K.; Rothe, M.; Mangat, P.K.; Garrett-Mayer, E.; Dib, E.G.; Duvivier, H.L.; Ahn, E.R.; Behl, D.; Borghaei, H.; Balmanoukian, A.S.; et al. Pertuzumab Plus Trastuzumab in Patients with Lung Cancer with *ERBB2* Mutation or Amplification: Results From the Targeted Agent and Profiling Utilization Registry Study. *JCO Precis. Oncol.* **2023**, *7*, e2300041. [CrossRef] [PubMed]
47. Mazieres, J.; Lafitte, C.; Ricordel, C.; Greillier, L.; Negre, E.; Zalcman, G.; Domblides, C.; Madelaine, J.; Bennouna, J.; Mascaux, C.; et al. Combination of Trastuzumab, Pertuzumab, and Docetaxel in Patients With Advanced Non–Small-Cell Lung Cancer Harboring *HER2* Mutations: Results From the IFCT-1703 R2D2 Trial. *J. Clin. Oncol.* **2022**, *40*, 719–728. [CrossRef] [PubMed]
48. Li, B.T.; Shen, R.; Buonocore, D.; Olah, Z.T.; Ni, A.; Ginsberg, M.S.; Ulaner, G.A.; Offin, M.; Feldman, D.; Hembrough, T.; et al. Ado-Trastuzumab Emtansine for Patients With *HER2*-Mutant Lung Cancers: Results From a Phase II Basket Trial. *J. Clin. Oncol.* **2018**, *36*, 2532–2537. [CrossRef] [PubMed]
49. Li, B.T.; Michelini, F.; Misale, S.; Cocco, E.; Baldino, L.; Cai, Y.; Shifman, S.; Tu, H.-Y.; Myers, M.L.; Xu, C.; et al. HER2-Mediated Internalization of Cytotoxic Agents in *ERBB2* Amplified or Mutant Lung Cancers. *Cancer Discov.* **2020**, *10*, 674–687. [CrossRef]
50. Peters, S.; Stahel, R.; Bubendorf, L.; Bonomi, P.; Villegas, A.; Kowalski, D.M.; Baik, C.S.; Isla, D.; Carpeno, J.D.C.; Garrido, P.; et al. Trastuzumab Emtansine (T-DM1) in Patients with Previously Treated HER2-Overexpressing Metastatic Non–Small Cell Lung Cancer: Efficacy, Safety, and Biomarkers. *Clin. Cancer Res.* **2019**, *25*, 64–72. [CrossRef]
51. Iwama, E.; Zenke, Y.; Sugawara, S.; Daga, H.; Morise, M.; Yanagitani, N.; Sakamoto, T.; Murakami, H.; Kishimoto, J.; Matsumoto, S.; et al. Trastuzumab Emtansine for Patients with Non–Small Cell Lung Cancer Positive for Human Epidermal Growth Factor Receptor 2 Exon-20 Insertion Mutations. *Eur. J. Cancer* **2022**, *162*, 99–106. [CrossRef] [PubMed]
52. Hurvitz, S.A.; Hegg, R.; Chung, W.-P.; Im, S.-A.; Jacot, W.; Ganju, V.; Chiu, J.W.Y.; Xu, B.; Hamilton, E.; Madhusudan, S.; et al. Trastuzumab Deruxtecan versus Trastuzumab Emtansine in Patients with HER2-Positive Metastatic Breast Cancer: Updated Results from DESTINY-Breast03, a Randomised, Open-Label, Phase 3 Trial. *Lancet* **2023**, *401*, 105–117. [CrossRef] [PubMed]
53. Tsurutani, J.; Iwata, H.; Krop, I.; Jänne, P.A.; Doi, T.; Takahashi, S.; Park, H.; Redfern, C.; Tamura, K.; Wise-Draper, T.M.; et al. Targeting HER2 with Trastuzumab Deruxtecan: A Dose-Expansion, Phase I Study in Multiple Advanced Solid Tumors. *Cancer Discov.* **2020**, *10*, 688–701. [CrossRef] [PubMed]
54. Li, B.T.; Smit, E.F.; Goto, Y.; Nakagawa, K.; Udagawa, H.; Mazières, J.; Nagasaka, M.; Bazhenova, L.; Saltos, A.N.; Felip, E.; et al. Trastuzumab Deruxtecan in *HER2*-Mutant Non–Small-Cell Lung Cancer. *N. Engl. J. Med.* **2022**, *386*, 241–251. [CrossRef] [PubMed]
55. Abuhelwa, Z.; Alloghbi, A.; Alqahtani, A.; Nagasaka, M. Trastuzumab Deruxtecan-Induced Interstitial Lung Disease/Pneumonitis in ERBB2-Positive Advanced Solid Malignancies: A Systematic Review. *Drugs* **2022**, *82*, 979–987. [CrossRef] [PubMed]
56. Goto, K.; Sang-We, K.; Kubo, T.; Goto, Y.; Ahn, M.-J.; Planchard, D.; Kim, D.-W.; Yang, J.C.-H.; Yang, T.-Y.; Pereira, K.M.C.; et al. LBA55 Trastuzumab Deruxtecan (T-DXd) in Patients (Pts) with HER2-Mutant Metastatic Non-Small Cell Lung Cancer (NSCLC): Interim Results from the Phase 2 DESTINY-Lung02 Trial. *Ann. Oncol.* **2022**, *33*, S1422. [CrossRef]
57. Patil, T.; Mushtaq, R.; Marsh, S.; Azelby, C.; Pujara, M.; Davies, K.D.; Aisner, D.L.; Purcell, W.T.; Schenk, E.L.; Pacheco, J.M.; et al. Clinicopathologic Characteristics, Treatment Outcomes, and Acquired Resistance Patterns of Atypical EGFR Mutations and HER2 Alterations in Stage IV Non–Small-Cell Lung Cancer. *Clin. Lung Cancer* **2020**, *21*, e191–e204. [CrossRef] [PubMed]
58. Mazières, J.; Peters, S.; Lepage, B.; Cortot, A.B.; Barlesi, F.; Beau-Faller, M.; Besse, B.; Blons, H.; Mansuet-Lupo, A.; Urban, T.; et al. Lung Cancer That Harbors an *HER2* Mutation: Epidemiologic Characteristics and Therapeutic Perspectives. *J. Clin. Oncol.* **2013**, *31*, 1997–2003. [CrossRef]
59. Cornelissen, R.; Prelaj, A.; Sun, S.; Baik, C.; Wollner, M.; Haura, E.B.; Mamdani, H.; Riess, J.W.; Cappuzzo, F.; Garassino, M.C.; et al. Poziotinib in Treatment-Naive NSCLC Harboring HER2 Exon 20 Mutations: ZENITH20-4, A Multicenter, Multicohort, Open-Label, Phase 2 Trial (Cohort 4). *J. Thorac. Oncol.* **2023**, *18*, 1031–1041. [CrossRef]
60. Jacobson, A. Trastuzumab Deruxtecan Improves Progression-Free Survival and Intracranial Response in Patients with HER2-Positive Metastatic Breast Cancer and Brain Metastases. *Oncologist* **2022**, *27*, S3–S4. [CrossRef]
61. Jerusalem, G.; Park, Y.H.; Yamashita, T.; Hurvitz, S.A.; Modi, S.; Andre, F.; Krop, I.E.; Gonzàlez Farré, X.; You, B.; Saura, C.; et al. Trastuzumab Deruxtecan in HER2-Positive Metastatic Breast Cancer Patients with Brain Metastases: A DESTINY-Breast01 Subgroup Analysis. *Cancer Discov.* **2022**, *12*, 2754–2762. [CrossRef] [PubMed]
62. Pérez-García, J.M.; Vaz Batista, M.; Cortez, P.; Ruiz-Borrego, M.; Cejalvo, J.M.; de la Haba-Rodriguez, J.; Garrigós, L.; Racca, F.; Servitja, S.; Blanch, S.; et al. Trastuzumab Deruxtecan in Patients with Central Nervous System Involvement from HER2-Positive Breast Cancer: The DEBBRAH Trial. *Neuro Oncol.* **2023**, *25*, 157–166. [CrossRef]
63. Bartsch, R.; Berghoff, A.S.; Furtner, J.; Marhold, M.; Bergen, E.S.; Roider-Schur, S.; Starzer, A.M.; Forstner, H.; Rottenmanner, B.; Dieckmann, K.; et al. Trastuzumab Deruxtecan in HER2-Positive Breast Cancer with Brain Metastases: A Single-Arm, Phase 2 Trial. *Nat. Med.* **2022**, *28*, 1840–1847. [CrossRef] [PubMed]

64. D'Amico, L.; Menzel, U.; Prummer, M.; Müller, P.; Buchi, M.; Kashyap, A.; Haessler, U.; Yermanos, A.; Gébleux, R.; Briendl, M.; et al. A Novel Anti-HER2 Anthracycline-Based Antibody-Drug Conjugate Induces Adaptive Anti-Tumor Immunity and Potentiates PD-1 Blockade in Breast Cancer. *J. Immunother. Cancer* **2019**, *7*, 16. [CrossRef] [PubMed]
65. Cheema, P.; Hartl, S.; Koczywas, M.; Hochmair, M.; Shepherd, F.A.; Chu, Q.; Galletti, G.; Gustavson, M.; Iyer, S.; Carl Barrett, J.; et al. 695 Efficacy and Safety of Trastuzumab Deruxtecan (T-DXd) with Durvalumab in Patients with Non-Small Cell Lung Cancer (HER2 Altered NSCLC) Who Progressed on Anti-PD1/PD-L1 Therapy (HUDSON). In *Proceedings of the Regular and Young Investigator Award Abstracts;* BMJ Publishing Group Ltd.: London, UK, 2023; p. A787.
66. Planchard, D.; Brahmer, J.R.; Yang, J.C.-H.; Kim, H.R.; Li, R.K.; Han, J.-Y.; Cortinovis, D.L.; Runglodvatana, Y.; Nakajima, E.; Ragone, A.; et al. 1507TiP Phase Ib Multicenter Study of Trastuzumab Deruxtecan (T-DXd) and Immunotherapy with or without Chemotherapy in First-Line Treatment of Patients (Pts) with Advanced or Metastatic Nonsquamous Non-Small Cell Lung Cancer (NSCLC) and HER2 Overexpression (OE): DESTINY-Lung03. *Ann. Oncol.* **2023**, *34*, S848–S849. [CrossRef]
67. Suh, K.J.; Sung, J.H.; Kim, J.W.; Han, S.-H.; Lee, H.S.; Min, A.; Kang, M.H.; Kim, J.E.; Kim, J.-W.; Kim, S.H.; et al. EGFR or HER2 Inhibition Modulates the Tumor Microenvironment by Suppression of PD-L1 and Cytokines Release. *Oncotarget* **2017**, *8*, 63901–63910. [CrossRef] [PubMed]
68. Nikanjam, M.; Kato, S.; Kurzrock, R. Liquid Biopsy: Current Technology and Clinical Applications. *J. Hematol. Oncol.* **2022**, *15*, 131. [CrossRef] [PubMed]
69. Mao, S.; Yang, S.; Liu, X.; Li, X.; Wang, Q.; Zhang, Y.; Chen, J.; Wang, Y.; Gao, G.; Wu, F.; et al. Molecular Correlation of Response to Pyrotinib in Advanced NSCLC with HER2 Mutation: Biomarker Analysis from Two Phase II Trials. *Exp. Hematol. Oncol.* **2023**, *12*, 53. [CrossRef]
70. Falk, M.; Willing, E.; Schmidt, S.; Schatz, S.; Galster, M.; Tiemann, M.; Ficker, J.H.; Brueckl, W.M. Response of an HER2-Mutated NSCLC Patient to Trastuzumab Deruxtecan and Monitoring of Plasma CtDNA Levels by Liquid Biopsy. *Curr. Oncol.* **2023**, *30*, 1692–1698. [CrossRef]

cancers

MDPI

Review

NRG1 Gene Fusions—What Promise Remains Behind These Rare Genetic Alterations? A Comprehensive Review of Biology, Diagnostic Approaches, and Clinical Implications

Tomasz Kucharczyk [1,*], Marcin Nicoś [1], Marek Kucharczyk [2] and Ewa Kalinka [3]

1 Department of Pneumonology, Oncology and Allergology, Medical University of Lublin, 20-059 Lublin, Poland; marcin.nicos@umlub.pl
2 Department of Zoology and Nature Conservation, Institute of Biology, Maria Curie-Sklodowska University in Lublin, 20-033 Lublin, Poland; marek.kucharczyk@mail.umcs.pl
3 Oncology Clinic, Institute of the Polish Mother's Health Center in Lodz, 93-338 Lodz, Poland; ewakalinka@wp.pl
* Correspondence: tomasz.kucharczyk@umlub.pl

Simple Summary: Neuregulin-1 (NRG1) is an important regulator of ErbB-mediated pathways involved in cancer development. Recently, there have been several studies analyzing *NRG-1* gene fusions engaged in altering the dimerization of HER family proteins and the consecutive results of their activation in different types of cancer. Non-small cell lung cancer (NSCLC) patients can benefit from pan-HER inhibitors, and knowledge of *NRG-1* fusions can help tailor the treatment to a specific group of patients. New drugs targeting cells with *NRG-1* fusions are under clinical trials and show effectiveness in NSCLC treatment.

Abstract: Non-small cell lung cancer (NSCLC) presents a variety of druggable genetic alterations that revolutionized the treatment approaches. However, identifying new alterations may broaden the group of patients benefitting from such novel treatment options. Recently, the interest focused on the neuregulin-1 gene (*NRG1*), whose fusions may have become a potential predictive factor. To date, the occurrence of *NRG1* fusions has been considered a negative prognostic marker in NSCLC treatment; however, many premises remain behind the targetability of signaling pathways affected by the *NRG1* gene. The role of *NRG1* fusions in ErbB-mediated cell proliferation especially seems to be considered as a main target of treatment. Hence, NSCLC patients harboring *NRG1* fusions may benefit from targeted therapies such as pan-HER family inhibitors, which have shown efficacy in previous studies in various cancers, and anti-HER monoclonal antibodies. Considering the increased interest in the *NRG1* gene as a potential clinical target, in the following review, we highlight its biology, as well as the potential clinical implications that were evaluated in clinics or remained under consideration in clinical trials.

Keywords: NSCLC; *NRG1* fusions; molecularly targeted

1. Introduction

In the era of precision therapy, novel driver alterations are extensively studied to qualify the patient for the best-fitting treatment. In 2020, lung cancer accounted for 11.9% of all new cancer diagnoses in Europe, constituting about 480,000 people [1]. Non-small cell lung cancer (NSCLC), which includes 85% of lung cancer cases, is one of the cancer types that presents a variety of actionable genetic changes, with quite a few available targeted drugs that have revolutionized the treatment approaches [2]. Nevertheless, the percentage of patients that receive already popular epidermal growth factor receptor (EGFR) tyrosine kinase inhibitors (TKIs), anaplastic lymphoma kinase (ALK) inhibitors, or less common molecules is still low. The search for new targets is still ongoing to allow a larger group of patients treatment tailored specifically for them [3]. Identifying such alterations or fusions may effectively select those benefiting from such novel treatment options. Apart from gene mutations, rearrangements and fusions of different genes are among the most commonly diagnosed cancer cell driver alterations [2]. The inhibitors of anaplastic lymphoma kinase (*ALK*), ROS proto-oncogene 1 (*ROS1*), rearranged during transfection (*RET*), and neurotrophic tyrosine receptor kinase (*NTRK*) rearrangements are already present in our everyday clinical practice [4,5]. Recently, interest has focused on neuregulin-1 (NRG1) as a potential oncogenic target.

The *NRG1* gene harbors several variants that have been classified on the Evidence for Sequence-variant Classification (ESCAT) scale as likely benign (rs3924999—intron region; rs7832768—promoter region) or uncertain significance (rs10503929 and rs16879552—both intronic), and their clinical relevance should be confirmed [6]. On the other hand, *NRG1* fusions might be considered the main oncogenic factor in solid tumors. The first description of such fusions (*CD74-NRG1*) in invasive mucinous adenocarcinoma of the lung (IMA) was in 2014. The targeted drugs for patients with *NRG1* fusions are still in clinical trials, and the search continues [7].

NRG1 rearrangements are uncommon compared to other more often described gene alterations found in NSCLC. In one of the studies, it was present in 0.5% of patients (2 of 404 analyzed cases) [8], in the other in 0.3% of patients (25 of 9252 analyzed samples) [9]. The prevalence of *NRG1* rearrangements in other types of cancers is similar to NSCLC and amounts to 0.5% in cholangiocarcinoma, pancreatic carcinoma, and renal cell carcinoma, 0.4% in ovarian cancer, and 0.2% in breast cancer and sarcoma [7,9,10]. Thus, the NRG1 fusions may be considered as biomarkers in various cancer types. Moreover, these fusions exclude the occurrence of other cancer-driving genetic changes. In some NSCLC cases, however, their presence was described along with mutations in *KRAS* and *BRAF* or *ALK* rearrangements [9,11].

To date, the occurrence of *NRG1* fusions was considered a negative prognostic marker in NSCLC treatment. Patients harboring such alterations presented reduced overall survival (OS) when treated with standard chemotherapy, chemoimmunotherapy, or immunotherapy alone. However, there are many premises behind the targetability of pathways affected by this alteration [9,12,13].

Considering the increased interest in the *NRG1* gene as a potential clinical target, this review discusses the structure and biology of the *NRG1* gene and the occurrence of potential genetic fusions. Furthermore, we indicate the *NRG1* fusion detection methods that are based on both high-throughput or single-gene approaches. In the end, we highlight the druggable applications of *NRG1* fusions as the first or secondary target in the treatment of NSCLC, which are already available in clinics or are still under consideration in clinical trials.

2. Structure and Biology of NRG1 Fusions

Neuregulin 1 is a protein, encoded by the *NRG1* gene located on the short arm of chromosome 8 (8p12), that is involved in various biological processes, including neural development, synaptic plasticity, myelination, and inter-cell signaling in the heart and breast [14]. The function of NRG1 is necessary for the early stages of development, and its absence, as was shown in mouse models, does not allow for proper embryonic development [15]. *NRG1* gene has many tissue-specific isoforms, created through alternative splicing, that differ structurally from each other. However, most isoforms contain the same extracellular epidermal growth factor-like (EGF-like) domain [14,16,17], which is crucial in the case of *NRG1* fusions to keep the functionality of the aberrant protein and drive cancer cell development. Most isoforms of NRG1 are bound to the cell membrane as a precursor. During proteolytic processes, the mature NRG1 is released, which can be transported further from the cell of origin and activate receptors on the surface of other cells. However, isoform III of neuregulin 1 retains the EGF-like domain in the membrane, which allows for the activation of mainly neighboring cells [18]. Moreover, there are some premises that epigenetic changes may also dysregulate the NRG1 expression, leading to its involvement in cancer development and progression [19]. The schematic localisation and structure of the *NRG1* gene is presented in Figure 1.

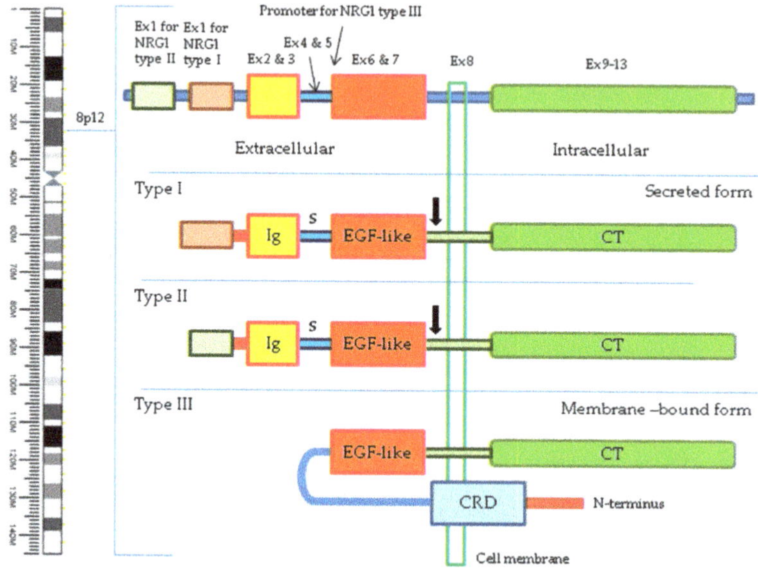

Figure 1. Schematic of *NRG1* gene structure showing position on chromosome 8, composition of exons, and structure of protein isoforms. Ig—immunoglobulin-like domain; S—stalk; EGF-like—EGF-like domain; CT—cytoplasmic tail; CRD—cysteine-rich domain. Black arrows indicate the location of cleavage in secreted types of NRG1.

The EGF-like domain of NRG1 protein is mainly an activator of Erb-B2 tyrosine kinase receptor 3 (ErbB3 also called HER3, human epidermal growth factor receptor 3), subsequently activating heterodimerization, most frequently ErbB2-ErbB3, but also EGFR or ErbB4, and further downstream signaling through mitogen-activated protein kinase (MAPK) and phosphoinositide-3-kinase (PI3K)/AKT/mammalian target of rapamycin (mTOR) pathways [20,21]. Although ErbB3 has seriously decreased kinase activity, its dimerization with other ErbB family receptors, after activation by NRG1, allows further downstream activation of the aforementioned pathways [22,23]. The NRG1 fusion proteins act as abnormal activators of ErbB-mediated cell proliferation pathways, and the result of such activation is the promotion of proliferation of molecularly altered cells.

It is postulated that *NRG1* fusions act similarly to neuregulin-1 isoform III, with a membrane-attached EGF-like domain. Different *NRG1* fusions can activate different homo- or heterodimers of ErbB [24], hence, they can activate diverse downstream pathways and result in alternative results from blockade attempts. The scheme of dimeric ErbB downstream signaling pathways regulated by the NRG1 protein is presented in Figure 2.

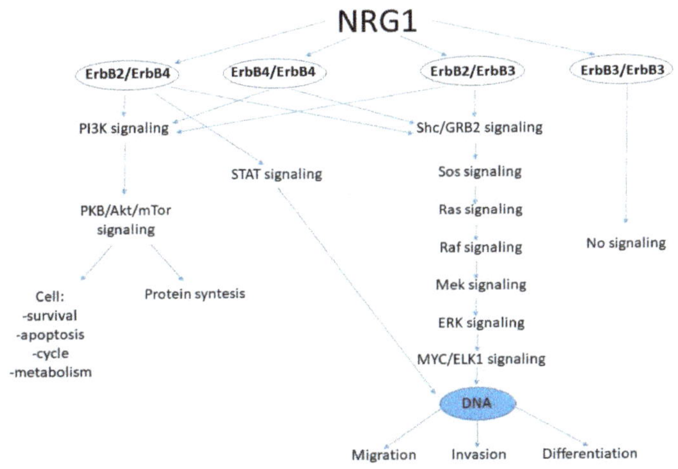

Figure 2. The scheme of dimeric ErbB downstream signaling pathways regulated by the NRG1 protein. The NRG1 protein, through different biological cascades, affects the ErbB dimers for protein synthesis, cell survival, cell apoptosis, control of cell cycle and metabolism, as well as cell migration, invasion, or differentiation.

3. Occurrence of *NRG1* Fusions

As previously stated, the first discovery of the *CD74-NRG1* fusion was described in 2014 in a study of 25 lung adenocarcinoma patients without *KRAS* or *EGFR* mutations. The described five cases were detected in non-smoking females with the IMA subtype [7]. Since that discovery, most of the *CD74-NRG1* fusion cases have been presented in this subtype of lung adenocarcinoma [25]. Subsequently, other groups of researchers identified different fusion partners of the *NRG1* gene: *SLC3A2* [9,12,26], *SDC4* [9,27], *RBPMS* [9,28], *VAMP2* [29], *WRN* [28], *ATP1B1* [27], *ROCK1* [25], *RALGAPA1* [30], *TNC* [9], *MDK* [9], *DIP2B* [9], *MRPL13* [9], *DPYSL2* [9], *PARP8* [9], *THAP7* [25], *SMAD4* [25], *KIF13B* [13], *ITGB1* [31], *UBXN8* [10], *NPTN* [32], *CADM1* [33], *F11R* [33], *FGFR1* [33], *FLYWCH1* [33], *KRAS* [33], *PLCG2* [33], and *VAPB* [33]. To date, the study by Jonna and co-workers is the most comprehensive analysis regarding *NRG1* fusions in solid tumors, where the incidence of the most common partners was as follows: *CD74* (29%), *ATP1B1* (10%), *SDC4* (7%), and *RBPMS* (5%). All the other fusions detected in the analyzed group of 21,858 solid tumor samples occurred with 2% frequency [9].

Fusions of *NRG1* and a few different partner genes have been described in other tumors as well, namely in ovarian cancer: *SETD4* [9], *TSHZ2* [9], *ZMYM2* [9], *RAB3IL1* [25], and *CLU* [25,34], and in pancreatic ductal adenocarcinoma: *VTCN1* [9], *CDH1* [9], *CDH6* [35], *SARAF* [10,35], *APP* [36], and *CDK1* [37]. Other described individual cases include breast cancer: *ADAM9* [9], *COX10-AS1* [9], *AKAP13* [25], *FOXA1* [25], *DDHD2* [10], *FUT10* [10], *BRE* [10], *CD9* [10], *ARHGEF39* [38], *FAM91A1* [38], and *ZNF704* [38], colorectal carcinoma: *IKBKB* [10], *ZCCHC7* [10], *TNRFSF10B* [10], *ERO1L* [10], and *KCTD9* [10], esophageal carcinoma: *BIN3* [10] and *CCAR2* [10], gallbladder carcinoma: *NOTCH2* [9], head and neck squamous carcinoma: *THBS1* [25] and *PDE7A* [25], bladder cancer: *GDF15* [9], renal cell carcinoma: *PCM1* [25], prostate carcinoma: *STMN2* [25] and *UNC5D* [39], neuroendocrine tumor of the nasopharynx: *HMBOX1* [9,40], spindle cell sarcoma: *WHSC1L1* [9] and *PPHLN1* [40], as well as uterine carcinosarcoma: *PMEPA1* [25]. Interestingly, in fusions of *NRG1* and *PCM1*, *STMN2*, and *PMEPA1*, the EGF-like domain was not observed; hence, the functionality and the activating ability of these fusions may not be relevant as oncogenic drivers [25]. The overview of locations of *NRG1* fusion gene partners in different tumors within chromosome 8 and other chromosomes is presented in Table 1 and Figures 3 and 4, respectively.

Table 1. Partner genes, with their chromosomal localization and translocation description, including the *NRG1* gene, in different cancer types.

Fusion Gene	Localization	Aberration	Cancer Type
ADAM9	8p11.22	t(8;8)(p12;p11)	Breast Cancer
AKAP13	15q25.3	t(8;15)(p12;q25)	
ARHGEF39	9p13.3	t(8;9)(p12;p13)	
BABAM2/BRE	2p23.2	t(8;2)(p12;p23)	
CD9	12p13.31	t(8;12)(p12;p13)	
COX10-AS1	17p12	t(8;17)(p12;p12)	
DDHD2	8p11.23	t(8;8)(p12;p11)	
FAM91A1	8q24.13	t(8;8)(p12;q24)	
FOXA1	14q21.1	t(8;14)(p12;q21)	
FUT10	8p12	t(8;8)(p12;p12)	
TENM4	11q14.1	t(8;11)(p12;q14)	
ZNF704	8q21.13	t(8;8)(p12;q21)	
ATP1B1	1q24.2	t(8;1)(p12;q24)	Breast Cancer/Cholangiocarcinoma/Pancreatic Ductal Adenocarcinoma
ERO1L	14q22.1	t(8;14)(p12;q22)	Colorectal Cancer
IKBKB	8p11.21	t(8;8)(p12;p11)	
KCTD9	8p21.2	t(8;8)(p12;q21)	
POMK	8p11.21	t(8;8)(p12;p11)	
TNFRSF10B	8p21.3	t(8;8)(p12;p21)	
ZCCHC7	9p13.2	t(8;9)(p12;p13)	
BIN3	8p21.3	t(8;8)(p12;p21)	Esophageal Carcinoma
CCAR2	8p21.3	t(8;8)(p12;p21)	
NOTCH2	1p12	t(8;1)(p12;p12)	Gallbladder Cancer
PDE7A	8q13.1	t(8;8)(p12;q13)	Head and Neck Squamous Cell Carcinoma
THBS1	15q14	t(8;15)(p12;q14)	
PCM1	8p22	t(8;8)(p12;p22)	Kidney Renal Clear Cell Carcinoma
CD74	5q33.1	t(8;5)(p12;q33)	Lung Adenocarcinoma/Pancreatic Adenocarcinoma
CADM1	11q23.3	t(8;11)(p12;q23)	Lung Cancer
DIP2B	12q13.12	t(8;12)(p12;q13)	
DPYSL2	8p21.2	t(8;8)(p12;p21)	
F11R	1q23.3	t(8;1)(p12;q23)	
FGFR1	8p11.23	t(8;8)(p12;q11)	
FLYWCH1	16p13.3	t(8;16)(p12;p13)	
ITGB1	10p11.22	t(8;10)(p12;p11)	
KIF13B	8p12	t(8;8)(p12;p12)	
KRAS	12p12.2	t(8;12)(p12;p12)	
MDK	11p11.2	t(8;11)(p12;p11)	
MRPL13	8q24.12	t(8;8)(p12;q24)	
NPTN	15q24.1	t(8;15)(p12;q24)	
PARP8	5q11.1	t(8;5)(p12;q11)	
PLCG2	16q23.3	t(8;16)(p12;q23)	
RALGAPA1	14q13.2	t(8;14)(p12;q13)	
SDC4	20q13.12	t(8;20)(p12;q13)	
SLC3A2	11q12.3	t(8;11)(p12;q12)	
SMAD4	18q21.2	t(8;18)(p12;q21)	
THAP7	22q11.21	t(8;22)(p12;q11)	
TNC	9q33.1	t(8;9)(p12;q33)	

Table 1. *Cont.*

Fusion Gene	Localization	Aberration	Cancer Type
WRN	8p12	t(8;8)(p12;p12)	Lung Cancer/Breast Cancer
RBPMS	8p12	t(8;8)(p12;p12)	Lung Cancer/Renal Cell Carcinoma
HMBOX1	8p21.1-p12	t(8;8)(p12;p21)	Neuroendocrine Tumor of the Nasopharynx/Spindle Cell Sarcoma
CLU RAB3IL1 SETD4 TSHZ2 ZMYM2	8p21.1 11q12.2-q12.3 21q22.12 20q13.2 13q12.11	t(8;8)(p12;p21) t(8;11)(p12;q12) t(8;21)(p12;q22) t(8;20)(p12;q13) t(8;13)(p12;q11)	Ovarian Cancer
APP CDH1 CDH6 CDK1 ROCK1 SARAF UNC5D VTCN1	21q21.3 16q22.1 5p13.3 10q21.2 18q11.1 8p12 8p12 1p13.1-p12	t(8;21)(p12;q21) t(8;16)(p12;q22) t(8;5)(p12;p13) t(8;10)(p12;q21) t(8;18)(p12;q11) t(8;8)(p12;p12) t(8;8)(p12;p12) t(8;1)(p12;p13)	Pancreatic Adenocarcinoma
STMN2	8q21.13	t(8;8)(p12;q21)	Prostate Cancer
WHSC1L1	8p11.23	t(8;8)(p12;p11)	Sarcoma
MTUS1 PPHLN1	8p22 12q12	t(8;8)(p12;p22) t(8;12)(p12;q12)	Spindle Cell Sarcoma
GDF15	19p13.11	t(8;19)(p12;p13)	Urothelial Bladder Cancer
PMEPA1	20q13.31	t(8;20)(p12;q13)	Uterine Carcinosarcoma

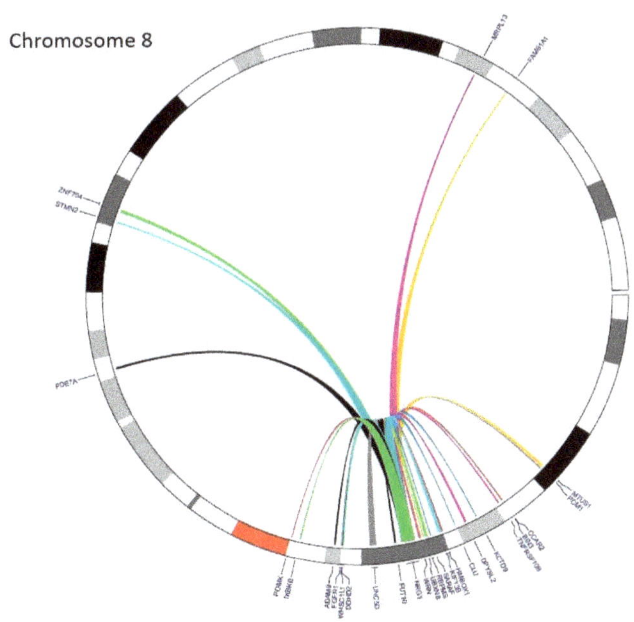

Figure 3. The circos plot presents the common fusion partners of *NRG1* within chromosome 8.

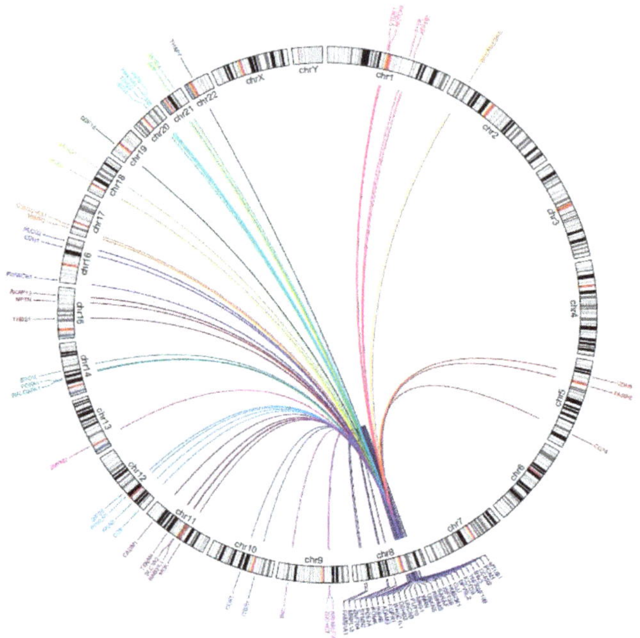

Figure 4. The circos plot presents the common fusion partners of NRG1 with genes localized within all chromosomes.

4. Detection of *NRG1* Fusions

The rare occurrence of *NRG1* fusions requires a robust detection method, especially in a wide range of partner genes. The main obstacle is to capture all probable fusions in a single sample using an economically viable tool.

The comprehensive identification of *NRG1* fusions may involve next-generation sequencing (NGS) technologies, both RNA and DNA-based, which allow for high-throughput genomic profiling of tumor samples [8,41]. Since the gene spans over 1Mb, NGS allows for full analysis of all *NRG1* alterations, those known and unknown as well [9]. RNA-based sequencing can be used to identify fusions located in-frame, allowing for detecting products of transcription from alternative splicing forms, which are common in the case of the *NRG1* gene [8,9]. The most useful method in such an approach would be whole transcriptome sequencing (WTS), as it allows the detection of all possible transcripts. However, the drawback of WTS is that the method needs high-quality RNA isolated from the sample [28,42]. The DNA-based NGS approaches, whole exome sequencing (WES) and targeted sequencing allow, on the other hand, for the description of exact sequences of breakpoints but do not tell if these sequences undergo translation [9]. Such methods also do not cover intronic sequences properly, which is a disadvantage, as the *NRG1* gene consists of large, non-coding fragments that might carry the possible breaking points [16]. Another drawback of the DNA-based NGS approach is the poor quality of DNA extracted from formalin-fixed paraffin-embedded (FFPE) tissue samples that, in the computational analysis, may deliver a high number of artifacts that may imitate the false positive results [43,44]. However, due to the mentioned limitations and high costs of NGS-based approaches, other single gene-based methods are still of great interest for *NRG1* gene detection.

The most common technique used for the detection of fusion genes and their protein products is immunohistochemistry (IHC). The method is relatively fast, cheap, and provides high sensitivity and specificity, although it needs a qualified pathologist for proper description [45]. It was postulated that phosphorylated ErbB3 (pErbB3) protein analysis could be the first step to identifying tumors carrying *NRG1* fusions [21,33]. The association between high pErbB3 expression detected with IHC in IMA and non-IMA lung cancer samples was shown by Trombetta et al. [46]. The main issue with IHC is that it might show false-positive results, as it presents fusion proteins that undergo full expression (transcription and translation) [47], hence, the method can be used mainly as the first step in screening and selection of samples for further, more complex analysis [9].

The third approach to the detection of *NRG1* fusions is the fluorescence in situ hybridization (FISH) technique [48]. It is commonly available in most molecular laboratories, but it requires more expertise and experience from the diagnostician when interpreting the results. It is more labor-consuming and works well with previously described fusions. Also, the technique cannot describe specific breakage points in fusion partners [36,46]. Besides IHC and FISH techniques, real-time PCR and Sanger sequencing also allow the detection of the exact known genomic breakpoints but remain underused and are very limiting [9]. On the other hand, Nanostring technology may become the RNA-based approach that will allow efficient estimation of the level of expression of all the exons in the region of interest within the *NRG1* gene [49].

5. *NRG1* Fusion as the Predictive Factor in Lung Cancer Treatment

The activation of signal transduction by binding of NRG1 ligand to ErbB family receptors or the process of ErbB family protein dimerization is considered the main target of treatment in patients harboring *NRG1* fusions [50,51]. Hence, NSCLC patients harboring *NRG1* fusions may benefit from targeted therapies such as HER family inhibitors, which have shown efficacy in previous studies in various cancers. The first choice in such an approach would be afatinib. This irreversible pan-HER inhibitor was proven effective in NSCLC patients harboring *EGFR* gene-activating mutations [52,53]. Several studies analyzed the effectiveness of afatinib in patients harboring *NRG1* fusions, although they were mainly case studies. Drilon et al. reported no response to afatinib treatment in four patients with IMA histology, although there were visible results in patient-derived xenograft mouse models [25]. Gay et al. presented two cases of lung cancer patients without *EGFR* mutations, carriers of *SLC3A2-NRG1* and *CD74-NRG1* fusions. The patients received afatinib with documented durable responses of 10 and 12 months, respectively [8]. Another case study of five lung cancer patients harboring *CD74-NRG1* or *SDC4-NRG1* fusions, treated with afatinib, resulted in four cases of partial response (PR) (5–27 months) and one stable disease (SD) (4 months) [54]. On the other hand, a single-patient case study with a *CD74-NRG1* fusion presented by Wu et al. indicated that afatinib showed PR for seven months until the progression of the disease [55]. A larger study by Liu et al. with different types of tumors included 29 NSCLC patients treated with afatinib, and it showed a 48.3% overall response rate (ORR), including three complete responses (CRs) and eleven PRs, with a median duration of response (DoR) of 6.8 months and median PFS of 6.1 months [56].

Tarloxotinib, another small molecule pan-ErbB inhibitor, in a hypoxic tumor environment decreased the phosphorylated ErbB-related process by targeting the membrane of reductase STEP4 protein, leading to tumor growth inhibition and cancer regression. The results were observed in patient-derived cell lines and multiple murine xenograft models harboring an *NRG1* fusion [9,57,58]. Apart from the small-molecule pan-ErbB inhibitors mentioned above, there are also many positive premises behind the inhibition of NRG1-related pathways by monoclonal antibodies binding to the ErbB receptors. Odintsov et al. reported that seribantumab (anti-ErbB3 antibody, MM-121/SAR256212) decreased activation of the PI3K-AKT, mTOR, and ERK pathways in *NRG1* fusion-positive patient-derived lung and breast cancer cell lines and patient-derived xenograft (PDX) models from lung and ovarian cancer patients. Moreover, seribantumab efficiently blocked other ErbB family members, indicating a similar to afatinib reduction of proliferation and induction of apoptosis [59]. In the end, Drilon et al. observed durable tumor regression in a PDX mouse model and anti-proliferative activity in the MDA-MB-175-VII cell line [25]. The summary of the effect of different drugs on the prognosis of lung cancer patients is presented in Table 2.

Table 2. A summary of the effect of different drugs on the prognosis of lung cancer patients.

Drug	Studied Material	Effect	Reference
Afatinib	NSCLC patients	10–12 months of durable response	Gay et al. [8]
		5–27 months of partial response 4 months of stable disease	Cadranel et al. [54]
		7 months of partial response	Wu et al. [55]
		48.3% of the overall response rate 6.8 months median duration of response 6.1 months median progression-free survival	Liu et al. [56]
Tarloxotinib	patient-derived cell lines, murine xenograft models	inhibition of tumor growth cancer regression	Bhandari et al. [57] Estrada-Bernal et al. [58]
Seribantumab	patient-derived cell lines, patient-derived xenograft models	reduction of proliferation induction of apoptosis	Odintsov et al. [59]
	patient-derived xenograft mouse model MDA-MB-175-VII cell line	durable tumor regression anti-proliferative activity	Drilon et al. [25]

6. *NRG1* Fusion as a Secondary Target in Lung Cancer Treatment

NRG1 gene fusions have the potential to affect the activity of ErbB-related pathways; thus, from one side, there is a potential treatment option for cancer patients harboring *NRG1* fusion-positive cancers by HER-targeted therapies [60]. However, some studies have indicated that *NRG1* fusion drives the primary resistance to molecularly targeted therapies by activation of the HER3 [61] and HER3/AKT [62] signaling pathway. Due to the complexity of the ErbB-related signal transduction pathways, the NRG1-driven resistance has the potential to be overcome by the application of treatment regimens based on multi-targeted agents. For instance, trastuzumab combined with anti-HER3 monoclonal antibody, pertuzumab, or poziotinib may revert the resistance process in cell lines [61–63].

In NSCLC, *NRG1* fusions are listed as acquired oncogenic alterations associated with the acquired resistance to EGFR-TKIs driven by activation of the NRG1/ErbB3 pathway [64,65]. Moreover, it was also shown that *ALK*-rearranged NSCLC cells acquire resistance to ALK inhibitors, losing the *EML4/ALK* fusion and activating the NRG1/ErbB3 pathway [9]. In such a situation, the sensitivity to crizotinib may be restored by pan-ErbB inhibitors, afatinib or dacomitinib, in the absence of other secondary *ALK* mutations [66]. In case of resistance to alectinib, the NRG1/ErbB3 activation maintains survival and stimulates mesenchymal activity, driving the epithelial–mesenchymal transition (EMT) that is the main hallmark of cancer dissemination [67,68]. This phenomenon may be confirmed by the observation that high expression of ErbB3 and NRG1 significantly correlated with brain metastases from primary lung tumors [69].

7. Clinical Trials Related to Patients with Solid Tumors Harboring *NRG1* Fusions

Besides the published results, there are also some clinical trials evaluating the targeted treatment possibilities in patients harboring *NRG1* fusions (Table 3). The clinical trial NCT03805841 [70] evaluated the ORR to tarloxotinib in NSCLC patients harboring insertion in exon 20 of the *EGFR* gene, activating mutation of *HER2* or *NRG1* fusion. However, the study was terminated, and the outcome has not been provided yet. The efficient blocking of ErbB family members by seribantumab was confirmed in metastatic cancer patients having high and low levels of NRG1 and ErbB2 expression, respectively (NCT01447706 [71], NCT01151046 [72], NCT00994123 [73]). Moreover, in the CRESTONE study (NCT04383210) in the cohort of NSCLC patients harboring *NRG1* fusions who received seribantumab, the ORR and the disease control rate were 39% and 94%, respectively. The overall duration of response ranged from 1.4 to 17.2 months [74,75].

Table 3. A summary of clinical trials dedicated to patients with solid tumors (including NSCLC) harboring *NRG1* fusions. Data were collected from the ClinalTrials.gov database (http://clinicaltrials.gov/ (accessed on 30 July 2024)).

Clinical Trial ID (Duration) Status	Tested Drug (Phase)	Genetic Eligibility	Conditions (Cohort)	Primary Measured Outcomes
NCT05919537 (09.2023–03.2031) Recruiting	HMBD-001 with/without chemotherapy (Phase I)	*NRG1* fusions Extracellular domain HER3 mutations	Advanced solid tumors (68)	1. Adverse events 2. Incidence and nature of dose-limiting toxicities (DLTs) 3. ORR
NCT02912949 (01.2015–12.2026) Recruiting	Zenocutuzumab (MCLA-128) (Phase 2)	*NRG1* fusions	Solid tumors (250)	1. ORR 2. Duration of response
NCT04383210 (09.2020–03-2025) Active, not-recruiting	Seribantumab (Phase 2)	*NRG1* fusions	Locally advanced or metastatic solid tumors (75)	ORR
NCT04750824 (10.2020–12.2021) Completed	Afatinib (Observational)	*NRG1* fusions	Solid tumors (110)	ORR
NCT05057013 (11.2021–09.2026) Recruiting	HMBD-001 (Phase 1)	*NRG1* fusions HER3 expression	Solid tumors (135)	1. Recommended dose 2. Adverse events 3. ORR
NCT03805841 (03.2019–04.2021) Terminated	Tarloxotinib (Phase 2)	*NRG1* fusions *ERBB* family fusions *EGFR* Exon 20 Insertion *HER2*-activating mutations	NSCLC or advanced solid tumors (41)	ORR

Further, Zenocutuzumab (MCLA-128), a bispecific monoclonal antibody against ErbB2 and ErbB3, in the eNRGy study (NCT02912949) demonstrated durable efficacy and a well-tolerated safety profile in patients with advanced solid tumors harboring *NRG1* fusion, regardless of tumor histology [76]. Moreover, the NCT01966445 trial showed that GSK2849330, an anti-ErbB3 monoclonal antibody, elicited a durable 19-month response in NSCLC patients harboring the *CD74-NRG1* fusion. In the end, poziotinib in the ZENITH20 clinical trial (NCT03318939) demonstrated antitumor activity with a durable response and manageable safety profile as the second-generation TKI in previously treated NSCLC patients with *HER2* exon 20 insertions [77–79].

8. Conclusions and Future Perspectives

The application of deep sequencing techniques, such as next-generation sequencing, has provided a wide array of data about the molecular background of NSCLC and opened routes for treatment personalization. This advancement revolutionized the management of therapy in these deadly conditions. Moreover, the studies shed light on how rare alterations affect the signaling pathways, indicating they impact treatment response or acquire resistance to targeted approaches. To date, the occurrence of *NRG1* fusions, which is a very rare alteration in solid tumors, was considered a negative prognostic marker in NSCLC treatment; however, gaining knowledge about its impact on ErbB signaling pathways has provided significant attention in recent scientific research, offering a potential avenue for targeted therapy. Recent and ongoing clinical trials and preclinical studies have explored the effectiveness of both already available and new agents in *NRG1* fusion-positive lung cancers, demonstrating promising results in terms of response rates and disease control. Thus, including such NSCLC cases in planning treatment regimens is reasonable. Moreover, re-evaluating standard approaches to NSCLC molecular analysis to detect possibly actionable, novel gene fusions or alterations that may affect well-known signaling pathways seems relevant and shows promise for further clinical improvement.

Author Contributions: Writing—review and editing, T.K., M.N., M.K. and E.K. Graphics—T.K. and M.K. All authors have read and agreed to the published version of the manuscript.

Funding: This research received no external funding.

Conflicts of Interest: The authors declare no conflicts of interest.

References

1. Dyba, T.; Randi, G.; Bray, F.; Martos, C.; Giusti, F.; Nicholson, N.; Gavin, A.; Flego, M.; Neamtiu, L.; Dimitrova, N.; et al. The European Cancer Burden in 2020: Incidence and Mortality Estimates for 40 Countries and 25 Major Cancers. *Eur. J. Cancer* **2021**, *157*, 308–347. [CrossRef]
2. Wang, M.; Herbst, R.S.; Boshoff, C. Toward Personalized Treatment Approaches for Non-Small-Cell Lung Cancer. *Nat. Med.* **2021**, *27*, 1345–1356. [CrossRef]
3. Xiao, Y.; Liu, P.; Wei, J.; Zhang, X.; Guo, J.; Lin, Y. Recent Progress in Targeted Therapy for Non-Small Cell Lung Cancer. *Front. Pharmacol.* **2023**, *14*, 1125547. [CrossRef]
4. Schram, A.M.; Chang, M.T.; Jonsson, P.; Drilon, A. Fusions in Solid Tumours: Diagnostic Strategies, Targeted Therapy, and Acquired Resistance. *Nat. Rev. Clin. Oncol.* **2017**, *14*, 735–748. [CrossRef]
5. Chen, J.; Xu, C.; Lv, J.; Lu, W.; Zhang, Y.; Wang, D.; Song, Y. Clinical Characteristics and Targeted Therapy of Different Gene Fusions in Non-Small Cell Lung Cancer: A Narrative Review. *Transl. Lung Cancer Res.* **2023**, *12*, 895–908. [CrossRef]
6. Horak, P.; Griffith, M.; Danos, A.M.; Pitel, B.A.; Madhavan, S.; Liu, X.; Chow, C.; Williams, H.; Carmody, L.; Barrow-Laing, L.; et al. Standards for the Classification of Pathogenicity of Somatic Variants in Cancer (Oncogenicity): Joint Recommendations of Clinical Genome Resource (ClinGen), Cancer Genomics Consortium (CGC), and Variant Interpretation for Cancer Consortium (VICC). *Genet. Med. Off. J. Am. Coll. Med. Genet.* **2022**, *24*, 986–998. [CrossRef]
7. Fernandez-Cuesta, L.; Plenker, D.; Osada, H.; Sun, R.; Menon, R.; Leenders, F.; Ortiz-Cuaran, S.; Peifer, M.; Bos, M.; Daßler, J.; et al. CD74-NRG1 Fusions in Lung Adenocarcinoma. *Cancer Discov.* **2014**, *4*, 415–422. [CrossRef]
8. Gay, N.D.; Wang, Y.; Beadling, C.; Warrick, A.; Neff, T.; Corless, C.L.; Tolba, K. Durable Response to Afatinib in Lung Adenocarcinoma Harboring NRG1 Gene Fusions. *J. Thorac. Oncol. Off. Publ. Int. Assoc. Study Lung Cancer* **2017**, *12*, e107–e110. [CrossRef]
9. Jonna, S.; Feldman, R.A.; Swensen, J.; Gatalica, Z.; Korn, W.M.; Borghaei, H.; Ma, P.C.; Nieva, J.J.; Spira, A.I.; Vanderwalde, A.M.; et al. Detection of NRG1 Gene Fusions in Solid Tumors. *Clin. Cancer Res. Off. J. Am. Assoc. Cancer Res.* **2019**, *25*, 4966–4972. [CrossRef]
10. Severson, E.; Achyut, B.R.; Nesline, M.; Pabla, S.; Previs, R.A.; Kannan, G.; Chenn, A.; Zhang, S.; Klein, R.; Conroy, J.; et al. RNA Sequencing Identifies Novel NRG1 Fusions in Solid Tumors That Lack Co-Occurring Oncogenic Drivers. *J. Mol. Diagn.* **2023**, *25*, 454–466. [CrossRef]
11. Muscarella, L.A.; Trombetta, D.; Fabrizio, F.P.; Scarpa, A.; Fazio, V.M.; Maiello, E.; Rossi, A.; Graziano, P. ALK and NRG1 Fusions Coexist in a Patient with Signet Ring Cell Lung Adenocarcinoma. *J. Thorac. Oncol. Off. Publ. Int. Assoc. Study Lung Cancer* **2017**, *12*, e161–e163. [CrossRef]
12. Shin, D.H.; Kim, S.H.; Choi, M.; Bae, Y.-K.; Han, C.; Choi, B.K.; Kim, S.S.; Han, J.-Y. Oncogenic KRAS Promotes Growth of Lung Cancer Cells Expressing SLC3A2-NRG1 Fusion via ADAM17-Mediated Shedding of NRG1. *Oncogene* **2022**, *41*, 280–292. [CrossRef]
13. Xia, D.; Le, L.P.; Iafrate, A.J.; Lennerz, J. KIF13B-NRG1 Gene Fusion and KRAS Amplification in a Case of Natural Progression of Lung Cancer. *Int. J. Surg. Pathol.* **2017**, *25*, 238–240. [CrossRef]
14. Falls, D.L. Neuregulins: Functions, Forms, and Signaling Strategies. *Exp. Cell Res.* **2003**, *284*, 14–30. [CrossRef]
15. Liu, S.V. NRG1 Fusions: Biology to Therapy. *Lung Cancer Amst. Neth.* **2021**, *158*, 25–28. [CrossRef]
16. Meyer, D.; Yamaai, T.; Garratt, A.; Riethmacher-Sonnenberg, E.; Kane, D.; Theill, L.E.; Birchmeier, C. Isoform-Specific Expression and Function of Neuregulin. *Dev. Camb. Engl.* **1997**, *124*, 3575–3586. [CrossRef]
17. Steinthorsdottir, V.; Stefansson, H.; Ghosh, S.; Birgisdottir, B.; Bjornsdottir, S.; Fasquel, A.C.; Olafsson, O.; Stefansson, K.; Gulcher, J.R. Multiple Novel Transcription Initiation Sites for NRG1. *Gene* **2004**, *342*, 97–105. [CrossRef]
18. Wang, J.Y.; Miller, S.J.; Falls, D.L. The N-Terminal Region of Neuregulin Isoforms Determines the Accumulation of Cell Surface and Released Neuregulin Ectodomain. *J. Biol. Chem.* **2001**, *276*, 2841–2851. [CrossRef]

19. Li, H.; Xu, L.; Cao, H.; Wang, T.; Yang, S.; Tong, Y.; Wang, L.; Liu, Q. Analysis on the Pathogenesis and Treatment Progress of NRG1 Fusion-Positive Non-Small Cell Lung Cancer. *Front. Oncol.* **2024**, *14*, 1405380. [CrossRef]
20. Gollamudi, M.; Nethery, D.; Liu, J.; Kern, J.A. Autocrine Activation of ErbB2/ErbB3 Receptor Complex by NRG-1 in Non-Small Cell Lung Cancer Cell Lines. *Lung Cancer* **2004**, *43*, 135–143. [CrossRef]
21. Fernandez-Cuesta, L.; Thomas, R.K. Molecular Pathways: Targeting NRG1 Fusions in Lung Cancer. *Clin. Cancer Res. Off. J. Am. Assoc. Cancer Res.* **2015**, *21*, 1989–1994. [CrossRef]
22. Guy, P.M.; Platko, J.V.; Cantley, L.C.; Cerione, R.A.; Carraway, K.L. Insect Cell-Expressed p180erbB3 Possesses an Impaired Tyrosine Kinase Activity. *Proc. Natl. Acad. Sci. USA* **1994**, *91*, 8132–8136. [CrossRef]
23. Kim, H.-G.; Lee, C.-K.; Cho, S.-M.; Whang, K.; Cha, B.-H.; Shin, J.-H.; Song, K.-H.; Jeong, S.-W. Neuregulin 1 Up-Regulates the Expression of Nicotinic Acetylcholine Receptors through the ErbB2/ErbB3-PI3K-MAPK Signaling Cascade in Adult Autonomic Ganglion Neurons. *J. Neurochem.* **2013**, *124*, 502–513. [CrossRef]
24. Muthuswamy, S.K.; Gilman, M.; Brugge, J.S. Controlled Dimerization of ErbB Receptors Provides Evidence for Differential Signaling by Homo- and Heterodimers. *Mol. Cell. Biol.* **1999**, *19*, 6845–6857. [CrossRef]
25. Drilon, A.; Somwar, R.; Mangatt, B.P.; Edgren, H.; Desmeules, P.; Ruusulehto, A.; Smith, R.S.; Delasos, L.; Vojnic, M.; Plodkowski, A.J.; et al. Response to ERBB3-Directed Targeted Therapy in NRG1-Rearranged Cancers. *Cancer Discov.* **2018**, *8*, 686–695. [CrossRef]
26. Nakaoku, T.; Tsuta, K.; Ichikawa, H.; Shiraishi, K.; Sakamoto, H.; Enari, M.; Furuta, K.; Shimada, Y.; Ogiwara, H.; Watanabe, S.; et al. Druggable Oncogene Fusions in Invasive Mucinous Lung Adenocarcinoma. *Clin. Cancer Res. Off. J. Am. Assoc. Cancer Res.* **2014**, *20*, 3087–3093. [CrossRef]
27. Jones, M.R.; Lim, H.; Shen, Y.; Pleasance, E.; Ch'ng, C.; Reisle, C.; Leelakumari, S.; Zhao, C.; Yip, S.; Ho, J.; et al. Successful Targeting of the NRG1 Pathway Indicates Novel Treatment Strategy for Metastatic Cancer. *Ann. Oncol. Off. J. Eur. Soc. Med. Oncol.* **2017**, *28*, 3092–3097. [CrossRef]
28. Dhanasekaran, S.M.; Balbin, O.A.; Chen, G.; Nadal, E.; Kalyana-Sundaram, S.; Pan, J.; Veeneman, B.; Cao, X.; Malik, R.; Vats, P.; et al. Transcriptome Meta-Analysis of Lung Cancer Reveals Recurrent Aberrations in NRG1 and Hippo Pathway Genes. *Nat. Commun.* **2014**, *5*, 5893. [CrossRef]
29. Jung, Y.; Yong, S.; Kim, P.; Lee, H.-Y.; Jung, Y.; Keum, J.; Lee, S.; Kim, J.; Kim, J. VAMP2-NRG1 Fusion Gene Is a Novel Oncogenic Driver of Non-Small-Cell Lung Adenocarcinoma. *J. Thorac. Oncol. Off. Publ. Int. Assoc. Study Lung Cancer* **2015**, *10*, 1107–1111. [CrossRef]
30. McCoach, C.E.; Le, A.T.; Gowan, K.; Jones, K.; Schubert, L.; Doak, A.; Estrada-Bernal, A.; Davies, K.D.; Merrick, D.T.; Paul, A.; et al. Resistance Mechanisms to Targeted Therapies in ROS1+ and ALK+ Non-Small Cell Lung Cancer. *Clin. Cancer Res. Off. J. Am. Assoc. Cancer Res.* **2018**, *24*, 3334. [CrossRef]
31. Pan, Y.; Zhang, Y.; Ye, T.; Zhao, Y.; Gao, Z.; Yuan, H.; Zheng, D.; Zheng, S.; Li, H.; Li, Y.; et al. Detection of Novel NRG1, EGFR, and MET Fusions in Lung Adenocarcinomas in the Chinese Population. *J. Thorac. Oncol. Off. Publ. Int. Assoc. Study Lung Cancer* **2019**, *14*, 2003–2008. [CrossRef]
32. Nie, X.; Zhang, P.; Bie, Z.; Song, C.; Zhang, M.; Ma, D.; Cui, D.; Cheng, G.; Li, H.; Lei, Y.; et al. Durable Response to Afatinib in Advanced Lung Adenocarcinoma Harboring a Novel NPTN-NRG1 Fusion: A Case Report. *World J. Surg. Oncol.* **2023**, *21*, 246. [CrossRef]
33. Drilon, A.; Duruisseaux, M.; Han, J.-Y.; Ito, M.; Falcon, C.; Yang, S.-R.; Murciano-Goroff, Y.R.; Chen, H.; Okada, M.; Molina, M.A.; et al. Clinicopathologic Features and Response to Therapy of NRG1 Fusion-Driven Lung Cancers: The eNRGy1 Global Multicenter Registry. *J. Clin. Oncol. Off. J. Am. Soc. Clin. Oncol.* **2021**, *39*, 2791–2802. [CrossRef]
34. Murumägi, A.; Ungureanu, D.; Khan, S.; Hirasawa, A.; Arjama, M.; Välimäki, K.; Mikkonen, P.; Niininen, W.; Kumar, A.; Eldfors, S.; et al. Abstract 2945: Clinical Implementation of Precision Systems Oncology in the Treatment of Ovarian Cancer Based on Ex-Vivo Drug Testing and Molecular Profiling. *Cancer Res.* **2019**, *79*, 2945. [CrossRef]
35. Heining, C.; Horak, P.; Uhrig, S.; Codo, P.L.; Klink, B.; Hutter, B.; Fröhlich, M.; Bonekamp, D.; Richter, D.; Steiger, K.; et al. NRG1 Fusions in KRAS Wild-Type Pancreatic Cancer. *Cancer Discov.* **2018**, *8*, 1087–1095. [CrossRef]
36. Laskin, J.; Liu, S.V.; Tolba, K.; Heining, C.; Schlenk, R.F.; Cheema, P.; Cadranel, J.; Jones, M.R.; Drilon, A.; Cseh, A.; et al. NRG1 Fusion-Driven Tumors: Biology, Detection, and the Therapeutic Role of Afatinib and Other ErbB-Targeting Agents. *Ann. Oncol.* **2020**, *31*, 1693–1703. [CrossRef]
37. Jones, M.R.; Williamson, L.M.; Topham, J.T.; Lee, M.K.C.; Goytain, A.; Ho, J.; Denroche, R.E.; Jang, G.; Pleasance, E.; Shen, Y.; et al. NRG1 Gene Fusions Are Recurrent, Clinically Actionable Gene Rearrangements in KRAS Wild-Type Pancreatic Ductal Adenocarcinoma. *Clin. Cancer Res. Off. J. Am. Assoc. Cancer Res.* **2019**, *25*, 4674–4681. [CrossRef]
38. Howarth, K.D.; Mirza, T.; Cooke, S.L.; Chin, S.-F.; Pole, J.C.; Turro, E.; Eldridge, M.D.; Garcia, R.M.; Rueda, O.M.; Boursnell, C.; et al. NRG1 Fusions in Breast Cancer. *Breast Cancer Res.* **2021**, *23*, 3. [CrossRef]
39. Ptáková, N.; Martínek, P.; Holubec, L.; Janovský, V.; Vančurová, J.; Grossmann, P.; Navarro, P.A.; Rodriguez Moreno, J.F.; Alaghehbandan, R.; Hes, O.; et al. Identification of Tumors with NRG1 Rearrangement, Including a Novel Putative Pathogenic UNC5D-NRG1 Gene Fusion in Prostate Cancer by Data-Drilling a de-Identified Tumor Database. *Genes. Chromosomes Cancer* **2021**, *60*, 474–481. [CrossRef]
40. Dermawan, J.K.; Zou, Y.; Antonescu, C.R. Neuregulin 1 (NRG1) Fusion-Positive High-Grade Spindle Cell Sarcoma: A Distinct Group of Soft Tissue Tumors with Metastatic Potential. *Genes Chromosomes Cancer* **2022**, *61*, 123–130. [CrossRef]

41. Sturgill, E.G.; Srivastava, J.; Correia, J.; Schumacher, C.; Luckett, D.; Perez, C.A.; Wang, J.S.; Divers, S.G.; Bashir, B.; Johnson, J.; et al. Abstract 921: Identification of NRG1 Fusions in Patients with Solid Tumors: Analysis from a Real-World Community Oncology Network. *Cancer Res.* **2023**, *83*, 921. [CrossRef]
42. Chang, J.C.; Offin, M.; Falcon, C.; Brown, D.; Houck-Loomis, B.R.; Meng, F.; Rudneva, V.A.; Won, H.H.; Amir, S.; Montecalvo, J.; et al. Comprehensive Molecular and Clinicopathologic Analysis of 200 Pulmonary Invasive Mucinous Adenocarcinomas Identifies Distinct Characteristics of Molecular Subtypes. *Clin. Cancer Res.* **2021**, *27*, 4066–4076. [CrossRef] [PubMed]
43. Cappello, F.; Angerilli, V.; Munari, G.; Ceccon, C.; Sabbadin, M.; Pagni, F.; Fusco, N.; Malapelle, U.; Fassan, M. FFPE-Based NGS Approaches into Clinical Practice: The Limits of Glory from a Pathologist Viewpoint. *J. Pers. Med.* **2022**, *12*, 750. [CrossRef] [PubMed]
44. Mathieson, W.; Thomas, G.A. Why Formalin-Fixed, Paraffin-Embedded Biospecimens Must Be Used in Genomic Medicine: An Evidence-Based Review and Conclusion. *J. Histochem. Cytochem. Off. J. Histochem. Soc.* **2020**, *68*, 543–552. [CrossRef]
45. Dixon, A.R.; Bathany, C.; Tsuei, M.; White, J.; Barald, K.F.; Takayama, S. Recent Developments in Multiplexing Techniques for Immunohistochemistry. *Expert Rev. Mol. Diagn.* **2015**, *15*, 1171–1186. [CrossRef] [PubMed]
46. Trombetta, D.; Graziano, P.; Scarpa, A.; Sparaneo, A.; Rossi, G.; Rossi, A.; Di Maio, M.; Antonello, D.; Mafficini, A.; Fabrizio, F.P.; et al. Frequent NRG1 Fusions in Caucasian Pulmonary Mucinous Adenocarcinoma Predicted by Phospho-ErbB3 Expression. *Oncotarget* **2018**, *9*, 9661–9671. [CrossRef] [PubMed]
47. O'Hurley, G.; Sjöstedt, E.; Rahman, A.; Li, B.; Kampf, C.; Pontén, F.; Gallagher, W.M.; Lindskog, C. Garbage in, Garbage out: A Critical Evaluation of Strategies Used for Validation of Immunohistochemical Biomarkers. *Mol. Oncol.* **2014**, *8*, 783–798. [CrossRef] [PubMed]
48. Adélaïde, J.; Huang, H.-E.; Murati, A.; Alsop, A.E.; Orsetti, B.; Mozziconacci, M.-J.; Popovici, C.; Ginestier, C.; Letessier, A.; Basset, C.; et al. A Recurrent Chromosome Translocation Breakpoint in Breast and Pancreatic Cancer Cell Lines Targets the Neuregulin/NRG1 Gene. *Genes Chromosomes Cancer* **2003**, *37*, 333–345. [CrossRef] [PubMed]
49. Alì, G.; Bruno, R.; Savino, M.; Giannini, R.; Pelliccioni, S.; Menghi, M.; Boldrini, L.; Proietti, A.; Chella, A.; Ribechini, A.; et al. Analysis of Fusion Genes by NanoString System: A Role in Lung Cytology? *Arch. Pathol. Lab. Med.* **2018**, *142*, 480–489. [CrossRef]
50. Shin, D.H.; Jo, J.Y.; Han, J.-Y. Dual Targeting of ERBB2/ERBB3 for the Treatment of SLC3A2-NRG1-Mediated Lung Cancer. *Mol. Cancer Ther.* **2018**, *17*, 2024–2033. [CrossRef]
51. Rosas, D.; Raez, L.E.; Russo, A.; Rolfo, C. Neuregulin 1 Gene (NRG1). A Potentially New Targetable Alteration for the Treatment of Lung Cancer. *Cancers* **2021**, *13*, 5038. [CrossRef]
52. Harvey, R.D.; Adams, V.R.; Beardslee, T.; Medina, P. Afatinib for the Treatment of EGFR Mutation-Positive NSCLC: A Review of Clinical Findings. *J. Oncol. Pharm. Pract. Off. Publ. Int. Soc. Oncol. Pharm. Pract.* **2020**, *26*, 1461–1474. [CrossRef]
53. Jiang, Y.; Fang, X.; Xiang, Y.; Fang, T.; Liu, J.; Lu, K. Afatinib for the Treatment of NSCLC with Uncommon EGFR Mutations: A Narrative Review. *Curr. Oncol.* **2023**, *30*, 5337–5349. [CrossRef]
54. Cadranel, J.; Liu, S.V.; Duruisseaux, M.; Branden, E.; Goto, Y.; Weinberg, B.A.; Heining, C.; Schlenk, R.F.; Cheema, P.; Jones, M.R.; et al. Therapeutic Potential of Afatinib in NRG1 Fusion-Driven Solid Tumors: A Case Series. *Oncologist* **2021**, *26*, 7–16. [CrossRef]
55. Wu, X.; Zhang, D.; Shi, M.; Wang, F.; Li, Y.; Lin, Q. Successful Targeting of the NRG1 Fusion Reveals Durable Response to Afatinib in Lung Adenocarcinoma: A Case Report. *Ann. Transl. Med.* **2021**, *9*, 1507. [CrossRef]
56. Liu, S.V.; Frohn, C.; Minasi, L.; Fernamberg, K.; Klink, A.J.; Gajra, A.; Savill, K.M.Z.; Jonna, S. Real-World Outcomes Associated with Afatinib Use in Patients with Solid Tumors Harboring NRG1 Gene Fusions. *Lung Cancer* **2024**, *188*, 107469. [CrossRef]
57. Bhandari, V.; Hoey, C.; Liu, L.Y.; Lalonde, E.; Ray, J.; Livingstone, J.; Lesurf, R.; Shiah, Y.-J.; Vujcic, T.; Huang, X.; et al. Molecular Landmarks of Tumor Hypoxia across Cancer Types. *Nat. Genet.* **2019**, *51*, 308–318. [CrossRef]
58. Estrada-Bernal, A.; Le, A.T.; Doak, A.E.; Tirunagaru, V.G.; Silva, S.; Bull, M.R.; Smaill, J.B.; Patterson, A.V.; Kim, C.; Liu, S.V.; et al. Tarloxotinib Is a Hypoxia-Activated Pan-HER Kinase Inhibitor Active against a Broad Range of HER-Family Oncogenes. *Clin. Cancer Res. Off. J. Am. Assoc. Cancer Res.* **2021**, *27*, 1463–1475. [CrossRef]
59. Odintsov, I.; Lui, A.J.W.; Sisso, W.J.; Gladstone, E.; Liu, Z.; Delasos, L.; Kurth, R.I.; Sisso, E.M.; Vojnic, M.; Khodos, I.; et al. The Anti-HER3 Monoclonal Antibody Seribantumab Effectively Inhibits Growth of Patient-Derived and Isogenic Cell Line and Xenograft Models with Oncogenic NRG1 Fusions. *Clin. Cancer Res. Off. J. Am. Assoc. Cancer Res.* **2021**, *27*, 3154–3166. [CrossRef]
60. Swain, S.M.; Shastry, M.; Hamilton, E. Targeting HER2-Positive Breast Cancer: Advances and Future Directions. *Nat. Rev. Drug Discov.* **2023**, *22*, 101–126. [CrossRef]
61. Yang, L.; Li, Y.; Shen, E.; Cao, F.; Li, L.; Li, X.; Wang, X.; Kariminia, S.; Chang, B.; Li, H.; et al. NRG1-Dependent Activation of HER3 Induces Primary Resistance to Trastuzumab in HER2-Overexpressing Breast Cancer Cells. *Int. J. Oncol.* **2017**, *51*, 1553–1562. [CrossRef]
62. Guardia, C.; Bianchini, G.; Arpí-LLucià, O.; Menendez, S.; Casadevall, D.; Galbardi, B.; Dugo, M.; Servitja, S.; Montero, J.C.; Soria-Jiménez, L.; et al. Preclinical and Clinical Characterization of Fibroblast-Derived Neuregulin-1 on Trastuzumab and Pertuzumab Activity in HER2-Positive Breast Cancer. *Clin. Cancer Res. Off. J. Am. Assoc. Cancer Res.* **2021**, *27*, 5096–5108. [CrossRef]
63. Udagawa, H.; Robichaux, J.P.; Elamin, Y.Y.; He, J.; Nilsson, M.B.; Heymach, J.V. Abstract 1182: Molecular Landscape and ErbB Family Signaling in NRG Fusion NSCLC: Therapeutic Implications for Pan-ErbB Family Inhibitors. *Cancer Res.* **2022**, *82*, 1182. [CrossRef]

64. Trombetta, D.; Sparaneo, A.; Fabrizio, F.P.; Di Micco, C.M.; Rossi, A.; Muscarella, L.A. NRG1 and NRG2 Fusions in Non-Small Cell Lung Cancer (NSCLC): Seven Years between Lights and Shadows. *Expert Opin. Ther. Targets* **2021**, *25*, 865–875. [CrossRef]
65. Taniguchi, H.; Yamada, T.; Wang, R.; Tanimura, K.; Adachi, Y.; Nishiyama, A.; Tanimoto, A.; Takeuchi, S.; Araujo, L.H.; Boroni, M.; et al. AXL Confers Intrinsic Resistance to Osimertinib and Advances the Emergence of Tolerant Cells. *Nat. Commun.* **2019**, *10*, 259. [CrossRef]
66. Taniguchi, H.; Akagi, K.; Dotsu, Y.; Yamada, T.; Ono, S.; Imamura, E.; Gyotoku, H.; Takemoto, S.; Yamaguchi, H.; Sen, T.; et al. Pan-HER Inhibitors Overcome Lorlatinib Resistance Caused by NRG1/HER3 Activation in ALK-Rearranged Lung Cancer. *Cancer Sci.* **2023**, *114*, 164–173. [CrossRef]
67. Tanimura, K.; Yamada, T.; Okada, K.; Nakai, K.; Horinaka, M.; Katayama, Y.; Morimoto, K.; Ogura, Y.; Takeda, T.; Shiotsu, S.; et al. HER3 Activation Contributes toward the Emergence of ALK Inhibitor-Tolerant Cells in ALK-Rearranged Lung Cancer with Mesenchymal Features. *NPJ Precis. Oncol.* **2022**, *6*, 5. [CrossRef]
68. Isozaki, H.; Ichihara, E.; Takigawa, N.; Ohashi, K.; Ochi, N.; Yasugi, M.; Ninomiya, T.; Yamane, H.; Hotta, K.; Sakai, K.; et al. Non-Small Cell Lung Cancer Cells Acquire Resistance to the ALK Inhibitor Alectinib by Activating Alternative Receptor Tyrosine Kinases. *Cancer Res.* **2016**, *76*, 1506–1516. [CrossRef]
69. Saunus, J.M.; Quinn, M.C.J.; Patch, A.-M.; Pearson, J.V.; Bailey, P.J.; Nones, K.; McCart Reed, A.E.; Miller, D.; Wilson, P.J.; Al-Ejeh, F.; et al. Integrated Genomic and Transcriptomic Analysis of Human Brain Metastases Identifies Alterations of Potential Clinical Significance. *J. Pathol.* **2015**, *237*, 363–378. [CrossRef]
70. Ye, L.; Chen, X.; Zhou, F. EGFR-Mutant NSCLC: Emerging Novel Drugs. *Curr. Opin. Oncol.* **2021**, *33*, 87–94. [CrossRef]
71. Liu, J.F.; Ray-Coquard, I.; Selle, F.; Poveda, A.M.; Cibula, D.; Hirte, H.; Hilpert, F.; Raspagliesi, F.; Gladieff, L.; Harter, P.; et al. Randomized Phase II Trial of Seribantumab in Combination with Paclitaxel in Patients with Advanced Platinum-Resistant or -Refractory Ovarian Cancer. *J. Clin. Oncol. Off. J. Am. Soc. Clin. Oncol.* **2016**, *34*, 4345–4353. [CrossRef]
72. Curley, M.D.; Sabnis, G.J.; Wille, L.; Adiwijaya, B.S.; Garcia, G.; Moyo, V.; Kazi, A.A.; Brodie, A.; MacBeath, G. Seribantumab, an Anti-ERBB3 Antibody, Delays the Onset of Resistance and Restores Sensitivity to Letrozole in an Estrogen Receptor-Positive Breast Cancer Model. *Mol. Cancer Ther.* **2015**, *14*, 2642–2652. [CrossRef]
73. Sequist, L.V.; Gray, J.E.; Harb, W.A.; Lopez-Chavez, A.; Doebele, R.C.; Modiano, M.R.; Jackman, D.M.; Baggstrom, M.Q.; Atmaca, A.; Felip, E.; et al. Randomized Phase II Trial of Seribantumab in Combination with Erlotinib in Patients with EGFR Wild-Type Non-Small Cell Lung Cancer. *Oncologist* **2019**, *24*, 1095–1102. [CrossRef]
74. Carrizosa, D.R.; Burkard, M.E.; Elamin, Y.Y.; Desai, J.; Gadgeel, S.M.; Lin, J.J.; Waqar, S.N.; Spigel, D.R.; Chae, Y.K.; Cheema, P.K.; et al. CRESTONE: Initial Efficacy and Safety of Seribantumab in Solid Tumors Harboring NRG1 Fusions. *J. Clin. Oncol.* **2022**, *40*, 3006. [CrossRef]
75. Patil, T.; Carrizosa, D.R.; Burkard, M.E.; Reckamp, K.L.; Desai, J.; Chae, Y.K.; Liu, S.V.; Konduri, K.; Gadgeel, S.M.; Lin, J.J.; et al. Abstract CT229: CRESTONE: A Phase 2 Study of Seribantumab in Adult Patients with Neuregulin-1 (NRG1) Fusion Positive Locally Advanced or Metastatic Solid Tumors. *Cancer Res.* **2023**, *83*, CT229. [CrossRef]
76. Schram, A.M.; Goto, K.; Kim, D.-W.; Martin-Romano, P.; Ou, S.-H.I.; O'Kane, G.M.; O'Reilly, E.M.; Umemoto, K.; Duruisseaux, M.; Neuzillet, C.; et al. Efficacy and Safety of Zenocutuzumab, a HER2 x HER3 Bispecific Antibody, across Advanced NRG1 Fusion (NRG1+) Cancers. *J. Clin. Oncol.* **2022**, *40*, 105. [CrossRef]
77. Le, X.; Cornelissen, R.; Garassino, M.; Clarke, J.M.; Tchekmedyian, N.; Goldman, J.W.; Leu, S.-Y.; Bhat, G.; Lebel, F.; Heymach, J.V.; et al. Poziotinib in Non-Small-Cell Lung Cancer Harboring HER2 Exon 20 Insertion Mutations After Prior Therapies: ZENITH20-2 Trial. *J. Clin. Oncol. Off. J. Am. Soc. Clin. Oncol.* **2022**, *40*, 710–718. [CrossRef]
78. Cornelissen, R.; Prelaj, A.; Sun, S.; Baik, C.; Wollner, M.; Haura, E.B.; Mamdani, H.; Riess, J.W.; Cappuzzo, F.; Garassino, M.C.; et al. Poziotinib in Treatment-Naive NSCLC Harboring HER2 Exon 20 Mutations: ZENITH20-4, A Multicenter, Multicohort, Open-Label, Phase 2 Trial (Cohort 4). *J. Thorac. Oncol. Off. Publ. Int. Assoc. Study Lung Cancer* **2023**, *18*, 1031–1041. [CrossRef]
79. Socinski, M.A.; Cornelissen, R.; Garassino, M.C.; Clarke, J.; Tchekmedyian, N.; Molina, J.; Goldman, J.W.; Bhat, G.; Lebel, F.; Le, X. LBA60 ZENITH20, a Multinational, Multi-Cohort Phase II Study of Poziotinib in NSCLC Patients with EGFR or HER2 Exon 20 Insertion Mutations. *Ann. Oncol.* **2020**, *31*, S1188. [CrossRef]

cancers

MDPI

Article

Pretreatment Tumor Growth Rate and Radiological Response as Predictive Markers of Pathological Response and Survival in Patients with Resectable Lung Cancer Treated by Neoadjuvant Treatment

Toulsie Ramtohul [1,*], Léa Challier [1], Vincent Servois [1] and Nicolas Girard [2,3]

[1] Department of Radiology, Institut Curie Paris, PSL Research University, 75005 Paris, France; lea.challier@curie.fr (L.C.); vincent.servois@curie.fr (V.S.)
[2] Institut du Thorax Curie Montsouris, Institut Curie, 75005 Paris, France; nicolas.girard2@curie.fr
[3] Paris Saclay Campus, Versailles Saint Quentin University, 78000 Versailles, France
* Correspondence: toulsie.ramtohul@curie.fr; Tel.: +33-14432-4200; Fax: +33-14329-0203

Simple Summary: For patients with resectable non-small cell lung carcinoma (NSCLC), neoadjuvant nivolumab and chemotherapy are associated with increased major pathological responses and better event-free survival. Identification of earlier biomarkers associated with progression precluding surgery or disease recurrence after surgery is of importance in this population. The aim of our retrospective study was to assess the potential added value of pretreatment tumor growth rate (TGR_0) using computed tomography (CT) and/or positron emission tomography (PET)-CT scans before and at baseline. We confirmed in 32 patients with resectable stage IB ($\geq$4 cm) to IIIA NSCLC that the assessment of TGR_0 helps identify patients who would benefit from neoadjuvant treatment and outperforms RECIST assessments for survival outcomes. TGR_0 may be an early noninvasive marker for more favorable genetic and/or biological profiles, leading to improved disease control and overall survival.

Abstract: Introduction: Predictive biomarkers associated with pathological response, progression precluding surgery, and/or recurrence after surgery are needed for patients with resectable non-small cell lung carcinoma (NSCLC) treated by neoadjuvant treatment. We evaluated the clinical impact of the pretreatment tumor growth rate (TGR_0) and radiological response for patients with resectable NSCLC treated with neoadjuvant therapies. Methods: Consecutive patients with resectable stage IB ($\geq$4 cm) to IIIA NSCLC treated by neoadjuvant platinum-doublet chemotherapy with or without nivolumab at our tertiary center were retrospectively analyzed. TGR_0 and RECIST objective responses were determined. Multivariable analyses identified independent predictors of event-free survival (EFS), overall survival (OS), and major pathological response (MPR). Results: Between November 2017 and December 2022, 32 patients (mean [SD] age, 63.8 [8.0] years) were included. At a median follow-up of 54.8 months (95% CI, 42.3–60.4 months), eleven patients (34%) experienced progression or recurrence, and twelve deaths (38%) were recorded. The TGR_0 cutoff of 30%/month remained the only independent factor associated with EFS (HR = 0.04; 95% CI, 0.01–0.3; p = 0.003) and OS (HR = 0.2; 95% CI, 0.03–0.7; p = 0.01). The TGR_0 cut-off had a mean time-dependent AUC of 0.83 (95% CI, 0.64–0.95) and 0.80 (95% CI, 0.62–0.97) for predicting EFS and OS, respectively. Fifteen of 26 resection cases (58%) showed MPR including nine with pathological complete responses (35%). Only the objective response of the primary tumor was associated with MPR (OR = 27.5; 95% CI, 2.6–289.1; p = 0.006). Conclusions: Assessment of TGR_0 can identify patients who should benefit from neoadjuvant treatment. A tumor objective response might be a predictor of MPR after neoadjuvant treatment, which will help to adapt surgical management.

Keywords: non-small cell lung cancer; pretreatment tumor growth rate; event-free survival; overall survival; major pathological response

1. Introduction

Lung cancer is the leading cause of cancer-related death worldwide [1]. Due to the increased use of screening CT scans in high-risk patients, the proportion of non-small cell lung cancer (NSCLC) diagnosed at an early-stage increases up to 30% [2–4]. Surgery remains the best treatment modality for curing patients diagnosed with resectable stage I-IIIA NSCLC [5]. However, more than 50% of patients with resectable NSCLC will experience recurrence after surgery alone [6–8]. Adjuvant chemotherapy is associated with an improved recurrence-free survival for stage IB-IIIA patients, resulting in absolute survival benefits of 5.4% to 6.9% at five years [9,10]. Neoadjuvant chemotherapy provides few significant pathological responses and a comparable level of risk reduction compared to adjuvant chemotherapy [11]. Recently, the CheckMate 816 trial, which evaluated the combination of neoadjuvant nivolumab and chemotherapy in 358 newly diagnosed patients with resectable stage IB to IIIA NSCLC, reported increased major and complete pathological responses and better event-free survival (EFS) compared to chemotherapy alone, without more adverse effects on surgical feasibility or surgical outcomes [12]. However, 15% to 20% of patients do not undergo definitive surgery, and predictive factors of the long-term benefits of neoadjuvant treatment are still under investigation [13]. Overall, predicting pathological complete response (pCR) and major pathological responses (MPR) is a challenge for optimizing surgical approaches.

The tumor growth rate (TGR) provides a means of quantitative evaluation of tumor volume changes over time that may be calculated before treatment onset, leading to a better understanding of natural growth kinetics [14]. TGR-derived parameters have been validated as radiological markers of progression-free survival and overall survival in different cancer types [15,16]. Pretreatment TGR is associated with inferior progression-free survival and distant control among patients with locally advanced NSCLC and helps in identifying hyperprogressive disease in patients treated with PD-1/PD-L1 [17,18]. The aim was to evaluate the clinical impact of pretreatment TGR on the survival outcomes and pathological responses of patients with resectable NSCLC treated with neoadjuvant chemotherapy with or without nivolumab.

2. Materials and Methods

2.1. Patients and Study Design

This single-center study was approved by the institutional ethics review boards, and written informed consent was obtained from all patients. Between November 2017 and December 2022, consecutive patients with resectable NSCLC treated by neoadjuvant treatment at our tertiary center were retrospectively analyzed. Key eligibility criteria included age older than or equal to 18 years, histologically confirmed resectable stage IB ($\geq$4 cm) to IIIA NSCLC (according to the staging criteria of the American Joint Committee on Cancer, 7th edition), Eastern Cooperative Oncology Group performance status 0 to 1, no previous anticancer therapy or measurable disease per Response Evaluation Criteria in Solid Tumors (RECIST) version 1.1. Patients with known ALK translocations or EGFR mutations were excluded. Patients received nivolumab (360 mg) plus platinum-doublet chemotherapy or platinum-doublet chemotherapy alone (every 3 weeks for three cycles) before undergoing definitive surgery. This study complied with the tenets of the Declaration of Helsinki.

2.2. Radiological Assessment

Tumors were assessed using computed tomography (CT) and/or positron emission tomography (PET)-CT scans before baseline, at baseline, and within 14 days prior to definitive surgery per response evaluation criteria in solid tumors (RECIST) v1.1. Examinations were centrally reviewed by two senior radiologists blinded to the clinical characteristics, treatment received, and outcomes. TGR_0 is expressed as the percentage change in tumor volume over 1 month (%/m): $TGR_0 = 100 \times [\exp(TG) - 1]$, where $TG = 3 \times \log(D2/D1)/\text{time}$ (months). D1 and D2 represent the largest diameter of the primary tumor according to

RECIST v1.1 on pretreatment and baseline imaging, respectively. Lymph nodes were not taken into account in the TGR_0 calculation. Detailed examples of TGR calculations are given in Figure 1. For each patient, the same imaging technique (CT/PET-CT scans) was preferred for each time point. The RECIST objective response was defined as the proportion of patients who experienced a complete response (CR) or partial response (PR) of the primary tumor after neoadjuvant treatment.

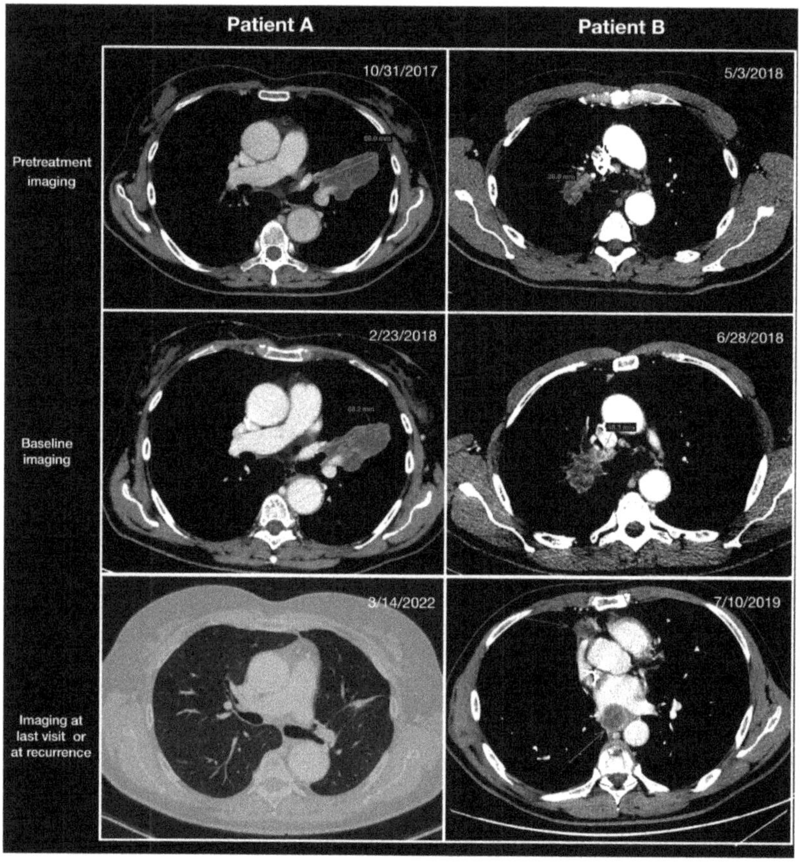

Figure 1. Illustration of the pretreatment tumor growth rate (TGR_0) in two patients treated with neoadjuvant therapies for resectable non-small cell lung cancer (NSCLC). Patient A: Lung CT scan of a 61-year-old male patient who received neoadjuvant chemoimmunotherapy for non-squamous NSCLC. The tumor manifested as a proximal mass in the left upper lobe and was classified (according to the staging criteria of the American Joint Committee on Cancer, 7th edition) as stage IIIa NSCLC. Pretreatment imaging examinations revealed no significant growth rate of the lung mass with a TGR_0 of 0.2%/month. The patient underwent a left upper lobectomy, with no evidence of recurrence after a follow-up of 46 months. Patient B: Lung CT scan of a 68-year-old male patient who received neoadjuvant chemoimmunotherapy for squamous NSCLC. The tumor manifested as a proximal mass in the right upper lobe and was classified (according to the staging criteria of the American Joint Committee on Cancer, 7th edition) as stage IIb NSCLC. Pretreatment imaging examinations revealed a significant growth rate of the lung mass with a TGR_0 of 52%/month. The patient underwent a right upper lobectomy. Mediastinal necrotic lymph nodes (arrows) appeared 9 months after surgery, and the patient died 13 months after surgery.

2.3. Survival Endpoints

Event-free survival (EFS) was calculated as the time from neoadjuvant treatment start to the occurrence of any radiologically identified disease progression precluding surgery, disease recurrence after surgery, disease progression in the absence of surgery, or death from any cause. Overall survival (OS) was calculated as the time from neoadjuvant treatment start to death from any cause.

2.4. Pathological Endpoints

An expert pathologist, unaware of patient characteristics and outcomes, reviewed hematoxylin and eosin-stained slides containing sections of the gross residual tumor post-surgery. The assessment involved comparing the estimated cross-sectional area of viable tumor with that of necrosis, inflammation, and fibrosis on each slide to determine the percentage of residual tumor. A major pathological response (MPR) was characterized as having a residual viable tumor of ≤10% within the primary tumor and sampled lymph nodes. Meanwhile, a complete pathological response (pCR) was identified when no residual viable tumor was present in either the primary tumor or lymph node tissue.

2.5. Statistical Analysis

Continuous variables were assessed utilizing the Wilcoxon–Mann–Whitney and Student tests, taking into account the normality of their distribution. Categorical variables underwent assessment through either the χ^2 test or Fisher's exact test. The inter-reader reliability of TGR_0 was assessed with the intraclass correlation coefficient (ICC). To illustrate time-to-event outcomes, the Kaplan–Meier product-limit approach was utilized, and a comparison between the curves was determined using the exact log-rank test. For patients who either did not experience an event or were alive on the specified date (2 August 2023), their data were censored at the most recent evaluation date. The identification of the optimal TGR_0 cutoff, differentiating patients based on EFS and OS, was carried out using a stepwise log-rank test. A multivariable Cox analysis was conducted, initially including all variables associated with EFS or OS from the univariate analysis at a significance level of $p < 0.05$. The assumption of proportional hazards was verified using Schoenfeld residuals. All computations were conducted using SAS software (version 9.4, SAS Institute, Cary, NC, USA). Statistical tests were two-tailed, and p-values below 0.05 were considered indicative of statistical significance.

3. Results

3.1. Patient Characteristics

Between November 2017 and December 2022, 32 patients (mean [SD] age, 63.8 [8.0] years) were identified in our database. All patients had at least one available imaging scan before baseline. The baseline characteristics are summarized in Table 1. Twenty-seven patients (84%) had stage IIIa disease. Neoadjuvant nivolumab-based therapy was delivered to 23 patients (72%). Twenty-six patients (81%) had low-TGR_0 (≤30%/month) and six (19%) high-TGR_0 (>30%/month). The median time between pretreatment and baseline imaging for the calculation of TGR_0 was 1.9 months (IQR: 1.3–2.3). Patients with low-TGR_0 had higher rates of non-squamous histology type (85% vs. 33%, $p = 0.009$) and disease stage IIIa (92% vs. 50%, $p = 0.01$) compared to those with high-TGR_0. The reproducibility of the assessment for TGR_0 among the two readers was very good with an ICC of 0.81 (95% CI: 0.76–0.95).

Table 1. Baseline characteristics. Data are numbers of patients with percentages in parentheses. Disease stage was based on TNM Classification of Malignant Tumors, 7th edition. Objective response was defined as the proportion of patients with a complete response or partial response of the primary tumor according to RECIST v1.1. A major pathological response was available for 26 patients. * Two patients with low-TGR_0 had missing data for PD-L1 status. *P*-values were obtained using the χ^2 test. Low-TGR_0: ≤30%/month; high-TGR_0: >30%/month.

Variables		Whole Cohort	Pretreatment TGR_0		*p*-Value
		(*n* = 32)	Low (*n* = 26)	High (*n* = 6)	
Age (years)	≤60	11 (34%)	7 (27%)	4 (67%)	0.07
	>60	21 (66%)	19 (73%)	2 (33%)	
Sex	Female	10 (31%)	8 (31%)	2 (33%)	0.90
	Male	22 (69%)	18 (69%)	4 (67%)	
ECOG performance status	0	27 (84%)	22 (85%)	5 (83%)	0.94
	1	5 (16%)	4 (15%)	1 (17%)	
Smoking status	Current or former smoker	29 (91%)	24 (92%)	5 (83%)	0.50
	Never smoked	3 (9%)	2 (8%)	1 (17%)	
Histologic type	Non-squamous	24 (75%)	22 (85%)	2 (33%)	0.009
	Squamous	8 (25%)	4 (15%)	4 (67%)	
PD-L1 status (%) *	≤10	11 (37%)	7 (29%)	4 (67%)	0.09
	>10	19 (63%)	17 (71%)	2 (33%)	
Disease stage	Ib or II	5 (16%)	2 (8%)	3 (50%)	0.01
	IIIa	27 (84%)	24 (92%)	3 (50%)	
Largest tumor size at baseline (mm)	≤50	19 (59%)	16 (62%)	3 (50%)	0.60
	>50	13 (41%)	10 (38%)	3 (50%)	
Nodal stage	N1/2	28 (88%)	23 (88%)	5 (83%)	0.73
	N0	4 (13%)	3 (12%)	1 (17%)	
Nivolumab-based neoadjuvant treatment	Present	23 (72%)	18 (69%)	5 (83%)	0.49
	Absent	9 (28%)	8 (31%)	1 (17%)	
RECIST objective response	Present	14 (44%)	12 (46%)	2 (33%)	0.57
	Absent	18 (56%)	14 (54%)	4 (67%)	
Major pathological response	Present	15 (58%)	15 (63%)	0 (0%)	0.09
	Absent	11 (42%)	9 (38%)	2 (100%)	

3.2. Prognostic Factors Associated with EFS and OS

The objective response rate of the primary tumor was 44% (14 of 32) (Figure 2). At the last database lock (2 August 2023), the median follow-up was 54.8 months (95% CI, 42.3–60.4 months). Overall, eleven patients (34%) experienced progression or recurrence, and twelve deaths (38%) were recorded. Seven patients (22%) had disease progression at the end of neoadjuvant treatment precluding surgery and four patients (13%) experienced lung cancer-related recurrence after surgery; two patients (6%) had a non-cancer death after surgery. The median EFS was not reached while the median OS was 54.1 months (95% CI, 23.7 months-not reached). There was a moderate difference in EFS ($p = 0.045$) but no significant difference in OS ($p = 0.13$) based on the RECIST objective response of the primary tumor during neoadjuvant treatment (Figure 3A,B). There was a larger difference in EFS ($p < 0.001$) and OS ($p < 0.001$) based on TGR_0 before neoadjuvant treatment (Figure 3C,D). For patients receiving nivolumab-based neoadjuvant treatment, similar EFS and OS patterns were observed according to the TGR_0 or objective response (Supplementary Figure S1).

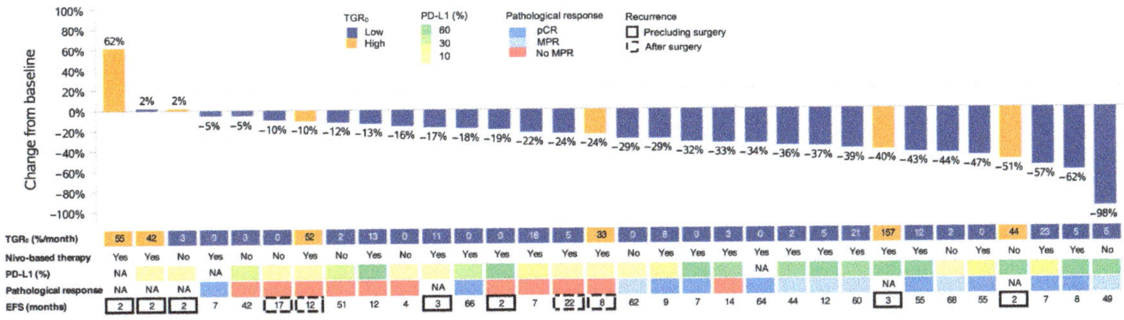

Figure 2. Change from baseline in tumor size, with patient data on pretreatment tumor growth rate (TGR_0), baseline PD-L1 status, type of neoadjuvant treatment, pathological response, and time to event-free survival (months). The change from baseline (%) is labeled on each bar. Low-TGR_0: ≤30%/month; high-TGR_0: >30%/month; NA: not available.

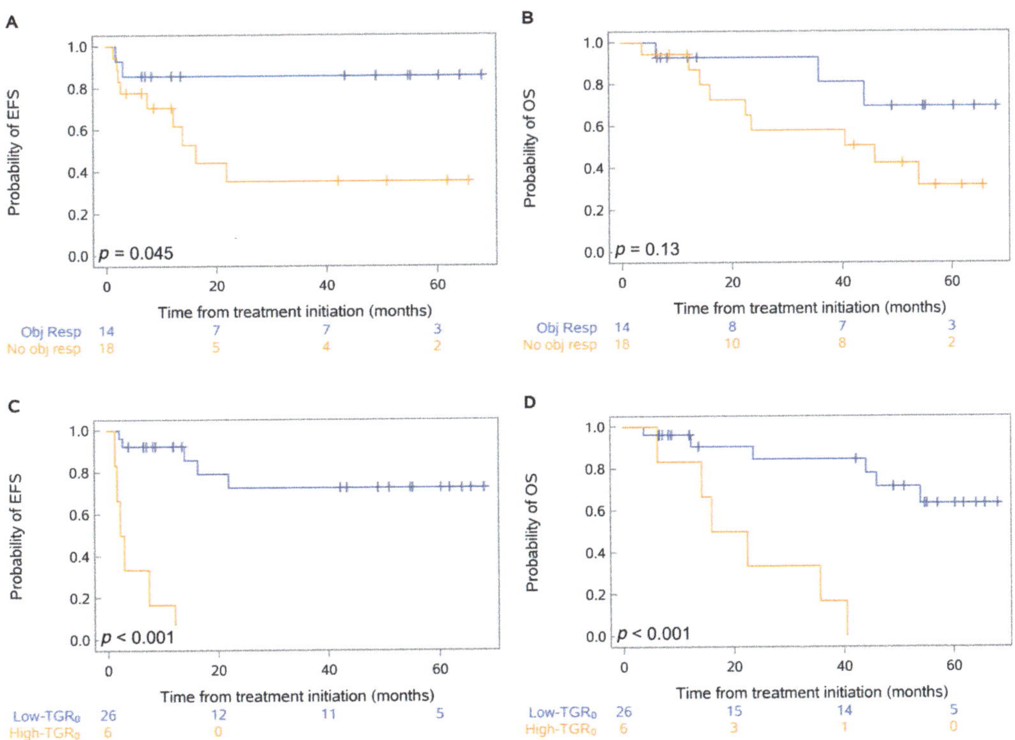

Figure 3. Kaplan-Meier analysis of event-free survival (EFS) and overall survival (OS) by RECIST objective response of the primary tumor (**A**,**B**) and pretreatment TGR_0 (**C**,**D**). *p*-values were obtained using the log-rank test.

For EFS, a non-squamous histology type (HR = 0.3; 95% CI, 0.1–0.09; $p = 0.047$), a PD-L1 > 10% (HR = 0.2; 95% CI, 0.1–0.7; $p = 0.01$), a disease stage IIIa (HR = 0.3; 95% CI, 0.1–0.9; $p = 0.04$) and a low-TGR_0 (HR = 0.04; 95% CI, 0.01–0.2; $p < 0.001$) were associated with higher EFS in univariable analyses and were included in the multivariable analysis (Table 2). Only the low-TGR_0 remained an independent factor associated with higher EFS (HR = 0.02; 95% CI, 0.01–0.3; $p = 0.003$).

Table 2. Multivariable analysis of EFS. Multivariable analysis was undertaken by entering all variables at the $p < 0.05$ level in the univariable analysis. HR: hazard ratio; EFS: event-free survival; CI: confidence interval.

Variable	Univariable Analysis of EFS		Multivariable Analysis of EFS	
	HR (95% CI)	*p*-Value	HR (95% CI)	*p*-Value
Age (years), >60 vs. ≤60	0.6 (0.2–1.8)	0.33		
Sex, female vs. male	0.8 (0.2–3.2)	0.80		
Smoking status, never smoked vs. current or former smoker	4.7 (0.9–25.9)	0.07		
Histologic type, non-squamous vs. squamous	0.3 (0.1–0.9)	0.047	0.6 (0.1–4.2)	0.58
PD-L1 (%), >10 vs. ≤10	0.2 (0.1–0.7)	0.01	0.3 (0.1–1.4)	0.13
Disease stage, IIIa vs. Ib/II	0.3 (0.1–0.9)	0.04	5.2 (0.5–58.9)	0.18
Largest tumor size at baseline (mm), >50 vs. ≤50	0.8 (0.2–2.9)	0.78		
Nodal stage, N1/2 vs. N0	1.2 (0.1–9.3)	0.88		
Nivolumab-based treatment, present vs. absent	1.3 (0.3–4.9)	0.72		
RECIST objective response, present vs. absent	0.2 (0.1–1.1)	0.07		
TGR_0 (%/month), ≤30 vs. >30	0.04 (0.01–0.2)	<0.001	0.04 (0.01–0.3)	0.003

For OS, univariable analyses showed that a low-TGR_0 (HR = 0.1; 95% CI, 0.02–0.4; $p = 0.001$) a PD-L1 > 10% (HR = 0.2; 95% CI, 0.1–0.8; $p = 0.08$) and disease stage IIIa (HR = 0.1; 95% CI, 0.02–0.4; $p = 0.002$) were associated with higher OS and were thus included in the multivariable analysis (Table 3). Only the low-TGR_0 remained an independent factor associated with higher OS (HR = 0.2; 95% CI, 0.03–0.7; $p = 0.01$). The TGR_0 cutoff had a mean time-dependent AUC of 0.83 (95% CI, 0.64–0.95) and 0.80 (95% CI, 0.62–0.97) for predicting EFS and OS, respectively (Figure 4).

Table 3. Multivariable analysis of OS. Multivariable analysis was undertaken by entering all variables at the $p < 0.05$ level in the univariable analysis. HR: hazard ratio; OS: overall survival; CI: confidence interval.

Variable	Univariable Analysis of OS		Multivariable Analysis of OS	
	HR (95% CI)	*p*-Value	HR (95% CI)	*p*-Value
Age (years), >60 vs. ≤60	0.9 (0.3–3.1)	0.90		
Sex, female vs. male	1.2 (0.4–4.0)	0.76		
Smoking status, never smoked vs. current or former smoker	-	>0.99		
Histologic type, non-squamous vs. squamous	0.4 (0.1–1.2)	0.10		
PD-L1 (%), >10 vs. ≤10	0.2 (0.1–0.8)	0.02	0.4 (0.1–1.6)	0.19
Disease stage, IIIa vs. Ib/II	0.2 (0.1–0.7)	0.02	0.5 (0.1–1.9)	0.29
Largest tumor size at baseline (mm), >50 vs. ≤50	1.0 (0.3–3.1)	0.97		
Nodal stage, N1/2 vs. N0	0.8 (0.1–6.2)	0.82		
Nivolumab-based treatment, present vs. absent	1.2 (0.4–4.1)	0.76		
RECIST objective response, present vs. absent	0.4 (0.1–1.4)	0.15		
TGR_0 (%/month), ≤30 vs. >30	0.1 (0.02–0.4)	0.001	0.2 (0.03–0.7)	0.01

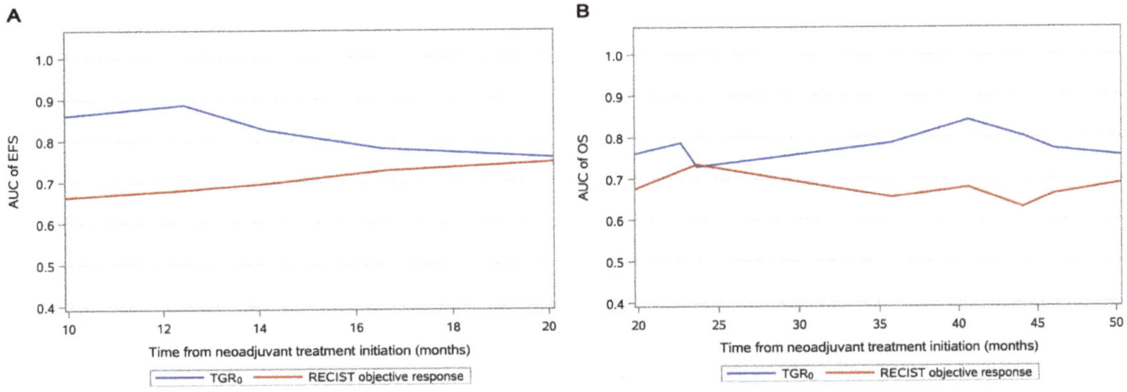

Figure 4. A time-dependent area under the TGR_0 curve compared with the RECIST objective response of the primary tumor for event-free survival (**A**) and overall survival (**B**). The TGR_0 model had a mean time-dependent AUC of 0.83 (95% CI, 0.64–0.95) and 0.80 (95% CI, 0.62–0.97) for predicting EFS and OS, respectively. The RECIST objective response model had a mean time-dependent AUC of 0.70 (95% CI, 0.46–0.94) and 0.66 (95% CI, 0.43–0.88) for predicting EFS and OS, respectively.

3.3. Predictors of Pathological Response after Neoadjuvant Treatment

Twenty-six patients (81%) underwent surgery after neoadjuvant treatment and all resected patients had an R0 resection. Among them, 15 (58%) had MPR, including nine with pCR (36%). MPR was observed for both PD-L1–positive and PD-L1–negative tumors and was associated with both EFS (log-rank $p = 0.002$) and OS (log-rank $p = 0.01$) (Supplementary Figure S2). Only an objective response of the primary tumor was associated with MPR (OR = 27.5; 95% CI, 2.6–289.1; $p = 0.006$) with a sensitivity and specificity of 0.73 (11/15 patients; [95% CI: 0.51, 0.96]) and 0.91 (10/11 patients; [95% CI: 0.74, 0.99]), respectively (Table 4 and Figure 5).

Table 4. Multivariable analysis of major pathological response. Multivariable analysis was undertaken by entering all variables at the $p < 0.05$ level in the univariable analysis. MPR: major pathological response; OR: odds ratio; CI: confidence interval.

Variable	Univariable Analysis of MPR		Multivariable Analysis of MPR	
	OR (95% CI)	*p*-Value	OR (95% CI)	*p*-Value
Age (years), >60 vs. ≤60	1.1 (0.2–5.8)	0.87		
Sex, female vs. Male	0.4 (0.1–2.3)	0.32		
Smoking status, never smoked vs. current or former smoker	-	>0.99		
Histologic type, non-squamous vs. squamous	1.5 (0.2, 9.4)	0.66		
PD-L1 (%), >10 vs. ≤10	3.1 (0.6–17.3)	0.20		
Disease stage, IIIa vs. Ib/II	5.2 (0.5–59.3)	0.18		
Largest tumor size at baseline (mm), >50 vs. ≤50	3.9 (0.6–24.7)	0.14		
Nodal stage, N1/2 vs. N0	0.7 (0.1–8.2)	0.74		
Nivolumab-based treatment, present vs. absent	2.3 (0.4–13.3)	0.36		
RECIST objective response, present vs. absent	27.5 (2.6–289.1)	0.006	27.5 (2.6–289.1)	0.006
TGR_0 (%/month), ≤30 vs. >30	-	>0.99		

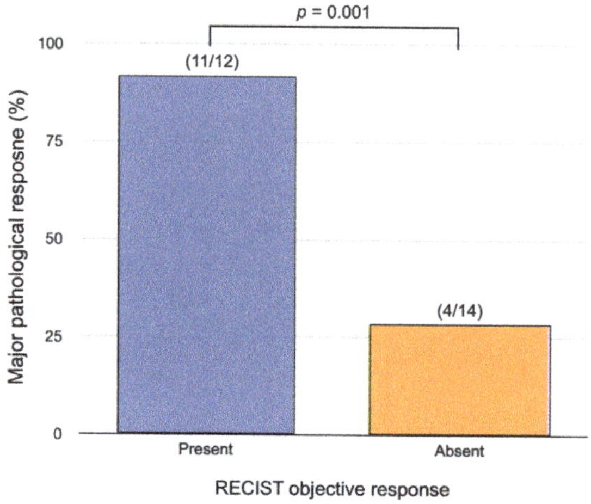

Figure 5. Correlation of major pathological response and RECIST objective response of the primary tumor.

4. Discussion

Identification of earlier biomarkers associated with progression precluding surgery or disease recurrence after surgery is of importance in patients with resectable NSCLC treated by neoadjuvant treatment [19]. In our study, a lower pretreatment tumor growth rate (TGR_0) was a strong factor associated with longer event-free and overall survivals after neoadjuvant treatment. The TGR_0 could provide an early, noninvasive, cost-effective, and time-efficient method to identify patients likely to benefit from a neoadjuvant strategy.

Among patients with advanced lung cancer, the tumor growth rate has been recognized as an important marker of tumor response and progression, especially in the setting of immune checkpoint inhibitor therapy [17,18,20]. He et al. found that patients with metastatic NSCLC undergoing immunotherapy and exhibiting a high pretreatment TGR_0 (> 25%/month) had lower PFS and a less durable clinical rate [20]. Interestingly, the threshold of TGR_0 that best separated the groups with distinct clinical outcomes was comparable with the result in our study and consistent with a tumor volume doubling time of 80 days. In our study, all patients exhibiting a high pretreatment tumor growth rate (> 30%/month) experienced progression or recurrence of the disease and lower OS, indicating that neoadjuvant strategies including anti-PD-1 antibodies are not capable of inhibiting rapidly growing tumors, eradicating micrometastatic disease, and preventing tumor relapse [21–23].

The RECIST objective response of the primary tumor correlated moderately with clinical outcomes. Among patients without progression before surgery or recurrence after surgery, almost half did not show an objective response. The TGR_0 model outperformed RECIST assessments with regard to survival prediction (time-dependent AUC of OS: 0.80 vs. 0.66) and helped to identify a large subset of patients (81%) with a low rate of progression or recurrence. In addition, we reported no significant correlation between smoking status, histologic type, disease stage or PD-L1 expression and survival outcomes. Our results are in accordance with the NADIM trial in which no strong association between the tumor response to treatment according to RECIST criteria or PD-L1 expression and survival was found [13]. In their trial, undetectable ctDNA at the end of neoadjuvant treatment has been proposed as a surrogate endpoint for long-term outcomes. However, the lack of standardization and the limited sensitivity of the current detection methods remained an obstacle for widespread clinical application [24]. TGR_0 may be more biologically and clinically relevant for predicting patient clinical outcomes. Uncontrolled tumor growth is associated with a larger tumor burden, aberrant vascularization, and an altered immune microenvironment unfavorable for the action of PD-1 axis inhibitors [25,26].

A major pathologic response defined as $\leq$ 10% residual viable tumor in the primary tumor and lymph nodes was seen in almost 60% of patients receiving neoadjuvant treatment, mainly with nivolumab, which was consistent with other studies [12,13,27,28]. A major pathologic response might represent a promising surrogate endpoint for survival outcomes for patients with resectable tumors after neoadjuvant treatment [29–31]. A link between MPR and survival outcomes was also observed in this study. Additionally, we found that the RECIST objective response was the only variable associated with MPR. Indeed, neither clinical variables nor PD-L1 staining predicted MPR. This may have significant implications for optimizing the approach and extent of surgical resection as well as radiological follow-up after surgery or for adjuvant decisions [32].

Our study has several limitations. First, the study was retrospectively conducted at a single institute with a small sample size and a substantial number of exploratory variables; thus, results from multivariable analyses should be taken with caution. Second, a limitation of assessment of tumor growth dynamics is requirement of one imaging scan before baseline. In our experience, treatment decision-making often requires consecutive examinations separated by at least one month to confirm or complete staging information of NSCLC (e.g., lung CT and PET-CT scans). Additionally, frequent delays between initial imaging and screening often lead to the necessity of performing an additional staging scan immediately before neoadjuvant treatment. Third, the TGR assessment was based on the largest axis of the primary tumor, like the time-efficient approach of RECIST v1.1, although it may not reflect the whole tumor burden as non-targets like lymph nodes were not considered. In addition, automatic segmentation tools could be a more robust approach with less variability for volume growth assessment but were not investigated.

5. Conclusions

Pretreatment tumor growth rate (TGR_0) provides information to select patients with slow-growing non-small cell lung cancer who should benefit from first-line neoadjuvant treatment. TGR_0 may be an early radiographic marker for more favorable genetic and/or biologic profiles that result in improved disease control and overall survival. The objective response of the primary tumor has the potential to serve as a surrogate of a major pathological response.

Supplementary Materials: The following supporting information can be downloaded at: https://www.mdpi.com/article/10.3390/cancers15164158/s1, Figure S1: Kaplan-Meier analysis of event-free survival (EFS) and overall survival (OS) by RECIST objective response of the primary tumor (**A**,**B**) and pretreatment TGR0 (**C**,**D**) in a subgroup of patients treated by neoadjuvant nivolumab-based therapy. p-values were obtained using the log-rank test.; Figure S2: Kaplan-Meier analysis of event-free survival (**A**) and overall survival (**B**) by major pathological response (MPR). *p*-values were obtained using the log-rank test. EFS: event-free survival; OS: overall survival.

Author Contributions: Conceptualization, T.R. and V.S.; methodology, T.R.; software, T.R.; validation, T.R., V.S. and N.G.; formal analysis, T.R.; investigation, V.S.; resources, N.G.; data curation, T.R. and L.C.; writing—original draft preparation, T.R. and L.C.; writing—review and editing, V.S. and N.G.; visualization, V.S.; supervision, V.S.; project administration, N.G.; funding acquisition, N.G. All authors have read and agreed to the published version of the manuscript.

Funding: This research has received no external funding.

Institutional Review Board Statement: The study was conducted in accordance with the guidelines of the Declaration of Helsinki, and approved by the Institutional Review Board of Institut Curie (protocol CRI-02-2023 approved on 2 April 2023).

Informed Consent Statement: Informed consent was obtained from all subjects involved in the study.

Data Availability Statement: The data presented in this study are available on request from the corresponding author.

Conflicts of Interest: Dr. Girard reports receiving research grants/support from AstraZeneca, Amgen, Boehringer Ingelheim, Eli Lilly, Hoffmann-La Roche, Janssen, Merck, Merck Sharp & Dohme, Novartis, Pfizer, Sivan, and Trizell; having consultative services for Bristol Myers Squibb, AstraZeneca, AbbVie, Amgen, Boehringer Ingelheim, Eli Lilly, Hoffmann-La Roche, Janssen, Merck, Merck Sharp & Dohme, Mirati, Novartis, Pfizer, Roche, Sanofi, and Sivan; receiving payment for expert testimony from AstraZeneca; having participation on a data safety monitoring board for Roche; having a leadership role in the International Thymic Malignancy Interest Group; and having employment of a family member with AstraZeneca. Other authors declare no conflict of interest.

References

1. Thai, A.A.; Solomon, B.J.; Sequist, L.V.; Gainor, J.F.; Heist, R.S. Lung Cancer. *Lancet* **2021**, *398*, 535–554. [CrossRef] [PubMed]
2. de Koning, H.J.; van der Aalst, C.M.; de Jong, P.A.; Scholten, E.T.; Nackaerts, K.; Heuvelmans, M.A.; Lammers, J.-W.J.; Weenink, C.; Yousaf-Khan, U.; Horeweg, N.; et al. Reduced Lung-Cancer Mortality with Volume CT Screening in a Randomized Trial. *N. Engl. J. Med.* **2020**, *382*, 503–513. [CrossRef] [PubMed]
3. Pastorino, U.; Silva, M.; Sestini, S.; Sabia, F.; Boeri, M.; Cantarutti, A.; Sverzellati, N.; Sozzi, G.; Corrao, G.; Marchianò, A. Prolonged Lung Cancer Screening Reduced 10-Year Mortality in the MILD Trial: New Confirmation of Lung Cancer Screening Efficacy. *Ann. Oncol.* **2019**, *30*, 1162–1169. [CrossRef]
4. Flores, R.; Patel, P.; Alpert, N.; Pyenson, B.; Taioli, E. Association of Stage Shift and Population Mortality Among Patients with Non–Small Cell Lung Cancer. *JAMA Netw. Open* **2021**, *4*, e2137508. [CrossRef]
5. Raman, V.; Yang, C.-F.J.; Deng, J.Z.; D'Amico, T.A. Surgical Treatment for Early Stage Non-Small Cell Lung Cancer. *J. Thorac. Dis.* **2018**, *10*, S898–S904. [CrossRef]
6. Cruz, C.; Afonso, M.; Oliveiros, B.; Pêgo, A. Recurrence and Risk Factors for Relapse in Patients with Non-Small Cell Lung Cancer Treated by Surgery with Curative Intent. *Oncology* **2017**, *92*, 347–352. [CrossRef] [PubMed]
7. Fedor, D.; Johnson, W.R.; Singhal, S. Local Recurrence Following Lung Cancer Surgery: Incidence, Risk Factors, and Outcomes. *Surg. Oncol.* **2013**, *22*, 156–161. [CrossRef]
8. Kelsey, C.R.; Marks, L.B.; Hollis, D.; Hubbs, J.L.; Ready, N.E.; D'Amico, T.A.; Boyd, J.A. Local Recurrence after Surgery for Early Stage Lung Cancer: An 11-Year Experience with 975 Patients. *Cancer* **2009**, *115*, 5218–5227. [CrossRef] [PubMed]
9. Bria, E.; Gralla, R.J.; Raftopoulos, H.; Cuppone, F.; Milella, M.; Sperduti, I.; Carlini, P.; Terzoli, E.; Cognetti, F.; Giannarelli, D. Magnitude of Benefit of Adjuvant Chemotherapy for Non-Small Cell Lung Cancer: Meta-Analysis of Randomized Clinical Trials. *Lung Cancer* **2009**, *63*, 50–57. [CrossRef] [PubMed]
10. NSCLC Meta-Analyses Collaborative Group. Adjuvant Chemotherapy, with or without Postoperative Radiotherapy, in Operable Non-Small-Cell Lung Cancer: Two Meta-Analyses of Individual Patient Data. *Lancet* **2010**, *375*, 1267–1277. [CrossRef]
11. NSCLC Meta-Analysis Collaborative Group. Preoperative Chemotherapy for Non-Small-Cell Lung Cancer: A Systematic Review and Meta-Analysis of Individual Participant Data. *Lancet* **2014**, *383*, 1561–1571. [CrossRef]
12. Forde, P.M.; Spicer, J.; Lu, S.; Provencio, M.; Mitsudomi, T.; Awad, M.M.; Felip, E.; Broderick, S.R.; Brahmer, J.R.; Swanson, S.J.; et al. Neoadjuvant Nivolumab plus Chemotherapy in Resectable Lung Cancer. *N. Engl. J. Med.* **2022**, *386*, 1973–1985. [CrossRef] [PubMed]
13. Provencio, M.; Serna-Blasco, R.; Nadal, E.; Insa, A.; García-Campelo, M.R.; Casal Rubio, J.; Dómine, M.; Majem, M.; Rodríguez-Abreu, D.; Martínez-Martí, A.; et al. Overall Survival and Biomarker Analysis of Neoadjuvant Nivolumab Plus Chemotherapy in Operable Stage IIIA Non–Small-Cell Lung Cancer (NADIM Phase II Trial). *J. Clin. Oncol.* **2022**, *40*, 2924–2933. [CrossRef]
14. Ferté, C.; Fernandez, M.; Hollebecque, A.; Koscielny, S.; Levy, A.; Massard, C.; Balheda, R.; Bot, B.; Gomez-Roca, C.; Dromain, C.; et al. Tumor Growth Rate Is an Early Indicator of Antitumor Drug Activity in Phase I Clinical Trials. *Clin. Cancer Res.* **2014**, *20*, 246–252. [CrossRef] [PubMed]
15. Lamarca, A.; Ronot, M.; Moalla, S.; Crona, J.; Opalinska, M.; Lopez Lopez, C.; Pezzutti, D.; Najran, P.; Carvhalo, L.; Bezerra, R.O.F.; et al. Tumor Growth Rate as a Validated Early Radiological Biomarker Able to Reflect Treatment-Induced Changes in Neuroendocrine Tumors: The GREPONET-2 Study. *Clin. Cancer Res.* **2019**, *25*, 6692–6699. [CrossRef] [PubMed]
16. Ramtohul, T.; Cohen, A.; Rodrigues, M.; Piperno-Neumann, S.; Cabel, L.; Cassoux, N.; Lumbroso-Le Rouic, L.; Malaise, D.; Gardrat, S.; Pierron, G.; et al. Tumor Growth Rate Improves Tumor Assessment and First-Line Systemic Treatment Decision-Making for Immunotherapy in Patients with Liver Metastatic Uveal Melanoma. *Br. J. Cancer* **2022**, *127*, 258–267. [CrossRef] [PubMed]
17. Osorio, B.; Yegya-Raman, N.; Kim, S.; Simone II, C.B.; Theodorou Ross, C.; Deek, M.P.; Gaines, D.; Zou, W.; Lin, L.; Malhotra, J.; et al. Clinical Significance of Pretreatment Tumor Growth Rate for Locally Advanced Non-Small Cell Lung Cancer. *Ann. Transl. Med.* **2019**, *7*, 95. [CrossRef]
18. Champiat, S.; Dercle, L.; Ammari, S.; Massard, C.; Hollebecque, A.; Postel-Vinay, S.; Chaput, N.; Eggermont, A.; Marabelle, A.; Soria, J.-C.; et al. Hyperprogressive Disease Is a New Pattern of Progression in Cancer Patients Treated by Anti-PD-1/PD-L1. *Clin. Cancer Res.* **2017**, *23*, 1920–1928. [CrossRef]
19. Uprety, D.; Mandrekar, S.J.; Wigle, D.; Roden, A.C.; Adjei, A.A. Neoadjuvant Immunotherapy for NSCLC: Current Concepts and Future Approaches. *J. Thorac. Oncol.* **2020**, *15*, 1281–1297. [CrossRef] [PubMed]

20. He, L.; Zhang, X.; Li, H.; Chen, T.; Chen, C.; Zhou, Y.; Lin, Z.; Du, W.; Fang, W.; Yang, Y.; et al. Pre-Treatment Tumor Growth Rate Predicts Clinical Outcomes of Patients With Advanced Non-Small Cell Lung Cancer Undergoing Anti-PD-1/PD-L1 Therapy. *Front. Oncol.* **2021**, *10*, 621329. [CrossRef]
21. Wagner, N.B.; Lenders, M.M.; Kühl, K.; Reinhardt, L.; André, F.; Dudda, M.; Ring, N.; Ebel, C.; Stäger, R.; Zellweger, C.; et al. Pretreatment Metastatic Growth Rate Determines Clinical Outcome of Advanced Melanoma Patients Treated with Anti-PD-1 Antibodies: A Multicenter Cohort Study. *J. Immunother. Cancer* **2021**, *9*, e002350. [CrossRef] [PubMed]
22. Matsumura, S.; Kato, T.; Kujime, Y.; Kitakaze, H.; Nakano, K.; Hongo, S.; Yoshioka, I.; Okumi, M.; Nonomura, N.; Takada, S. Pre-Treatment Metastatic Growth Rate Is Associated with Clinical Outcome in Patients with Metastatic Renal Cell Carcinoma Treated with Nivolumab. *BMC Urol.* **2023**, *23*, 107. [CrossRef]
23. Zhang, L.; Wu, L.; Chen, Q.; Zhang, B.; Liu, J.; Liu, S.; Mo, X.; Li, M.; Chen, Z.; Chen, L.; et al. Predicting Hyperprogressive Disease in Patients with Advanced Hepatocellular Carcinoma Treated with Anti-Programmed Cell Death 1 Therapy. *EClinicalMedicine* **2021**, *31*, 100673. [CrossRef]
24. Peng, Y.; Mei, W.; Ma, K.; Zeng, C. Circulating Tumor DNA and Minimal Residual Disease (MRD) in Solid Tumors: Current Horizons and Future Perspectives. *Front. Oncol.* **2021**, *11*, 763790. [CrossRef] [PubMed]
25. Pitt, J.M.; Marabelle, A.; Eggermont, A.; Soria, J.-C.; Kroemer, G.; Zitvogel, L. Targeting the Tumor Microenvironment: Removing Obstruction to Anticancer Immune Responses and Immunotherapy. *Ann. Oncol.* **2016**, *27*, 1482–1492. [CrossRef]
26. Altorki, N.K.; Markowitz, G.J.; Gao, D.; Port, J.L.; Saxena, A.; Stiles, B.; McGraw, T.; Mittal, V. The Lung Microenvironment: An Important Regulator of Tumour Growth and Metastasis. *Nat. Rev. Cancer* **2019**, *19*, 9–31. [CrossRef] [PubMed]
27. Rothschild, S.I.; Zippelius, A.; Eboulet, E.I.; Savic Prince, S.; Betticher, D.; Bettini, A.; Früh, M.; Joerger, M.; Lardinois, D.; Gelpke, H.; et al. SAKK 16/14: Durvalumab in Addition to Neoadjuvant Chemotherapy in Patients with Stage IIIA(N2) Non–Small-Cell Lung Cancer—A Multicenter Single-Arm Phase II Trial. *J. Clin. Oncol.* **2021**, *39*, 2872–2880. [CrossRef]
28. Provencio, M.; Nadal, E.; González-Larriba, J.L.; Martínez-Martí, A.; Bernabé, R.; Bosch-Barrera, J.; Casal-Rubio, J.; Calvo, V.; Insa, A.; Ponce, S.; et al. Perioperative Nivolumab and Chemotherapy in Stage III Non-Small-Cell Lung Cancer. *N. Engl. J. Med.* **2023**, *389*, 504–513. [CrossRef] [PubMed]
29. Hellmann, M.D.; Chaft, J.E.; William, W.N.; Rusch, V.; Pisters, K.M.W.; Kalhor, N.; Pataer, A.; Travis, W.D.; Swisher, S.G.; Kris, M.G. Pathological Response after Neoadjuvant Chemotherapy in Resectable Non-Small-Cell Lung Cancers: Proposal for the Use of Major Pathological Response as a Surrogate Endpoint. *Lancet Oncol.* **2014**, *15*, e42–e50. [CrossRef]
30. Rosner, S.; Liu, C.; Forde, P.M.; Hu, C. Association of Pathologic Complete Response and Long-Term Survival Outcomes Among Patients Treated with Neoadjuvant Chemotherapy or Chemoradiotherapy for NSCLC: A Meta-Analysis. *JTO Clin. Res. Rep.* **2022**, *3*, 100384. [CrossRef]
31. Pataer, A.; Weissferdt, A.; Vaporciyan, A.A.; Correa, A.M.; Sepesi, B.; Wistuba, I.I.; Heymach, J.V.; Cascone, T.; Swisher, S.G. Evaluation of Pathologic Response in Lymph Nodes of Patients with Lung Cancer Receiving Neoadjuvant Chemotherapy. *J. Thorac. Oncol.* **2021**, *16*, 1289–1297. [CrossRef] [PubMed]
32. Huynh, C.; Walsh, L.A.; Spicer, J.D. Surgery after Neoadjuvant Immunotherapy in Patients with Resectable Non-Small Cell Lung Cancer. *Transl. Lung Cancer Res.* **2021**, *10*, 563–580. [CrossRef] [PubMed]

cancers

MDPI

Article

Combining Classic and Novel Neutrophil-Related Biomarkers to Identify Non-Small-Cell Lung Cancer

Yunzhao Ren [1,2], Qinchuan Wang [1,2,3], Chenyang Xu [1,2], Qian Guo [1,2], Ruoqi Dai [1,2], Xiaohang Xu [1,2], Yuhao Zhang [1,2], Ming Wu [4], Xifeng Wu [1,2,5,*] and Huakang Tu [1,2,*]

1 Department of Big Data in Health Science, School of Public Health, Center of Clinical Big Data and Analytics, The Second Affiliated Hospital, Zhejiang University School of Medicine, 866 Yuhangtang Rd., Hangzhou 310058, China; 11918143@zju.edu.cn (Y.R.); wangqinchuan@zju.edu.cn (Q.W.); 22118855@zju.edu.cn (C.X.); 22118920@zju.edu.cn (Q.G.); 22118915@zju.edu.cn (R.D.); 12018344@zju.edu.cn (X.X.); zhangyuhao0580@outlook.com (Y.Z.)

2 The Key Laboratory of Intelligent Preventive Medicine of Zhejiang Province, 866 Yuhangtang Rd., Hangzhou 310058, China

3 Department of Surgical Oncology, The Affiliated Sir Run Run Shaw Hospital, Zhejiang University School of Medicine, 3 East Qingchun Rd., Hangzhou 310016, China

4 Department of Thoracic Surgery, The Second Affiliated Hospital, Zhejiang University School of Medicine, 88 Jiefang Rd., Hangzhou 310009, China; iwuming22@zju.edu.cn

5 Cancer Center, Zhejiang University, 866 Yuhangtang Rd., Hangzhou 310058, China

* Correspondence: xifengw@zju.edu.cn (X.W.); huakangtu@zju.edu.cn (H.T.); Tel.: +86-13016839096 (X.W.); +86-0571-88981319 (H.T.)

Simple Summary: This study explored the predictive value of neutrophils and neutrophil-related biomarkers as auxiliary diagnosis biomarkers of NSCLC in an ongoing large cohort. IL-6 and IL-1RA were identified as independent risk factors for NSCLC. These findings can improve the predictive performance beyond epidemiological variables and classic neutrophil-related biomarkers in identifying NSCLC.

Abstract: Background: Recent studies have revealed that neutrophils play a crucial role in cancer progression. This study aimed to explore the diagnostic value of neutrophil-related biomarkers for non-small-cell lung cancer (NSCLC). Methods: We initially assessed the associations between classic neutrophil-related biomarkers (neutrophil-to-lymphocyte ratio (NLR), absolute neutrophil counts (NEU), absolute lymphocyte counts (LYM)) and NSCLC in 3942 cases and 6791 controls. Then, we measured 11 novel neutrophil-related biomarkers via Luminex Assays in 132 cases and 66 controls, individually matching on sex and age (±5 years), and evaluated their associations with NSCLC risk. We also developed the predictive models by sequentially adding variables of interest and assessed model improvement. Results: Interleukin-6 (IL-6) (odds ratio (OR) = 10.687, 95% confidence interval (CI): 3.875, 29.473) and Interleukin 1 Receptor Antagonist (IL-1RA) (OR = 8.113, 95% CI: 3.182, 20.689) shows strong associations with NSCLC risk after adjusting for body mass index, smoking status, NLR, and carcinoembryonic antigen. Adding the two identified biomarkers to the predictive model significantly elevated the model performance from an area under the receiver operating characteristic curve of 0.716 to 0.851 with a net reclassification improvement of 97.73%. Conclusions: IL-6 and IL-1RA were recognized as independent risk factors for NSCLC, improving the predictive performance of the model in identifying disease.

Keywords: non-small-cell lung cancer; neutrophils; biomarkers; Interleukin-6; Interleukin 1 receptor antagonist; diagnosis

1. Introduction

Worldwide, lung cancer stands as the principal cause of cancer-related deaths. In 2020, approximately 2.2 million new lung cancer cases were diagnosed, with 1.8 million fatalities. Over 85% of all lung cancer patients have non-small-cell lung cancer (NSCLC) [1].

Neutrophils, the most abundant cells in human blood circulation, are recently identified as a crucial player during carcinogenesis [2]. Neutrophils are the primary cell type during the acute inflammatory response, rapidly recruited to the affected tissue through a multi-step cascade [3], and capable of eliminating pathogens through diverse mechanisms, including phagocytosis, release of antimicrobial proteins, and formation of neutrophil extracellular traps (NETs) [4]. During the resolution of inflammation or in an anti-inflammatory state, the involvement of neutrophils is also significant. Their phagocytic activity aids in the clearance of dead cells and bacteria, thereby contributing to the elimination and reconstruction of the affected area. This can be attributed to the essential functions of several proteases expressed by neutrophils, including MMP9 and VEGFA, in tissue repair, remodeling, and angiogenesis [5,6]. The persistent infiltration of neutrophils causes chronic inflammation, which in turn leads to tissue damage and plays a significant role in the onset of cancer. This lasting and unresolved tissue inflammation is a characteristic feature of the tumor microenvironment [6]. It has been proven that neutrophils can modulate tumor progression during the onset and growth of cancer, possessing both pro-tumoral and anti-tumoral functions [5,7]. Based on the different mediators of cancer cells and the tumor microenvironment, neutrophils can be polarized into different activation states, thereby playing distinct functions in alteration of tumor progression [8]. For instance, neutrophils could promote tumorigenesis via reactive oxygen species (ROS) induced DNA damage in a lung cancer model [9], whereas it could also attack tumor cells by a neutrophil-dependent cytotoxic effect via a phagocytosis signaling of signal regulatory protein-α (SIRPα)–CD47 interaction [10,11]. However, given the multifaceted roles and varied phenotypes of neutrophils, the current research on the connection between neutrophils and lung cancer is limited.

Neutrophils and tumor-associated neutrophils (TANs) are associated with key features of resistance to immune checkpoint inhibition, such as adaptive immune cell polarization and suppression, tumor neoangiogenesis, immune exclusion, and cancer-cell-intrinsic characteristics [11–15]. Also, multiple studies have shown that neutrophil-to-lymphocyte ratio (NLR) can predict the clinical response of ICI treatment [12,13]. In clinical practice, the balance of inflammatory and immune responses is frequently reflected by the NLR in peripheral blood [14]. It has emerged as a prognostic factor for the survival and treatment responses in several cancers [15]. NLR is also reported as a promising predictive biomarker for immune checkpoint inhibition in NSCLC patients [14]. However, the diagnostic value of NLR in NSCLC and its underlying mechanisms are yet to be extensively studied.

Depending on the context, neutrophils play a dual role in tumor development. They promote inflammation through the release of ROS or proteases, and promote tumor dissemination and metastasis by facilitating immune suppression, angiogenesis, cancer cell motility, and epithelial-to-mesenchymal transition (EMT) [8,16]. Recent investigations have highlighted the critical role of NETs in tumor initiation and metastasis [16]. Meanwhile, neutrophils can restrict cancer growth through cytotoxic activities, such as the release of iNOS, which exerts cytotoxic effects on cancer cells. Moreover, they can inhibit tumor metastasis through mechanisms mediated by H202 or TSP1 [5,16]. TANs could impact anti-tumor immunity via secreting cytokines crosstalk with $CD8^+$ T cells. TANs could produce proinflammatory factors such as monocyte chemoattractant protein-1, interleukin-8, macrophage inflammatory protein-1 alpha, interleukin-6 (IL-6), and anti-inflammatory interleukin 1 receptor antagonist (IL-1RA), thereby bolstering anti-tumor immunity in early-stage lung cancer [17]. Also, other cytokines, including interleukins (ILs), colony-stimulating factor (CSF), interferon (IFN), and chemokines, demonstrated significant associations with tumorigenesis in terms of modulating intercellular interactions and regulating immune responses [18–20]. Understanding the multifunctionality of neutrophils, their diverse phenotypes in different environments, and the potential for reprogramming has significant implications for understanding cancer initiation and progression [5,21]. Monitoring neutrophil-associated cytokines throughout disease progression may serve as a predictive tool for disease onset and development. However, until now, no comprehensive study has been implemented to systematically illustrate their impact on lung cancer risk.

We carried out a multi-phase study to investigate the potential diagnostic value of neutrophil-related biomarkers in NSCLC development. In the first phase, we explored the predictive efficacy of NLR as an auxiliary biomarker in diagnosing NSCLC in a large cohort encompassing lung cancer patients and healthy controls. In the second phase, we further included 132 patients from the lung cancer cohort and 66 matched healthy individuals and investigated the association between eleven novel blood neutrophil-related biomarkers and the risk of NSCLC. In the third phase, we developed predictive models incorporating classic and novel neutrophil-related biomarkers and clinical variables for the diagnosis of NSCLC.

2. Materials and Methods

2.1. Study Population and Data Collection

In the first phase of this study, a case–control design was adopted to explore the associations between classic blood neutrophil-related biomarkers (NLR, absolute neutrophil counts (NEU), absolute lymphocyte counts (LYM)), and the risk of NSCLC. From an ongoing cohort study begun in 2020 at The Second Affiliated Hospital Zhejiang University School of Medicine (SAHZU), a total of 3942 NSCLC patients were recruited. And 6791 healthy controls were drawn from the concurrently recruited healthy controls who had health check-up examinations in the general practice clinic at the SAHZU. In the second phase of the study, we further measured 11 novel blood neutrophil-related biomarkers in 132 cases and 66 healthy controls, which were individually matched on sex and age ($\pm$5 years).

The inclusion criteria for the case group are as follows: (1) with clinically and histopathologically confirmed NSCLC; (2) the patient has provided informed consent or waived consent; and (3) with data on classic blood neutrophil-related biomarkers. The exclusion criteria for the case group are as follows: (1) multiple cancers; and (2) any previous treatment undertaken by the patient at the point of enrollment.

The inclusion criteria for the healthy control group are as follows: (1) regular health examination participants from the SAHZU; and (2) participants have provided informed consent or waived consent. The exclusion criteria are as follows: (1) suffering from severe lung disease; and (2) diagnosis of any malignant neoplasm.

This study has received approval from the Institutional Review Board of SAHZU. Clinical pathological information was derived from detailed chart reviews. Concentrations of NEU, LYM, and carcinoembryonic antigen (CEA) in the blood were quantified at NSCLC diagnosis for cases or during routine health examinations for the controls. The staging of lung cancer patients was carried out by the attending physicians and pathologists in accordance with the NCCN Clinical Practice Guidelines Non-small-cell lung cancer v1, 2022.

Staff members gathered epidemiological information through face-to-face interviews. The participants' weight, height, history of hypertension (yes or no), history of diabetes (yes or no), and smoking status were recorded upon enrollment. The body mass index (BMI) was derived by taking the ratio of weight to the square of the height (kg/m^2). According to the WHO guidelines, we divided BMI into two categories: underweight/normal (<25 kg/m^2) and overweight/obese ($\geq$25 kg/m^2). The categorization of smoking status depends on whether the subject had ever smoked (defined as having smoked at least 100 cigarettes in their lifetime) [22,23].

2.2. Detection of Novel Neutrophil-Related Biomarkers via Luminex Assays

In stage 2, venous blood samples of 20 mL were collected from 198 participants using ethylenediaminetetraacetic acid tubes and promptly delivered to the SAHZU laboratory. Prior to the initiation of the experiment, plasma was isolated, divided into aliquots, and preserved at −80 °C. Plasma samples were defrosted on ice [23]. The concentrations of 11 novel neutrophil-related biomarkers (IL-1α, IL-1β, IL-1RA, IL-6, IL-17, G-SCF, GM-CSF, CXCL2, CXCL5, IFN-α, and S100B) in the plasma samples were quantitatively determined utilizing the Luminex Discovery Assay—Human Premixed Multi-Analyte Kit (R&D Systems, Minneapolis, MN, USA, LXSAHM-11), strictly adhering to the protocol provided by the manufacturer [24]. To ensure the reliability and accuracy of the measurements, each sample was assayed in duplicates on a 96-well plate using a Luminex FLEXMAP 3D system (Luminex Corp, Austin, TX, USA), utilizing undiluted plasma. Each plate incorporated both positive and negative controls, as well as samples for the purpose of generating the standard curve [23,25]. Laboratory personnel were completely blinded to the case and control status. The assay was performed in alignment with the manufacturer's instructions.

We selected 11 novel blood neutrophil-related biomarkers based on comprehensive literature reviews and the feasibility of using Luminex assays. G-CSF is a critical regulatory agent in the biological genesis of neutrophils, with its receptors being expressed throughout the entire bone marrow lineage, ranging from early stem cells and progenitor cells to the mature status of neutrophils [26]. GM-CSF and IL-6 are both acknowledged as cytokines involved in granulocyte formation and neutrophil proliferation in various types of cancer [26]. IL-1, once activated, functions as a robust pro-inflammatory cytokine locally, instigating vasodilation and recruiting monocytes and neutrophils to the stress location [27]. The generation of active IL-1β is facilitated by inflammasomes or neutrophil proteases through cleaving pro-IL-1β, a process mediated by caspase-1 [27]. IL-1α triggers sterile inflammation through the induction of neutrophil mobilization in reaction to cell death. IL-1 RA, which can be released by neutrophils, binds and blocks IL-1 Receptor Type 1, competitively inhibiting the pro-inflammatory action of IL-1 [28]. IL-17, generated by neutrophils, T cells, innate lymphoid cells, natural killer cells, macrophages, and so on, exerts a crucial role in the recruitment of neutrophils [29]. CXCL2 plays a crucial role in neutrophil recruitment by interacting with CXCR2 on neutrophils [30]. CXCL5 can enhance the immunosuppressive features of the tumor microenvironment by stimulating immune cell migration to the tumor and recruiting vascular endothelial cells for angiogenesis, thus promoting tumor progression [31]. By targeting both tumor cells as well as immune cells, type I interferons have demonstrated a pivotal role in inhibiting tumor growth [32,33]. The S100 family proteins also hold vital value in natural immunity and act as mediators in inflammatory responses. Neutrophils, among other immune cells, can produce considerable amounts of S100 A8/A9, which control inflammation by triggering the discharge of cytokines and ROS. S100B is one of the most active members of the S100 family [34,35]. Among the 11 biomarkers, 10 (except S100B) are produced by neutrophils under certain circumstances. In general, IL-1RA, IFN-α, G-CSF, and GM-CSF are considered anti-tumor biomarkers, while the remaining factors are considered pro-tumor biomarkers.

2.3. Statistical Analysis

Categorical variables are characterized by frequencies with percentages. Statistical differences between groups were compared using chi-square tests or Fisher's exact probability method. Continuous variables are depicted using mean ± standard deviation (SD) or median [25th and 75th percentiles (Q1–Q3)] depending on the distribution type. A comparison of groups for statistical differences was conducted using *t*-tests (for normal distributions), Kruskal–Wallis tests or Wilcoxon rank-sum tests (for non-normal distributions). For the matched paired samples in stage 2, the comparison of measurement data between the two groups was conducted using the paired samples *t*-test or non-parametric test, while the comparison of categorical data was performed using the paired chi-square test or non-parametric test. Furthermore, we conducted a sensitivity analysis in stage 1 with cases and controls being individually matched on sex and age (±5 years).

All biomarkers were further processed as categorical variables to minimize skewness. Using the median value in the control group as the cutoff, the values of NLR, NEU, LYM, IL-6, CXCL2, IL-1RA, IL-1α, and CXCL5 were classified into low and high groups. Meanwhile, the values of S100B and GM-CSF were divided into low and high groups, with the experimental detection limit serving as the cutoff because a substantial of individuals had levels under the experimental detection limit. The CEA value was divided into normal and abnormal groups using the threshold of 5 ng/mL.

In stage 2, given the matched case–control design, the associations between novel neutrophil-related biomarkers and NSCLC risk were examined by conditional logistic regression. We first performed a univariate analysis for the 11 measured novel markers, followed by a multivariate analysis adjusting for epidemiological variables, NLR, and CEA, founded on the univariate analysis results and prior knowledge. The model outputs the odds ratio (OR) and its 95% confidence interval (CI) to estimate the strength of the association between the novel biomarkers and NSCLC risk.

In stage 3, we developed the predictive models by sequentially adding variables of interest in the study population from stage 2. The risk prediction model was initially built based on health history (model 1: BMI + smoking status). Then, we included relevant clinical biomarkers (model 2: model 1 + NLR + CEA). Further, two newly identified novel blood neutrophil-related biomarkers were incorporated (model 3: model 2 + IL-6 + IL-1RA). We applied the receiver operating characteristic (ROC) curve and the area under the curve (AUC) to evaluate the discriminative capacity of different models for NSCLC risk. Delong's test was used to test whether the differences in model performance across different models. Additionally, the true positive rate (TPR) and false positive rate (FPR) were calculated. To evaluate whether the predictive performance of models was enhanced after the addition of more predictors, we calculated the net reclassification improvement (NRI) and integrated discrimination improvement (IDI) metrics.

The collection of data and the presentation of tables were conducted in Excel (Microsoft Office 2021 version). R software (v4.2.1) was utilized for all data analysis and visualization. Specifically, the R software packages "readxl", "readr", "plyr", "dplyr", "data.table", "tableone", "table1", "stringr", "forcats", "reshape2", "broom", "tidyverse", and "tidyr" were used for data preparation (including reading, cleaning, data transformation, etc.). The R package "Hmisc" was applied for the correlation analysis and the "pROC" for the ROC curve. The restricted cubic spline analysis used the R packages "rms", "Hmisc", "car", and "smoothHR". The comparative analysis was carried out by the R package "rstatix". The values of NRI and IDI were calculated using the R packages "nricens" and "PredictABEL". In order to visualize our results, the R packages "pheatmap", "ggpubr", "ggplot2", "ggthemes", "grid", "gridExtra", and "forestploter" were employed. The R packages used for the logistic regression analysis were "pubh", "rms", "survival", "car". All R packages and their instructions used in our study can be found in the link (https://cran.r-project.org/web/packages/available_packages_by_name.html (accessed on 25 September 2023). All statistical tests were two sided, setting the significance level at 0.05.

3. Results

3.1. Stage 1

Table S1 presents the baseline characteristics of all subjects in stage 1. More than half of the healthy participants were female, while nearly two-thirds of patients were male. The age of the subjects increased with disease status and severity. The proportion of smokers was higher among patients with invasive adenocarcinoma (IAC), which includes early-stage and late-stage NSCLC. The majority of patients did not present with either lymph node metastasis or distant metastasis at the time of enrollment.

The restricted cubic spline analysis revealed a positive association between NEU and NSCLC risk, while an inverse association was seen with LYM (Figure 1B,C). NLR showed a strong non-linear positive association with the risk of NSCLC (Figure 1A).

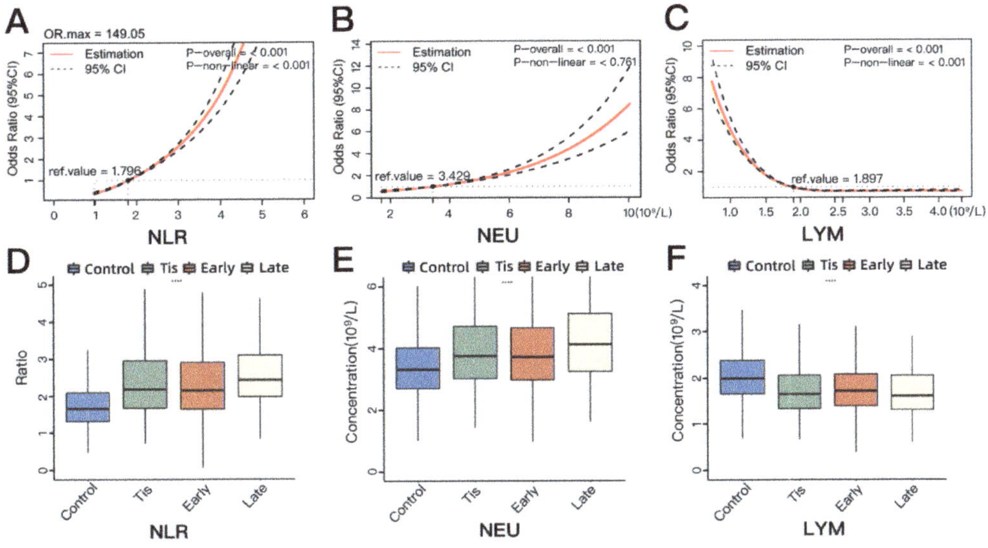

Figure 1. The associations between classic neutrophil-related blood biomarkers (NLR, NEU, and LYM) and NSCLC risk based on restricted cubic splines (**A**–**C**) and the inter-group comparisons of classic biomarkers' distribution across four groups (control, Tis, early-stage NSCLC, and late-stage NSCLC) (**D**–**F**). (**A**) The associations between the plasma level of NLR and NSCLC risk. (**B**) The associations between the plasma concentration of NEU and NSCLC risk. (**C**) The associations between the plasma concentration of LYM and NSCLC risk. (**D**) The distribution of plasma concentration of NLR across four groups. (**E**) The distribution of plasma concentration of NEU across four groups. (**F**) The distribution of plasma concentration of LYM across four groups. Tis, carcinoma in situ; NSCLC, non-small-cell lung cancer; NLR, neutrophil-to-lymphocyte ratio; NEU, absolute neutrophil counts; LYM, absolute lymphocyte counts; OR, odds ratio; CI, confidence interval. Early stage included stage 1 and 2 diseases. Late stage included stage 3 and 4 diseases. The reference value (ref. value) means the level of the biomarker when the corresponding OR is 1 (the horizontal dotted line). The *p*-overall indicates the statistical significance of the association between the biomarker and NSCLC risk, with *p*-overall < 0.05 indicating a statistically significant association. The *p*-non-linear value indicates whether there is a nonlinear relationship between the biomarker and NSCLC risk, with *p*-non-linear < 0.05 indicating that the association between the biomarker and NSCLC risk could not be fully explained by a linear relationship; in other words, there was a non-linear association. **** $p < 0.0001$.

The unconditional univariate logistic regression analysis indicated significant correlations between classic blood biomarkers and NSCLC risk (Table S2 and Figure 2). Elevated levels of NLR and NEU were associated with increased risk of NSCLC (OR_{NLR} = 2.983 (95% CI: 2.737, 3.252), OR_{NEU} = 1.755 (95% CI: 1.620, 1.902)), whereas a high blood concentration of LYM was associated with decreased risk of NSCLC (OR = 0.434, 95% CI: 0.399, 0.471). After adjusting for age, sex, BMI, and smoking status, compared with the low-level groups, the high-level groups of NLR and NEU were associated with higher risks of NSCLC (OR_{NLR} = 2.608 (95% CI: 2.333, 2.918), OR_{NEU} = 2.222 (95% CI: 1.994, 2.477)), while the higher LYM was associated with a lower NSCLC risk (OR_{LYM} = 0.650 (95% CI: 0.583, 0.724)) (all *p* values were less than 0.05). Stratified analysis by age, sex, BMI, and smoking status also revealed significant and highly consistent associations between these three indicators and NSCLC risks (Figure 2). In addition, our sensitivity analysis indicated the results were notably consistent after matching the cases and controls on age and sex.

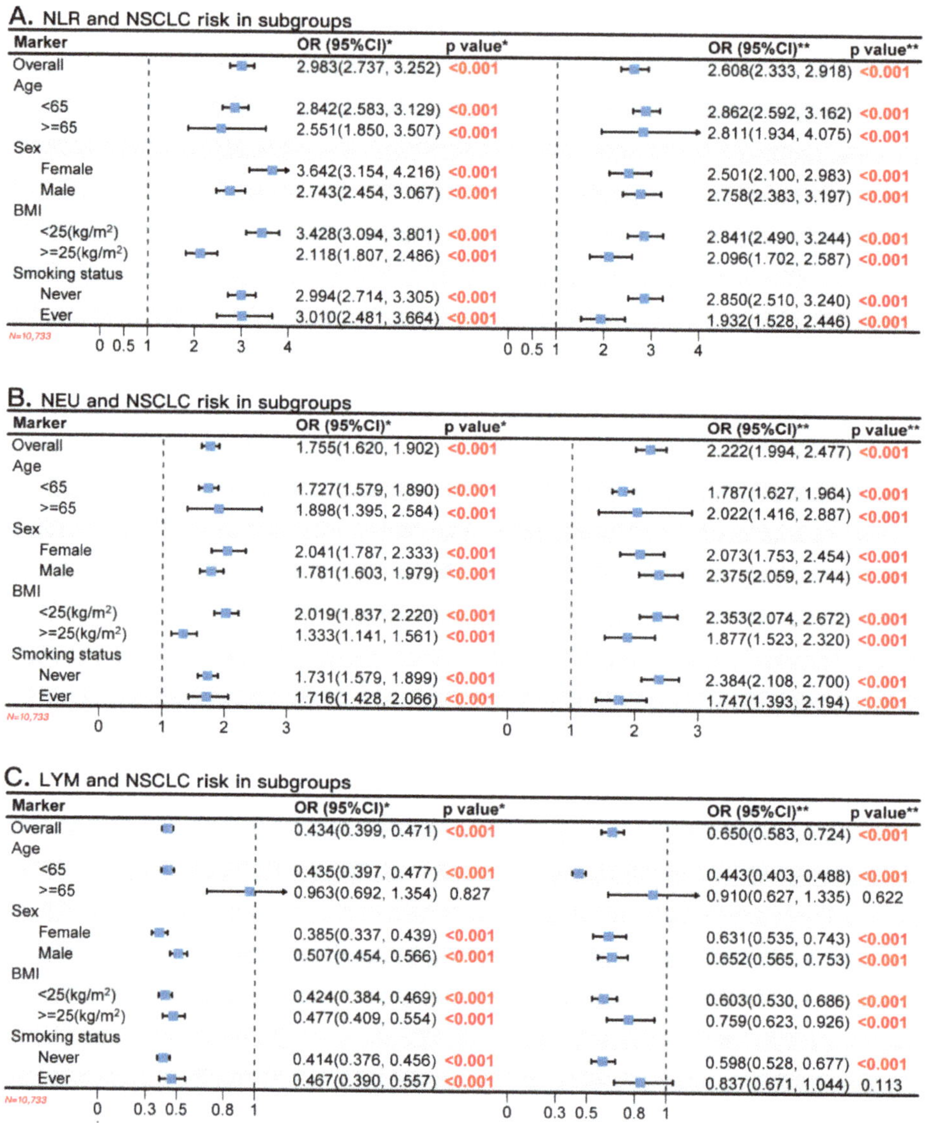

Figure 2. The overall and stratified associations between classic blood neutrophil-related biomarkers (NLR, NEU, and LYM) and NSCLC risk in a large clinical cohort. (**A**) The univariate (**left**) and multiple (**right**) logistic regression analysis of NLR and the stratified analysis. (**B**) The univariate (**left**) and multiple (**right**) logistic regression analysis of NEU and the stratified analysis. (**C**) The univariate (**left**) and multiple (**right**) logistic regression analysis of LYM and the stratified analysis. NLR, neutrophil-to-lymphocyte ratio; NEU, absolute neutrophil counts; LYM, absolute lymphocyte counts; BMI, body mass index; OR, odds ratio; CI, confidence interval. * shows the OR (95%CI) and p values of the univariate conditional logistic regression analysis. ** shows the OR (95%CI) and p values of the multiple conditional logistic regression analysis.

We further analyzed the differences in NEU, LYM, and NLR across the control, carcinoma in situ (Tis), early-stage, and late-stage groups (Table 1). The concentrations of NEU in cancer groups were significantly higher while LYM was lower in comparison to the control group. The NLR was observed to be higher in the NSCLC groups. However, the difference between the Tis and early-stage groups among these three indicators was not significant. Compared to early-stage patients, late-stage patients had higher NLR and NEU levels, while the difference was not significant in LYM level. Figure 1D–F graphically illustrates the differences across those groups.

Table 1. The distribution of classic blood neutrophil-related biomarkers by disease status in stage 1.

Markers	Control (n = 6791) Median [Q1–Q3]	NSCLC (n = 3942) Median [Q1–Q3]			* p	** p	*** p	**** p
		Tis (n = 450)	Early (n = 3376)	Late (n = 116)				
NLR	1.657 [1.315–2.090]	2.179 [1.670–2.962]	2.142 [1.644–2.911]	2.436 [1.972–3.102]	<0.001	<0.001	0.332	<0.001
NEU (10^9/L)	3.310 [2.690–4.020]	3.750 [3.020–4.710]	3.715 [2.970–4.660]	4.120 [3.248–5.115]	<0.001	<0.001	0.670	0.005
LYM (10^9/L)	1.990 [1.650–2.380]	1.645 [1.333–2.060]	1.710 [1.388–2.080]	1.600 [1.305–2.053]	<0.001	<0.001	0.360	0.142

Early stage indicates stage 1 and 2 diseases, late stage indicates stage 3 and 4 diseases, the staging criteria according to NCCN Clinical Practice Guidelines Non-small-cell lung cancer v1, 2022. NSCLC, non-small-cell lung cancer; Tis, carcinoma in situ; NLR, neutrophil-to-lymphocyte ratio; NEU, absolute neutrophil counts; LYM, absolute lymphocyte counts. * p indicates control vs. NSCLC. ** p indicates control vs. Tis. *** p indicates Tis vs. early stage. **** p indicates early stage vs. late stage.

Figure S1 depicts the comparison of classic blood biomarkers across various clinical features among the healthy controls.

3.2. Stage 2

Table S3 shows the host characteristics of the subset of the study participants with measurements of 11 novel blood neutrophil-related biomarkers. This phase included a total of 198 participants, 132 of whom were NSCLC patients and 66 were healthy controls. In the NSCLC group, there were 90 cases of invasive cancer (68.18%). Subsequent analyses did not include IL-1β, IFN-α, IL-17, and G-CSF due to excessive missing assay values.

The differences in NLR, NEU, and LYM between cases and controls in stage 2 were consistent with those in stage 1 (Table S4). In terms of novel blood neutrophil-related biomarkers, the IL-6 and IL-1RA levels in the controls were significantly lower than in the cases ($p < 0.001$) (Table S4 and Figure S2).

Figure S3 presents the pairwise correlations among the novel blood neutrophil-related biomarkers. IL-1α and CXCL5 have a correlation coefficient of 0.9, indicating a strong correlation. The correlation coefficients for the remaining factors were all below 0.5, indicating weaker correlations.

In univariate analysis (Table 2), among the analyzed novel blood neutrophil-related biomarkers, higher levels of IL-6 and IL-1RA were significantly associated with increased risk of NSCLC, with OR values of 9.339 (95% CI: 3.882, 22.646) and 7.535 (95% CI: 3.293, 17.244), respectively. In multivariate conditional logistic regression analysis conducted on IL-6, CXCL2, IL-1RA, IL-1α, CXCL5, S100B, and GM-CSF, we found that after adjusting for BMI, smoking status, NLR, and CEA, higher plasma levels of IL-6 and IL-1RA were associated with substantially elevated risk of NSCLC. The risk of NSCLC in individuals with higher plasma levels of IL-6 was 10.687 times (95% CI: 3.875, 29.473) that of the low-level group, and similarly, the risk was 8.113 times (95% CI: 3.182, 20.689) higher for those with high levels of IL-1RA. The forest plots of the univariate and multiple conditional analysis are shown in Figure S4.

Table 2. The associations between novel blood neutrophil-related biomarkers and NSCLC in stage 2.

Markers	Control (n = 66)	NSCLC (n = 132)	OR (95% CI) *	p Value *	OR (95% CI) **, a	p Value **, a
IL-6						
Low	34 (51.52)	15 (11.36)	1 (ref)	<0.001	1 (ref)	<0.001
High	32 (48.48)	117 (88.64)	9.339 (3.882, 22.464)		10.687 (3.875, 29.473)	
CXCL2						
Low	33 (50.00)	52 (39.39)	1 (ref)	0.151	1 (ref)	0.196
High	33 (50.00)	80 (6.61)	1.570 (0.848, 2.907)		1.824 (0.903, 3.683)	
IL-1RA						
Low	33 (50.00)	17 (12.88)	1 (ref)	<0.001	1 (ref)	<0.001
High	33 (50.00)	115 (87.12)	7.535 (3.293, 17.244)		8.113 (3.182, 20.689)	
IL-1α						
Low	33 (50.00)	64 (48.48)	1 (ref)	0.849	1 (ref)	0.615
High	33 (50.00)	68 (51.52)	1.056 (0.603, 1.850)		1.314 (0.700, 2.466)	
CXCL5						
Low	33 (50.00)	62 (46.97)	1 (ref)	0.703	1 (ref)	0.594
High	33 (50.00)	70 (53.3)	1.116 (0.636, 1.958)		1.371 (0.719, 2.613)	
S100B						
Low	42 (63.64)	70 (53.3)	1 (ref)	0.184	1 (ref)	0.531
High	24 (36.36)	62 (46.97)	1.481 (0.830, 2.645)		1.287 (0.663, 2.501)	
GM-CSF						
Low	52 (78.79)	100 (75.76)	1 (ref)	0.633	1 (ref)	0.611
High	14 (21.21)	32 (24.24)	1.191 (0.582, 2.436)		1.294 (0.563, 2.977)	

The values of IL-6, CXCL2, IL-1RA, IL-1α, and CXCL5 were divided into low and high groups based on the median value in the control group as the cutoff. The values of S100B and GM-CSF were divided into low and high groups, with the experimental detection limit serving as the cutoff. [a] Adjusted factors: BMI + smoking status + NLR + CEA. * shows the OR (95% CI) and p values of the univariate logistic regression analysis. ** shows the OR (95% CI) and p values of the multiple logistic regression analysis. NSCLC, non-small-cell lung cancer; BMI, body mass index; IL-6, Interleukin 6; IL-1α, Interleukin 1 alpha; IL-1RA, Interleukin-1 receptor antagonist; GM-CSF, Granulocyte-macrophage colony-stimulating factor; CXCL2, C-X-C Motif Chemokine Ligand 2; CXCL5, C-X-C Motif Chemokine Ligand 5; S100B, S100 Calcium-binding Protein B; NLR, neutrophil-to-lymphocyte ratio; CEA, carcinoembryonic antigen; OR, odds ratio; CI, confidence interval. Values are presented as n (%) unless otherwise specified.

3.3. Stage 3

Table S5 shows the comparison of the discrimination ability of different predictive models. In Model 1, which only included the epidemiological predictors (BMI and smoking status), the AUC value was 0.603 (95% CI: 0.527, 0.678). After adding NLR and CEA, the AUC value of Model 2 increased to 0.716 (95% CI: 0.637, 0.794). Upon the addition of IL-6 and IL-1RA to Model 2 to create Model 3, the best performance was achieved with an AUC value of 0.851 (95% CI: 0.793, 0.908), a TPR of 0.856, and a FPR of 0.333. The differences in AUCs from the three models were statistically significant. The visualized ROC curves for Models 1, 2, and 3 are shown in Figure 3.

We employed NRI and IDI to measure the improvement of the predictive performance of models after the inclusion of new risk factors, enabling the comparison among models (Table 3). Initially, we compared Model 2 (Model 1 + NLR + CEA) with Model 1 (baseline model incorporating only epidemiological indicators: BMI + smoking status). The results indicated a noteworthy enhancement in the predictive performance of Model 2 in contrast to Model 1. The NRI reached 59.85% (95% CI: 0.331, 0.935), and the IDI was 0.128 (95% CI: 0.079, 0.176), both differences being statistically significant. By incorporating IL-6 and IL-1RA into Model 2, the NRI for Model 3 increased to 97.73%, signifying that the integration of IL-6 and IL-1RA could escalate the accurate reclassification proportion of the model by 97.73% (95% CI: 0.667, 1.279). The extent of improvement was also significantly increased, with an IDI of 0.198 (95% CI: 0.137, 0.259).

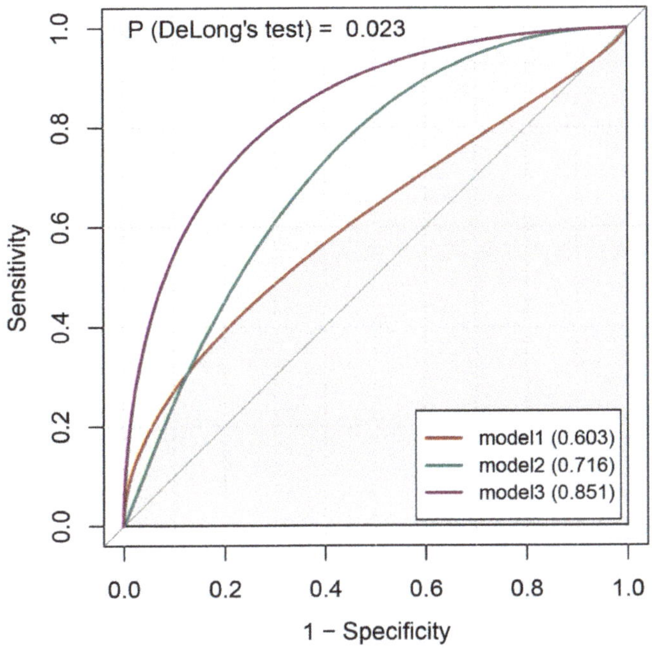

Figure 3. Comparison of ROC curves for different models. The AUC value of Model 1 was only 0.603 based on epidemiological variables (BMI and smoking status) data. In Model 2 (added NLR and CEA), the AUC value increased to 0.716. Model 3 (added IL-6 and IL-1RA) showed the best performance, with an AUC value of 0.851. IL-6, Interleukin 6; IL-1RA, Interleukin-1 receptor antagonist; NLR, neutrophil-to-lymphocyte ratio; CEA, carcinoembryonic antigen; BMI, body mass index; ROC, receiver operating characteristic; AUC, area under the curve.

Table 3. The comparative efficacy of predictive models.

Models	NRI	95% CI [a]	* *p* Value	IDI	95% CI [b]	** *p* Value
Model 1 vs. Model 2	59.85%	0.331, 0.935	<0.001	0.128	0.079, 0.176	<0.001
Model 2 vs. Model 2 + IL-6	75.00%	0.319, 1.022	<0.001	0.148	0.092, 0.203	<0.001
Model 2 vs. Model 2 + IL-1RA	61.36%	0.122, 0.962	0.003	0.122	0.068, 0.176	<0.001
Model 2 + IL-6 vs. Model 3	16.67%	−0.055, 0.779	0.423	0.050	0.012, 0.088	0.010
Model 2 + IL-1RA vs. Model 3	29.55%	−0.001, 0.866	0.227	0.076	0.033, 0.119	0.001
Model 2 vs. Model 3	97.73%	0.667, 1.279	<0.001	0.198	0.137, 0.259	<0.001

Model 1: epidemiology variables: BMI + smoking status. Model 2: epidemiology variables + NLR + CEA. Model 3: epidemiology variables + NLR + CEA + IL-6 + IL-1RA. BMI, body mass index; IL-6, Interleukin 6; IL-1RA, Interleukin-1 receptor antagonist; NLR, neutrophil to lymphocyte ratio; CEA, carcinoembryonic antigen; NRI, net reclassification improvement; IDI, integrated discrimination improvement; CI, confidence interval. [a] shows the 95% CI on NRI. [b] shows the 95% CI on IDI. * indicates the *p* value on the difference of NRI. ** indicates the *p* value on the difference of IDI.

4. Discussion

Our findings indicated a significant association between NLR and NSCLC risk. Plasma IL-6 and IL-1RA also emerged as independent risk factors for NSCLC. These mean high levels of NLR, IL-6, and IL-1RA signaling a dramatically increased NSCLC risk. Moreover, we established a model incorporating NLR and novel blood neutrophil-related biomarkers to aid in predicting NSCLC diagnosis in the clinic, outperforming models reliant on the classic cancer marker CEA. Thus, our research underscores the importance of neutrophil-related biomarkers in predicting NSCLC risk, offering valuable assistance in clinical diagnosis.

NLR, a parameter derived from the NEU divided by the LYM, is one of the most widely investigated features based on blood cell counts and serves as an indicator of systemic inflammation [36]. NLR captures the balance between the detrimental effects of increased neutrophils and the beneficial roles of adaptive immunity mediated by lymphocytes [37]. In the field of cancer research, NLR serves as an effective indicator of the dynamic balance between pro-tumor and anti-tumor responses in the body. Most published studies have mainly focused on the correlation between NLR and the prognosis of NSCLC [36–40], with little attention paid to its association with disease risk. Therese Haugdahl Nøst et al. conducted a study on approximately 440,000 participants based on data from the UK Biobank, evaluating the longitudinal relationships between four systemic inflammation indicators (NLR, systemic immune-inflammation index, platelet-to-lymphocyte ratio, lymphocyte-to-monocyte ratio) and the risk of 17 cancer sites diagnosed clinically in the years preceding the study [41]. The study found that NSCLC patients exhibited higher NLR values, which was particularly apparent in the year before diagnosis, possibly driven by an elevated neutrophil count.

IL-6, a multifunctional cytokine, exhibits both pro-inflammatory and anti-inflammatory properties [42]. Under cancerous conditions, IL-6 can participate in various processes such as tumor formation, cancer cell proliferation, epithelial–mesenchymal transition, interactions between tumor cells and the matrix environment, tumor dissemination, and drug resistance. By releasing chemokine receptors including CXCR3/4 and CCR5/7, and secreting pro-inflammatory cytokines, notably IL-6, TANs play a pivotal role in the pathogenesis of lung cancer [43]. Currently, the predictive capabilities evaluation of IL-6 in cancer mostly focuses on treatment outcomes or survival status [44–50], making it one of the most discussed prognostic markers for NSCLC patients. However, studies dedicated to evaluating associated risks remain sparse. In a case–control study conducted at the National Cancer Institute in Maryland, Sharon R. Pine and colleagues analyzed the association between IL-6 and lung cancer in six pairs of patients and controls. They discovered a notable association between lung cancer and the highest quartile of serum IL-6 levels (OR = 3.29, 95% CI: 1.88–5.77) [51]. In a subsequent verification within the prospective Prostate, Lung, Colorectal, and Ovarian Cancer Screening Trial, they found that an increase in IL-6 levels was exclusively connected to lung cancer cases diagnosed within two years of blood sampling. A nested case–control study including 224 cases and 644 controls indicated higher blood IL-6 was associated with rising hepatocellular carcinoma (HCC) risk [52]. The relative risk was 5.12 (95% CI: 1.54–20.1) for HCC in the top tertile of IL-6 levels. This result was not influenced by variables such as hepatitis virus infection, lifestyle-related elements, and radiation exposure. Scholars have underscored the importance of early monitoring of IL-6 levels. Further gene expression analysis showed an increased expression of IL-6 in patients compared to controls. The latest research on the Lung Cancer Cohort Consortium, an international cohort with over two million participants from North America, Europe, Asia, and Australia, identified 36 proteins independently and reproducibly associated with the imminent risk of being diagnosed with lung cancer. IL-6 was found a robust correlation with lung cancer within the first year following diagnosis (OR = 2.56, 95% CI: 1.92–3.41) [34]. Therefore, the measurement of IL-6 in plasma can serve as an early auxiliary diagnostic indicator for NSCLC, providing predictive value for clinical work.

IL-1RA, a component of the IL-1 family, operates as a competitive binding factor that can inhibit the signal cascade response and suppress the pro-inflammatory signal transduction activated by IL-1α and IL-1β [53]. IL-1RA is commonly generated by the cells that concurrently produce IL-1α or IL-1β, notably monocytes, macrophages, dendritic cells, neutrophils, and so on. An increase in IL-1 production is often paired with a raised IL-1RA level. Even though IL-1RA does not trigger a biological response, it is described as an anti-inflammatory molecule due to its capacity to impede the pro-inflammatory activities of IL-1α and IL-1β [53]. The expression of IL-1RA has been investigated in many human diseases, such as inflammatory diseases, immune-related diseases, and numerous kinds of cancer. Generally, IL-1RA in cancer is considered to exert a tumor-suppressive effect due to its ability to inhibit pro-tumor cytokines [54]. Post myocardial infarction, animals with high expression of IL-1RA displayed decreased symptoms of inflammation, less neutrophil infiltration, and reduced ventricular expansion [55]. In a range of diseases, increased levels of circulating IL-1RA have been reported, such as chronic arthritis, inflammatory bowel disease, rheumatoid arthritis, and Acute Respiratory Distress Syndrome [56]. Synthesized research suggested that the balance between IL-1RA and the pro-inflammatory cytokine IL-1 was associated with increased risks of various cancers, including NSCLC [57]. Elevated serum IL-1RA concentrations had been observed in patients with Hodgkin's disease, lung cancer, colorectal cancer, cervical cancer, and endometrial cancer, underscoring the critical function of IL-1RA throughout the formation and advancement of tumors [54]. In 2013, a nested case–control study was performed on 526 lung cancer patients and 592 control subjects [20]. Using the Luminex beads-based experimental method, more than 70 serum inflammation markers were examined. The researchers found a correlation between increased serum IL-1RA levels and a decreased risk of lung cancer (OR = 0.71, 95% CI: 0.51–1.00). Genetic research in humans also showed that the genetic variability of IL-1α, IL-1β, and IL-1RA correlated with elevated risks of tumors, including NSCLC [58]. Although IL-1RA was discovered almost concurrently with IL-1α and IL-1β, its value in cancer remains relatively uncertain compared to the other two. Current research on IL-1RA in tumors mostly reports its anti-inflammatory role. However, some scholars have reported that the role of IL-1RA in cancer is not limited to suppressing inflammation, but it can also promote the growth of malignant tumors [53].

We acknowledge that there are several limitations. First, several novel factors that were to be tested in the samples inherently had low expression levels, which made their detection difficult. They were excluded from data analysis due to numerous undetected values, despite our meticulous handling of the samples, including storage at −80 degrees Celsius, prevention of repeated freeze–thaw cycles, and strict adherence to the instructions during the procedures. Second, as a cross-sectional study, we were unable to determine causality. Further rigorous validation work along with an increase in sample size is needed. Third, our research focused on studying the NSCLC population. However, the significant variations among different cancers and subtypes limit the generalizability of our findings.

5. Conclusions

Our study suggests that IL-6 and IL-1RA play key roles in lung carcinogenesis and progression. NLR, IL-6, and IL-1RA in the blood can serve as biomarkers for diagnosis of NSCLC. Combining these with patients' clinical features and tumor markers (for example, CEA) may enhance the effectiveness of diagnosing NSCLC, potentially providing heightened early warning at pre-diagnosis and diagnosis.

Supplementary Materials: The following supporting information can be downloaded at: https://www.mdpi.com/article/10.3390/cancers16030513/s1, Figure S1: the distribution of classic blood biomarkers (NLR, NEU, and LYM) across different epidemiological features (age, sex, BMI, and smoking status) among the healthy controls in stage 1; Figure S2: the differences analysis of blood neutrophil-related biomarkers in paired samples; Figure S3: the pairwise correlations among the novel blood neutrophil-related biomarkers; Figure S4: the associations between novel blood neutrophil-related biomarkers and NSCLC risk in stage 2; Table S1: host characteristics of the study participants in stage 1; Table S2: the associations between classic blood neutrophil-related biomarkers and NSCLC risk in stage 1; Table S3: host characteristics of the subset of the study participants with measurements of novel blood neutrophil-related biomarkers; Table S4: the difference in novel neutrophil-related biomarkers in stage 2; Table S5: the comparison of model performance of models with different predictors.

Author Contributions: Conceptualization: H.T., Q.W. and X.W.; Data Curation: Y.R., X.X., C.X., R.D., Q.G., M.W. and X.W.; Formal Analysis: Y.R., C.X., R.D., Q.G., H.T., Q.W., X.W. and Y.Z.; Funding Acquisition: H.T. and X.W.; Investigation: X.X., C.X., R.D., Q.G., H.T., M.W. and X.W.; Methodology: Y.R., X.X., H.T., Q.W., X.W. and Y.Z.; Project Administration: H.T. and X.W.; Resources: H.T., M.W. and X.W.; Software: Y.R., H.T. and Y.Z.; Supervision: H.T. and X.W.; Validation: Y.Z., Q.W. and H.T.; Visualization: Y.R. and Y.Z.; Writing—Original Draft: Y.R., H.T. and X.W.; Writing—Review and

Editing: Y.R., X.X., C.X., R.D., Q.G., M.W., X.W., H.T., Q.W. and Y.Z. All authors have read and agreed to the published version of the manuscript.

Funding: This study was supported by the Key Laboratory of Intelligent Preventive Medicine of Zhejiang Province (2020E10004), the Leading Innovative and Entrepreneur Team Introduction Program of Zhejiang (2019R01007), the Key Research and Development Program of Zhejiang Province (2020C03002), and the Healthy Zhejiang One Million Cohort (all to Prof. Wu, K20230085) and Faculty Startup Funds. The funding source had no role in study design, data collection, data analysis, data interpretation, writing of the report, or the decision to submit the article for publication.

Institutional Review Board Statement: This research has been approved by the Institutional Review Board of The Second Affiliated Hospital of Zhejiang University School of Medicine on 14 October 2019 (ethics code: 2019LSYD338).

Informed Consent Statement: Written informed consent has been obtained from the patients to publish this paper.

Data Availability Statement: The data that support the findings of this study are available from the corresponding author upon reasonable request.

Acknowledgments: We extend our appreciation to the field team and laboratory staff for their exceptional efforts in the recruitment of patients, as well as their outstanding work in the collection and/or handling of clinical/epidemiological information and biological samples.

Conflicts of Interest: The authors have no conflicts of interest.

Abbreviations

AUC: area under the curve; CEA, carcinoembryonic antigen; CI, confidence interval; CXCL2, C-X-C Motif Chemokine Ligand 2; CXCL5, C-X-C Motif Chemokine Ligand 5; FPR, false positive rate; G-SCF, Granulocyte colony-stimulating factor; GM-CSF, Granulocyte-macrophage colony-stimulating factor; IAC, invasive adenocarcinoma; IDI, integrated discrimination improvement; IFN-α, Interferon-alpha; IL-1α, Interleukin 1 alpha; IL-1β, Interleukin 1 beta; IL-1RA, Interleukin-1 receptor antagonist; IL-6, Interleukin 6; IL-17, Interleukin 17; LYM, absolute lymphocyte counts; NEU, absolute neutrophil counts; NLR, neutrophil to lymphocyte ratio; NRI, net reclassification improvement; NSCLC, non-small-cell lung cancer; OR, odds ratio; ROC, receiver operating characteristic; ROS, reactive oxygen species; S100B, S100 Calcium-binding Protein B; TANs, tumor-associated neutrophils; Tis, carcinoma in situ; TPR, true positive rate.

References

1. Leiter, A.; Veluswamy, R.R.; Wisnivesky, J.P. The global burden of lung cancer: Current status and future trends. *Nat. Rev. Clin. Oncol.* **2023**, *20*, 624–639. [CrossRef] [PubMed]
2. Shaul, M.E.; Fridlender, Z.G. Tumour-associated neutrophils in patients with cancer. *Nat. Rev. Clin. Oncol.* **2019**, *16*, 601–620. [CrossRef] [PubMed]
3. Silvestre-Roig, C.; Braster, Q.; Ortega-Gomez, A.; Soehnlein, O. Neutrophils as regulators of cardiovascular inflammation. *Nat. Rev. Cardiol.* **2020**, *17*, 327–340. [CrossRef] [PubMed]
4. Papayannopoulos, V. Neutrophil extracellular traps in immunity and disease. *Nat. Rev. Immunol.* **2018**, *18*, 134–147. [CrossRef] [PubMed]
5. Coffelt, S.B.; Wellenstein, M.D.; de Visser, K.E. Neutrophils in cancer: Neutral no more. *Nat. Rev. Cancer* **2016**, *16*, 431–446. [CrossRef] [PubMed]
6. Kolaczkowska, E.; Kubes, P. Neutrophil recruitment and function in health and inflammation. *Nat. Rev. Immunol.* **2013**, *13*, 159–175. [CrossRef] [PubMed]
7. Mantovani, A.; Cassatella, M.A.; Costantini, C.; Jaillon, S. Neutrophils in the activation and regulation of innate and adaptive immunity. *Nat. Rev. Immunol.* **2011**, *11*, 519–531. [CrossRef] [PubMed]
8. Hedrick, C.C.; Malanchi, I. Neutrophils in cancer: Heterogeneous and multifaceted. *Nat. Rev. Immunol.* **2022**, *22*, 173–187. [CrossRef]
9. Wculek, S.K.; Bridgeman, V.L.; Peakman, F.; Malanchi, I. Early Neutrophil Responses to Chemical Carcinogenesis Shape Long-Term Lung Cancer Susceptibility. *iScience* **2020**, *23*, 101277. [CrossRef]
10. Matlung, H.L.; Babes, L.; Zhao, X.W.; van Houdt, M.; Treffers, L.W.; van Rees, D.J.; Franke, K.; Schornagel, K.; Verkuijlen, P.; Janssen, H.; et al. Neutrophils Kill Antibody-Opsonized Cancer Cells by Trogoptosis. *Cell Rep.* **2018**, *23*, 3946–3959.e6. [CrossRef]

11. Faget, J.; Peters, S.; Quantin, X.; Meylan, E.; Bonnefoy, N. Neutrophils in the era of immune checkpoint blockade. *J. Immunother. Cancer* **2021**, *9*, e002242. [CrossRef] [PubMed]
12. Valero, C.; Lee, M.; Hoen, D.; Weiss, K.; Kelly, D.W.; Adusumilli, P.S.; Paik, P.K.; Plitas, G.; Ladanyi, M.; Postow, M.A.; et al. Pretreatment neutrophil-to-lymphocyte ratio and mutational burden as biomarkers of tumor response to immune checkpoint inhibitors. *Nat. Commun.* **2021**, *12*, 729. [CrossRef] [PubMed]
13. Sui, Q.; Zhang, X.; Chen, C.; Tang, J.; Yu, J.; Li, W.; Han, K.; Jiang, W.; Liao, L.; Kong, L.; et al. Inflammation promotes resistance to immune checkpoint inhibitors in high microsatellite instability colorectal cancer. *Nat. Commun.* **2022**, *13*, 7316. [CrossRef] [PubMed]
14. Rebuzzi, S.E.; Prelaj, A.; Friedlaender, A.; Cortellini, A.; Addeo, A.; Genova, C.; Naqash, A.R.; Auclin, E.; Mezquita, L.; Banna, G.L. Prognostic scores including peripheral blood-derived inflammatory indices in patients with advanced non-small-cell lung cancer treated with immune checkpoint inhibitors. *Crit. Rev. Oncol. Hematol.* **2022**, *179*, 103806. [CrossRef] [PubMed]
15. Que, H.; Fu, Q.; Lan, T.; Tian, X.; Wei, X. Tumor-associated neutrophils and neutrophil-targeted cancer therapies. *Biochim. Biophys. Acta Rev. Cancer* **2022**, *1877*, 188762. [CrossRef] [PubMed]
16. Adrover, J.M.; McDowell, S.A.C.; He, X.Y.; Quail, D.F.; Egeblad, M. NETworking with cancer: The bidirectional interplay between cancer and neutrophil extracellular traps. *Cancer Cell* **2023**, *41*, 505–526. [CrossRef] [PubMed]
17. Eruslanov, E.B.; Bhojnagarwala, P.S.; Quatromoni, J.G.; Stephen, T.L.; Ranganathan, A.; Deshpande, C.; Akimova, T.; Vachani, A.; Litzky, L.; Hancock, W.W.; et al. Tumor-associated neutrophils stimulate T cell responses in early-stage human lung cancer. *J. Clin. Investig.* **2014**, *124*, 5466–5480. [CrossRef]
18. Akbay, E.A.; Koyama, S.; Liu, Y.; Dries, R.; Bufe, L.E.; Silkes, M.; Alam, M.M.; Magee, D.M.; Jones, R.; Jinushi, M.; et al. Interleukin-17A Promotes Lung Tumor Progression through Neutrophil Attraction to Tumor Sites and Mediating Resistance to PD-1 Blockade. *J. Thorac. Oncol.* **2017**, *12*, 1268–1279. [CrossRef]
19. Schalper, K.A.; Carleton, M.; Zhou, M.; Chen, T.; Feng, Y.; Huang, S.P.; Walsh, A.M.; Baxi, V.; Pandya, D.; Baradet, T.; et al. Elevated serum interleukin-8 is associated with enhanced intratumor neutrophils and reduced clinical benefit of immune-checkpoint inhibitors. *Nat. Med.* **2020**, *26*, 688–692. [CrossRef]
20. Shiels, M.S.; Pfeiffer, R.M.; Hildesheim, A.; Engels, E.A.; Kemp, T.J.; Park, J.H.; Katki, H.A.; Koshiol, J.; Shelton, G.; Caporaso, N.E.; et al. Circulating inflammation markers and prospective risk for lung cancer. *J. Natl. Cancer Inst.* **2013**, *105*, 1871–1880. [CrossRef]
21. Xie, X.; Shi, Q.; Wu, P.; Zhang, X.; Kambara, H.; Su, J.; Yu, H.; Park, S.Y.; Guo, R.; Ren, Q.; et al. Single-cell transcriptome profiling reveals neutrophil heterogeneity in homeostasis and infection. *Nat. Immunol.* **2020**, *21*, 1119–1133. [CrossRef] [PubMed]
22. Pan, S.; Zhao, W.; Li, Y.; Ying, Z.; Luo, Y.; Wang, Q.; Li, X.; Lu, W.; Dong, X.; Wu, Y.; et al. Prediction of risk and overall survival of pancreatic cancer from blood soluble immune checkpoint-related proteins. *Front. Immunol.* **2023**, *14*, 1189161. [CrossRef] [PubMed]
23. Wang, Q.; He, Y.; Li, W.; Xu, X.; Hu, Q.; Bian, Z.; Xu, A.; Tu, H.; Wu, M.; Wu, X. Soluble Immune Checkpoint-Related Proteins in Blood Are Associated With Invasion and Progression in Non-Small Cell Lung Cancer. *Front. Immunol.* **2022**, *13*, 887916. [CrossRef] [PubMed]
24. Muri, J.; Cecchinato, V.; Cavalli, A.; Shanbhag, A.A.; Matkovic, M.; Biggiogero, M.; Maida, P.A.; Moritz, J.; Toscano, C.; Ghovehoud, E.; et al. Autoantibodies against chemokines post-SARS-CoV-2 infection correlate with disease course. *Nat. Immunol.* **2023**, *24*, 604–611. [CrossRef]
25. Genchi, A.; Brambilla, E.; Sangalli, F.; Radaelli, M.; Bacigaluppi, M.; Furlan, R.; Andolfo, A.; Drago, D.; Magagnotti, C.; Scotti, G.M.; et al. Neural stem cell transplantation in patients with progressive multiple sclerosis: An open-label, phase 1 study. *Nat. Med.* **2023**, *29*, 75–85. [CrossRef]
26. Zhao, Y.; Rahmy, S.; Liu, Z.; Zhang, C.; Lu, X. Rational targeting of immunosuppressive neutrophils in cancer. *Pharmacol. Ther.* **2020**, *212*, 107556. [CrossRef]
27. Schett, G.; Dayer, J.M.; Manger, B. Interleukin-1 function and role in rheumatic disease. *Nat. Rev. Rheumatol.* **2016**, *12*, 14–24. [CrossRef]
28. Afonina, I.S.; Müller, C.; Martin, S.J.; Beyaert, R. Proteolytic Processing of Interleukin-1 Family Cytokines: Variations on a Common Theme. *Immunity* **2015**, *42*, 991–1004. [CrossRef]
29. Isailovic, N.; Daigo, K.; Mantovani, A.; Selmi, C. Interleukin-17 and innate immunity in infections and chronic inflammation. *J. Autoimmun.* **2015**, *60*, 1–11. [CrossRef]
30. De Filippo, K.; Dudeck, A.; Hasenberg, M.; Nye, E.; van Rooijen, N.; Hartmann, K.; Gunzer, M.; Roers, A.; Hogg, N. Mast cell and macrophage chemokines CXCL1/CXCL2 control the early stage of neutrophil recruitment during tissue inflammation. *Blood* **2013**, *121*, 4930–4937. [CrossRef]
31. Zhou, S.L.; Dai, Z.; Zhou, Z.J.; Wang, X.Y.; Yang, G.H.; Wang, Z.; Huang, X.W.; Fan, J.; Zhou, J. Overexpression of CXCL5 mediates neutrophil infiltration and indicates poor prognosis for hepatocellular carcinoma. *Hepatology* **2012**, *56*, 2242–2254. [CrossRef] [PubMed]
32. Pylaeva, E.; Lang, S.; Jablonska, J. The Essential Role of Type I Interferons in Differentiation and Activation of Tumor-Associated Neutrophils. *Front. Immunol.* **2016**, *7*, 629. [CrossRef] [PubMed]
33. Yu, R.; Zhu, B.; Chen, D. Type I interferon-mediated tumor immunity and its role in immunotherapy. *Cell Mol. Life Sci.* **2022**, *79*, 191. [CrossRef] [PubMed]

34. The blood proteome of imminent lung cancer diagnosis. *Nat. Commun.* **2023**, *14*, 3042. [CrossRef] [PubMed]
35. Wang, T.; Du, G.; Wang, D. The S100 protein family in lung cancer. *Clin. Chim. Acta* **2021**, *520*, 67–70. [CrossRef] [PubMed]
36. Ancel, J.; Dormoy, V.; Raby, B.N.; Dalstein, V.; Durlach, A.; Dewolf, M.; Gilles, C.; Polette, M.; Deslée, G. Soluble biomarkers to predict clinical outcomes in non-small cell lung cancer treated by immune checkpoints inhibitors. *Front. Immunol.* **2023**, *14*, 1171649. [CrossRef] [PubMed]
37. Smith, D.; Raices, M.; Cayol, F.; Corvatta, F.; Caram, L.; Dietrich, A. Is the neutrophil-to-lymphocyte ratio a prognostic factor in non-small cell lung cancer patients who receive adjuvant chemotherapy? *Semin. Oncol.* **2022**, *49*, 482–489. [CrossRef] [PubMed]
38. Templeton, A.J.; McNamara, M.G.; Šeruga, B.; Vera-Badillo, F.E.; Aneja, P.; Ocaña, A.; Leibowitz-Amit, R.; Sonpavde, G.; Knox, J.J.; Tran, B.; et al. Prognostic role of neutrophil-to-lymphocyte ratio in solid tumors: A systematic review and meta-analysis. *J. Natl. Cancer Inst.* **2014**, *106*, dju124. [CrossRef]
39. Wang, F.; Chen, L.; Wang, Z.; Xu, Q.; Huang, H.; Wang, H.; Li, X.; Yu, M.; Chen, J.; Lin, F.; et al. Prognostic value of the modified systemic inflammation score in non-small-cell lung cancer with brain metastasis. *Cancer Cell Int.* **2022**, *22*, 320. [CrossRef]
40. Liu, W.; Ren, S.; Yang, L.; Xiao, Y.; Zeng, C.; Chen, C.; Wu, F.; Hu, Y. The predictive role of hematologic markers in resectable nsclc patients treated with neoadjuvant chemoimmunotherapy: A retrospective cohort study. *Int. J. Surg.* **2023**, *109*, 3519–3526. [CrossRef]
41. Nøst, T.H.; Alcala, K.; Urbarova, I.; Byrne, K.S.; Guida, F.; Sandanger, T.M.; Johansson, M. Systemic inflammation markers and cancer incidence in the UK Biobank. *Eur. J. Epidemiol.* **2021**, *36*, 841–848. [CrossRef] [PubMed]
42. Morris, E.C.; Neelapu, S.S.; Giavridis, T.; Sadelain, M. Cytokine release syndrome and associated neurotoxicity in cancer immunotherapy. *Nat. Rev. Immunol.* **2022**, *22*, 85–96. [CrossRef] [PubMed]
43. Ahmad, S.; Manzoor, S.; Siddiqui, S.; Mariappan, N.; Zafar, I.; Ahmad, A.; Ahmad, A. Epigenetic underpinnings of inflammation: Connecting the dots between pulmonary diseases, lung cancer and COVID-19. *Semin. Cancer Biol.* **2022**, *83*, 384–398. [CrossRef] [PubMed]
44. Taniguchi, K.; Karin, M. IL-6 and related cytokines as the critical lynchpins between inflammation and cancer. *Semin. Immunol.* **2014**, *26*, 54–74. [CrossRef]
45. Zhou, B.; Liu, J.; Wang, Z.M.; Xi, T. C-reactive protein, interleukin 6 and lung cancer risk: A meta-analysis. *PLoS ONE* **2012**, *7*, e43075. [CrossRef] [PubMed]
46. Mao, X.C.; Yang, C.C.; Yang, Y.F.; Yan, L.J.; Ding, Z.N.; Liu, H.; Yan, Y.C.; Dong, Z.R.; Wang, D.X.; Li, T. Peripheral cytokine levels as novel predictors of survival in cancer patients treated with immune checkpoint inhibitors: A systematic review and meta-analysis. *Front. Immunol.* **2022**, *13*, 884592. [CrossRef] [PubMed]
47. Liu, C.; Yang, L.; Xu, H.; Zheng, S.; Wang, Z.; Wang, S.; Yang, Y.; Zhang, S.; Feng, X.; Sun, N.; et al. Systematic analysis of IL-6 as a predictive biomarker and desensitizer of immunotherapy responses in patients with non-small cell lung cancer. *BMC Med.* **2022**, *20*, 187. [CrossRef]
48. Barrera, L.; Montes-Servín, E.; Barrera, A.; Ramírez-Tirado, L.A.; Salinas-Parra, F.; Bañales-Méndez, J.L.; Sandoval-Ríos, M.; Arrieta, Ó. Cytokine profile determined by data-mining analysis set into clusters of non-small-cell lung cancer patients according to prognosis. *Ann. Oncol.* **2015**, *26*, 428–435. [CrossRef]
49. Chang, C.H.; Hsiao, C.F.; Yeh, Y.M.; Chang, G.C.; Tsai, Y.H.; Chen, Y.M.; Huang, M.S.; Chen, H.L.; Li, Y.J.; Yang, P.C.; et al. Circulating interleukin-6 level is a prognostic marker for survival in advanced nonsmall cell lung cancer patients treated with chemotherapy. *Int. J. Cancer* **2013**, *132*, 1977–1985. [CrossRef]
50. Enewold, L.; Mechanic, L.E.; Bowman, E.D.; Zheng, Y.L.; Yu, Z.; Trivers, G.; Alberg, A.J.; Harris, C.C. Serum concentrations of cytokines and lung cancer survival in African Americans and Caucasians. *Cancer Epidemiol. Biomark. Prev.* **2009**, *18*, 215–222. [CrossRef]
51. Pine, S.R.; Mechanic, L.E.; Enewold, L.; Chaturvedi, A.K.; Katki, H.A.; Zheng, Y.L.; Bowman, E.D.; Engels, E.A.; Caporaso, N.E.; Harris, C.C. Increased levels of circulating interleukin 6, interleukin 8, C-reactive protein, and risk of lung cancer. *J. Natl. Cancer Inst.* **2011**, *103*, 1112–1122. [CrossRef] [PubMed]
52. Ohishi, W.; Cologne, J.B.; Fujiwara, S.; Suzuki, G.; Hayashi, T.; Niwa, Y.; Akahoshi, M.; Ueda, K.; Tsuge, M.; Chayama, K. Serum interleukin-6 associated with hepatocellular carcinoma risk: A nested case-control study. *Int. J. Cancer* **2014**, *134*, 154–163. [CrossRef] [PubMed]
53. Boersma, B.; Jiskoot, W.; Lowe, P.; Bourquin, C. The interleukin-1 cytokine family members: Role in cancer pathogenesis and potential therapeutic applications in cancer immunotherapy. *Cytokine Growth Factor. Rev.* **2021**, *62*, 1–14. [CrossRef] [PubMed]
54. Hu, Z.; Shao, M.; Chen, Y.; Zhou, J.; Qian, J.; Xu, L.; Ma, H.; Wang, X.; Xu, Y.; Lu, D.; et al. Allele 2 of the interleukin-1 receptor antagonist gene (IL1RN*2) is associated with a decreased risk of primary lung cancer. *Cancer Lett.* **2006**, *236*, 269–275. [CrossRef] [PubMed]
55. Cavalli, G.; Colafrancesco, S.; Emmi, G.; Imazio, M.; Lopalco, G.; Maggio, M.C.; Sota, J.; Dinarello, C.A. Interleukin 1α: A comprehensive review on the role of IL-1α in the pathogenesis and treatment of autoimmune and inflammatory diseases. *Autoimmun. Rev.* **2021**, *20*, 102763. [CrossRef] [PubMed]
56. Palomo, J.; Dietrich, D.; Martin, P.; Palmer, G.; Gabay, C. The interleukin (IL)-1 cytokine family--Balance between agonists and antagonists in inflammatory diseases. *Cytokine* **2015**, *76*, 25–37. [CrossRef]

57. Wu, T.C.; Xu, K.; Martinek, J.; Young, R.R.; Banchereau, R.; George, J.; Turner, J.; Kim, K.I.; Zurawski, S.; Wang, X.; et al. IL1 Receptor Antagonist Controls Transcriptional Signature of Inflammation in Patients with Metastatic Breast Cancer. *Cancer Res.* **2018**, *78*, 5243–5258. [CrossRef]
58. Mantovani, A.; Dinarello, C.A.; Molgora, M.; Garlanda, C. Interleukin-1 and Related Cytokines in the Regulation of Inflammation and Immunity. *Immunity* **2019**, *50*, 778–795. [CrossRef]

cancers

MDPI

Article

Baseline Blood CD8⁺ T Cell Activation Potency Discriminates Responders from Non-Responders to Immune Checkpoint Inhibition Combined with Stereotactic Radiotherapy in Non-Small-Cell Lung Cancer

Hanneke Kievit [1], M. Benthe Muntinghe-Wagenaar [1], Wayel H. Abdulahad [2], Abraham Rutgers [2], Lucie B. M. Hijmering-Kappelle [1], Birgitta I. Hiddinga [1], J. Fred Ubbels [3], Robin Wijsman [3], Marcel J. van der Leij [4], Johan Bijzet [2], Harry J. M. Groen [1], Huib A. M. Kerstjens [1], Anthonie J. van der Wekken [1], Bart-Jan Kroesen [4] and T. Jeroen N. Hiltermann [1,*]

[1] Department of Pulmonology, University of Groningen, University Medical Center Groningen, 9713 GZ Groningen, The Netherlands; h.a.m.kerstjens@umcg.nl (H.A.M.K.)
[2] Department of Rheumatology and Clinical Immunology, University of Groningen, University Medical Center Groningen, 9713 GZ Groningen, The Netherlands
[3] Department of Radiation Oncology, University of Groningen, University Medical Center Groningen, 9713 GZ Groningen, The Netherlands
[4] Department of Laboratory Medicine, Medical Immunology Laboratory, University of Groningen, University Medical Center Groningen, 9713 GZ Groningen, The Netherlands
* Correspondence: t.j.n.hiltermann@umcg.nl

Simple Summary: Cancer may be recognized by the immune system. For patients with non-small-cell lung cancer (NSCLC), immune checkpoint inhibitors (ICI) are a first-line treatment in most patients. However, most of these patients do not respond to ICI, implying that the treatment is ineffective, where it may have relevant side effects. Currently, there is no solid biomarker to predict response to ICI. Thus, there is an urgent need for a new biomarker to predict this response, preferably via minimally invasive techniques. We tested the potency of T cells to be activated ex vivo in the peripheral blood of patients with advanced NSCLC. We found an increased ability to activate $CD8^+$ T cells and produce intracellular IL-2 in peripheral $CD8^+$ T cells in patients that respond to ICI compared to non-responders and healthy controls before the start of ICI. The potency of peripheral T cells to be activated before treatment seems a promising biomarker.

Abstract: Background: Tumor-infiltrating immune cells have been correlated with prognosis for patients treated with immune checkpoint inhibitor (ICI) treatment of various cancers. However, no robust biomarker has been described to predict treatment response yet. We hypothesized that the activation potency of circulating T cells may predict response to ICI treatment. Methods: An exploratory analysis was conducted to investigate the association between the response to immune checkpoint inhibition (ICI) combined with stereotactic radiotherapy (SBRT) and the potency of circulating T cells to be activated. Blood-derived lymphocytes from 14 patients were stimulated ex vivo with, among others, Staphylococcal enterotoxin B (SEB) and compared to healthy controls (HCs). Patients were grouped into responders (>median progression free survival (PFS)) and non-responders (<median PFS). The expression of the T cell activation marker CD69 and intracellular cytokines (IL-2, IFNγ, TNFα) in both $CD4^+$ and $CD8^+$ T cells in response to stimulation was measured using flow cytometry. In addition, serum levels of BAFF, IFNγ, and IL-2 receptor (sIL-2R) were measured by Luminex. Results: At baseline, a higher percentage of activated $CD8^+$ T cells (15.8% vs. 3.5% ($p = <0.01$)) and $IL\text{-}2^+CD69^+CD8^+$ T cells (8.8% vs. 2.9% ($p = 0.02$)) was observed in responders compared to non-responders upon ex vivo stimulation with SEB. The concurrently measured serum cytokine levels were not different between responders and non-responders. Conclusion: Baseline blood $CD8^+$ T cell activation potency, measured by intracellular cytokine production after ex vivo stimulation, is a potential biomarker to discriminate responders from non-responders to SBRT combined with ICI.

Keywords: NSCLC; immune checkpoint inhibition; immunotherapy; biomarker; T cell function assay; cytokines; liquid biopsy

1. Introduction

Cancers can be recognized by the immune system, and immune checkpoint inhibition (ICI) presents an important treatment option, particularly for Non-Small-Cell Lung Cancer (NSCLC) [1–4]. However, not all tumors exhibit a favorable outcome to ICI. "Hot" tumors, characterized by a heightened T cell inflammatory profile, demonstrate a markedly superior response to ICI compared to "cold" tumors, which are defined by a low or absent T cell inflammatory profile. Resected tumors from individuals who respond to neoadjuvant ICI show a dense presence of immune activation, including infiltrating lymphocytes, macrophages, and tertiary lymphoid structures (TLS) [5–7]. The abundance of tumor-infiltrating $CD8^+$ T cells in both the TLS and the tumor microenvironment has been correlated with a more favorable prognosis in various cancers, including lung cancer [8–10].

To this date, the only biomarker in clinical practice used to predict the responsiveness of NSCLC to ICI is the expression of PD-L1 on tumor cells. In advanced lung cancer, tissue biopsy is essential for diagnosis and treatment decisions [11,12]. However, tissue availability is usually limited owing to small biopsies or cytological specimens. Consequently, there is ongoing exploration of less invasive techniques to obtain biomarkers, particularly those from peripheral blood sampling, to predict tumor response. In patients with NSCLC who received second-line nivolumab, a pre-treatment neutrophil-to-lymphocyte ratio (NLR) below 5 was associated with an extended progression free survival (PFS) and overall survival (OS) compared to a higher NLR ($\geq$5) [13,14]. Furthermore, elevated baseline levels of serum Interleukin 6 (IL-6) levels were associated with a shortened PFS and OS [15]. Flow cytometric analysis of peripheral blood lymphocytes of NSCLC patients receiving ICI treatment revealed that response to treatment was associated with a higher percentage of $CD62lowCD4^+$ T cells, whereas non-responders were characterized by a higher percentage of $CD25^+FOXP3^+CD4^+$ T cells prior to treatment [16]. Additionally, higher levels of $CD4^+$ T cells, $CD4^+/CD8^+$cell ratios, and absolute numbers of natural killer cells (NK cells) were associated with response to ICI [17].

Beyond the conventional assessment of immune status through flow cytometry, various aspects of T cell function, such as cytokine production following mitogenic ex vivo stimulation of T cells, can be assessed. Cytokines, such as tumor necrosis factor alpha (TNFα), interferon gamma (IFNγ), and interleukin 2 (IL-2) play an important role in modulating the immune response. Both IFNγ and IL-2 are known to promote the cytotoxicity of $CD8^+$ cells and NK cells and to stimulate the differentiation of $CD4^+$ T cells into Th1 helper cells [10,18]. In patients with NSCLC, the IFNγ signature is associated with prolonged survival [19].

We hypothesized that peripheral blood lymphocytes from NSCLC patients at baseline who respond to ICI exhibit increased sensitivity to mitogenic stimulation, leading to an augmented production of intracellular cytokines. If substantiated, this could potentially serve as a viable biomarker for selecting patients for ICI treatment. In this study we evaluated the baseline expression of the T cell activation marker CD69 and intracellular cytokines (IL-2, IFNγ, TNFα) in both $CD4^+$ and $CD8^+$ T cells after ex vivo stimulation and their association with response to combined ICI and SBRT treatment.

2. Materials and Methods

This exploratory analysis was part of a phase 1 study on safety and tolerability in patients with stage IIIB/IV NSCLC treated with stereotactic radiotherapy (SBRT) on a part of the primary tumor, combined with ICI treatment. ICI was administered in three different regimes in sequential cohorts as $\geq$2nd line treatment after platinum-based chemotherapy. The 1st cohort (n = 3) received durvalumab (PD-L1 inhibitor), while the 2nd and 3rd cohorts (both $n = 6$) received a combination of durvalumab and tremelimumab (CTLA-4 inhibitor) followed by durvalumab monotherapy. All patients were irradiated on the primary tumor (1 × 20 Gy on 9 cc) one week after the 1st dose of ICI. For further details regarding inclusion and exclusion criteria, patient characteristics, and full trial design of this phase 1 study, the SICI (stereotactic radiotherapy and immune checkpoint inhibition) trial, we refer to Kievit et al. [20]. Patients were grouped into responders (>median progression free survival (PFS)) and non-responders (≤median PFS).

Exploratory endpoints in this study were the correlation of response and survival with lymphocyte count, T cell activation potency, and the correlation between intracellular and serum cytokines at baseline just before the start of the combined ICI and SBRT treatment. PFS was defined from date of start of the treatment to the date of the first documented date of progression or death by any cause. OS was defined from date of start of the treatment to date of death. If a patient had not died, the OS was censored at the date of last follow-up. Cutoff date was 1 July 2024.

2.1. Lymphocyte Count

Blood lymphocyte counts were differentiated into $CD45^+CD3^+$ T cells, $CD45^+CD3^+CD4^+$ T helper cells, $CD45^+CD3^+CD8^+$ cytotoxic T cells, $CD45^+CD19^+$ B cells and $CD45^+CD16^+CD56^+$ NK cells. The NLR was calculated by dividing the absolute neutrophil count by absolute lymphocyte count and marked as low (<5) or high (≥5).

2.2. T Cell Activation Analysis

Potency of T cell activation was assessed using flow cytometric analysis of peripheral blood T cellT cells and compared to simultaneously included healthy controls, a standard assay that is used for the diagnosis of immune deficiencies as described by Stam et al. [21]. In short, whole blood was activated ex vivo using one of the following mitogens or antigens: 5 ug/mL Staphylococcal enterotoxin B (SEB; Sigma, Deisenhofen, Germany), anti-CD3 (aCD3; 10% *v*/*v* WT32 hybridoma culture supernatant), 5 ug/mL Phyto haemaglutine (PHA; Remel, Lenexa, KS, USA), a cocktail of 15 Lf/mL tetanus toxoid and diphtheria toxoid (cocktail; Netherlands Vaccine Institute; NVI, Bilthoven, The Netherlands), and 0.5 TE/mL purified protein derivative Tuberculosis (PPD; NVI, Bilthoven, The Netherlands). To block cytokine release from cells, 10 μg/mL brefeldin A (BFA; Sigma-Aldrich, St. Louis, MI, USA) was added to each sample. Next, samples were incubated for 24 h at 37 °C with 5% CO_2. Post-incubation, samples were treated with 10 μL of 40 mM EDTA in PBS for 10 min to inhibit activated cell adhesion. Following this, red blood cells were lysed, and white blood cells were fixed by adding 2 mL of FACS lysing solution (Becton Dickinson, Franklin Lakes, NJ, USA) for 10 min at room temperature, followed by a PBS wash with 1% bovine serum albumin (BSA). Next, each sample was permeabilized with 500 μL of Perm II (Becton Dickinson) containing varying concentrations and/or combinations of Pacific Blue (PB) and/or Pacific Orange (PO) dyes (Invitrogen, Carlsbad, CA, USA) to facilitate fluorescent cell barcoding (FCB) as depicted in Figure S1. Unstimulated samples were stained with 1.25 μg PB and 10 μg PO, while samples stimulated with PPD, SEB, aCD3 (WT32), cocktail, and PHA were stained with 0 μg PB and 0 μg PO, 1,25 μg PB and 0 μg PO, 10 μg PB and 0 μg PO, 0 μg PB and 10 μg PO, and 10 μg PB and 10 μg PO, respectively. After 10 min of incubation at room temperature in the dark, samples were washed and resuspended in PBS with 20% fetal calf serum (FCS). T cell activation potency was assessed by evaluating the expression of CD69 and intracellular cytokines (IL-2, IFNγ, TNFα). Alongside each patient's sample, we included a healthy control sample in our test. Flow cytometric analysis was done on 50,000 recorded events per analysis using a FACSCanto-II flow cytometer (Beckton Dickinson Biosciences, San Jose, CA, USA). Diva v9.3.1 (Beckton Dickinson) and Kaluza v2.2 (Beckman Coulter, Brea, CA, USA) software was used for gating and flow cytometry data analysis. The gating strategy is shown in Figure S1. Patients were divided into above median PFS (responders; 0 R) and below median PFS (non-responders; NR). We compared intracellular cytokine production in both

CD4$^+$ and CD8$^+$ T cells between both groups and with healthy controls (HC) using the Mann–Whitney U test.

2.3. Serum Cytokines

The intracellular cytokine production of IL-2, IFNγ and TNFα was correlated to serum concentrations of BAFF/BLyS/TNFSF13B, CD25/IL-2R alpha and IFNγ measured by multiplex Luminex as described by the protocol of the manufacturer (R&D, Austin, TX, USA). For an overall indication of T cell and B cell activation, the Luminex panel was expanded with CCL17/TARC, IL-6 and CXCL13/BLC/BCA-1.

2.4. Statistics

Relationships between lymphocyte count and response, as well as the relationship between serum cytokines and response were analyzed with the Mann–Whitney U test. The same test was used to compare differences in T cell activation assay between responders, non-responders, and healthy controls. Spearman's correlations were calculated between survival and NLR and the lymphocyte count and to assess relations between intracellular and serum cytokines.

2.5. Ethical and Regulatory Requirements

This study was performed in accordance with ethical principles that have their origin in the Declaration of Helsinki and are consistent with ICH-Good Clinical Practice, and applicable regulatory requirements concerning subject data protection. The study protocol was approved by the local medical ethic committee (ICTRP: NL-OMON44296; EudraCT: 2017–002797-39). All patients gave their written informed consent prior to the start of any study related procedures.

3. Results

Fifteen patients were included in the phase 1 SICI trial. In one patient, blood storage failed due to technical issues. Therefore, the data of 14 patients (8 non-responders and 6 responders) were available for the exploratory analysis at baseline. Patients categorized as responder by the mean PFS had either stable disease or partial response as best response to treatment and all non-responders had progressive disease as best response. As an internal control of the T cell activation assay, 14 healthy controls served as reference. Healthy controls were younger compared to the NSCLC patients (median age 48 years vs. 64 years, $p < 0.001$) and were more often female (71% vs. 14%, $p = 0.002$). None of the patients used corticosteroids at the start of treatment. Overall survival (OS) was 12 months (range 1—not reached) and superior for responders compared to non-responders (Figure S2 and phase 1 SICI trial [20]).

3.1. Lymphocyte Count

There was no significant difference in blood lymphocytes at baseline between responders and non-responders (Figure 1, Table S1). For all patients, OS showed a significant negative correlation with the neutrophil-to-lymphocyte ratio NLR (spearman's rho -0.55, $p = 0.04$). Furthermore, OS was positively correlated with all lymphocytes (0.71, $p = 0.005$), CD3$^+$ T cells (0.73, $p = 0.003$) and CD8$^+$ T cells (0.63, $p = 0.016$) (Table S2). All other correlations between the lymphocyte count and PFS or OS were not significantly different.

3.2. T Cell Activation Assay at Baseline

Before the start of treatment, responders had significantly higher percentages of activated (CD69$^+$)CD8$^+$ T cells, especially those producing IL-2 in response to SEB stimulation, compared to both non-responders (CD69$^+$CD8$^+$ 15.8% vs. 3.5% ($p = 0.008$); IL-2$^+$CD69$^+$CD8$^+$ 8.8% vs. 2.9% ($p = 0.02$)) and healthy controls (5.6% ($p = 0.009$) and 4.7%, respectively ($p = 0.001$)). No difference in the expression of IL-2 was observed between T cells of non-responders and healthy controls. In addition, significantly higher expression of TNFα and IFNγ was observed in activated CD8$^+$ T cells of responders when compared to healthy controls (resp 19.8% vs. 10.4%; $p = 0.01$ and 21.4% vs. 9.3%; $p = 0.03$). The expression of TNFα and IFNγ by T cells was not different between responders and non-responders (TNFα and IFNγ of non-responders resp. 8.0% and 5.5%) (Figure 2). No difference was observed in the percentage of activated CD8$^+$ T cells after stimulation with anti-CD3 between responders, non-responders, and healthy controls. Stimulation with the strong mitogen PHA resulted in strong T cell activation, evident from significant CD69 upregulation without observable differences between responders, non-responders, and healthy controls in activated T cells and intracellular cytokine production. Stimulation with the weak antigenic cocktail or PPD resulted in almost no stimulation of both CD8$^+$ and CD4$^+$ T cells with no differences between groups (Table S3).

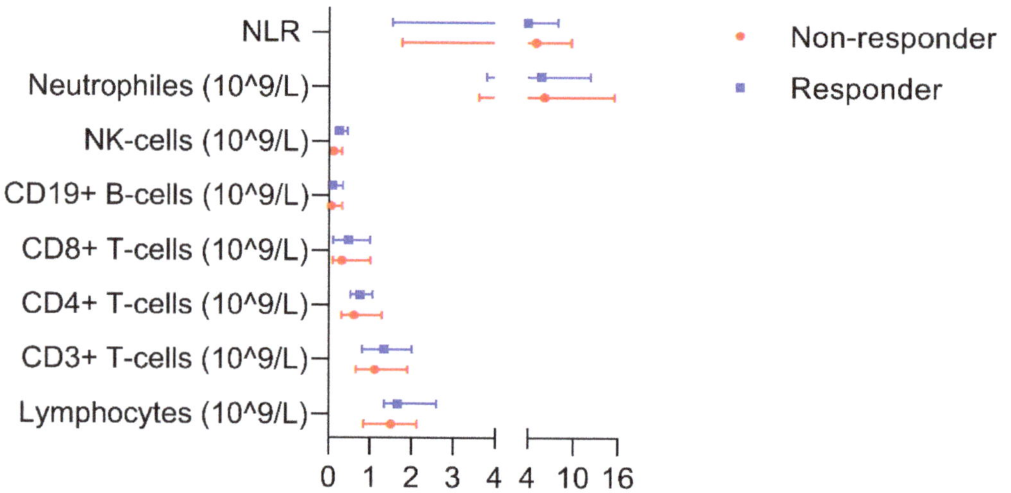

Figure 1. Blood lymphocytes between responders and non-responders at baseline. Data are presented as median with its range. No significant differences were observed, $p > 0.14$. NK cells: natural killer cells. NLR: neutrophile-to-lymphocyte ratio.

In contrast to CD8$^+$ T cells, in activated (CD69$^+$) CD4$^+$ T cells, no significant differences were observed between responders and non-responders after stimulation with SEB (CD69$^+$CD4$^+$ 18% vs. 12.6% ($p = 0.23$), IL-2$^+$CD69$^+$CD4$^+$ 17.1% vs. 12.5% ($p = 0.41$), TNFα$^+$CD69$^+$CD4$^+$ 22.5% vs. 15.4% ($p = 0.49$) and IFNγ$^+$CD69$^+$CD4$^+$ resp 9.8% and 6.3% ($p = 0.10$). Compared to healthy controls, only responders had significant higher percentages of activated (CD69$^+$) CD4$^+$ T cells when stimulated with SEB, including higher expression of IL-2 and TNFα (CD69$^+$CD4$^+$ 18% vs. 7.3% ($p = 0.03$), IL-2$^+$CD69$^+$CD4$^+$ 17.1% vs. 7.4% ($p = 0.009$), and TNFα$^+$CD69$^+$CD4$^+$ 22.5% vs. 9.6% ($p = 0.02$)). No significant difference was observed in IFNγ expression between responders and healthy controls (IFNγ$^+$CD69$^+$CD4$^+$ 9.8% vs. 3.1% ($p = 0.72$)). TNFα expression only, was significantly different between non-responders and healthy controls (TNFα$^+$CD69$^+$CD4$^+$15.4% vs. 9.4% ($p = 0.03$), Figure 3). For CD4$^+$ T cells stimulated with anti-CD3, a similar cytokine response as with SEB was observed (Table S3).

3.3. Luminex Assay

At baseline, serum cytokines (IFNγ, IL-2 (CD25/IL-2R alpha), TNFα (BAFF/BLyS/TNFSF13B), CCL17 (CCL17/TARC), IL-6, and CXCL13 (CXCL13/BLC/BCA-1)), which were selected based on complementarity, appearance, and function in immune activation, were not different between responders and non-responders (Table S4). In addition, there was no correlation between intracellular IL-2, IFNγ and TNFα expression by T cells and serum levels of three selected cytokines (BAFF/BLyS/TNFSF13B, CD25/IL-2R alpha, IFNγ) (Figure 4).

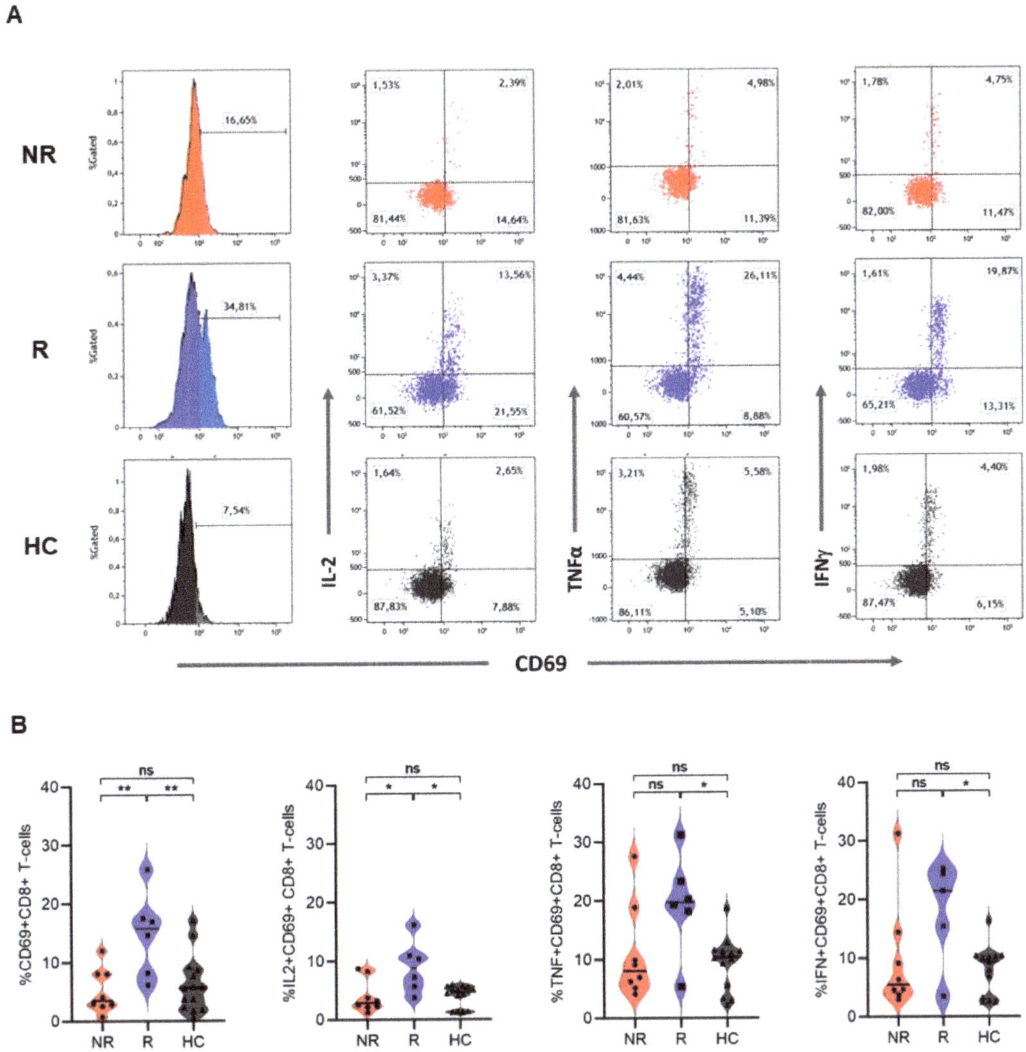

Figure 2. Activation of intracellular cytokine expression in CD8⁺ T cells after stimulation with SEB. (**A**) Representative flow cytometry plots of SEB-stimulated CD8⁺ T cells showing CD69 expression alone (histograms) or CD69 expression in combination with expression of intracellular cytokines IL-2, TNFα, IFNγ (dot-plots) in a non-responder (NR; upper plots), a responder (R; middle plots), and an age- and sex-matched healthy control (HC; lower plots). Values in each gate represent the percentage of positive cells; (**B**) frequencies of CD69, IL-2, TNFα, and IFNγ expression within responding CD8⁺ T cells from NR (n = 8), R (n = 6) and HC (n = 14) after in vitro stimulation with SEB. Horizontal lines represent the median percentage. p-values were calculated using the Mann–Whitney U test. *: p <0.05. **: p <0.001. ns: non-significant.

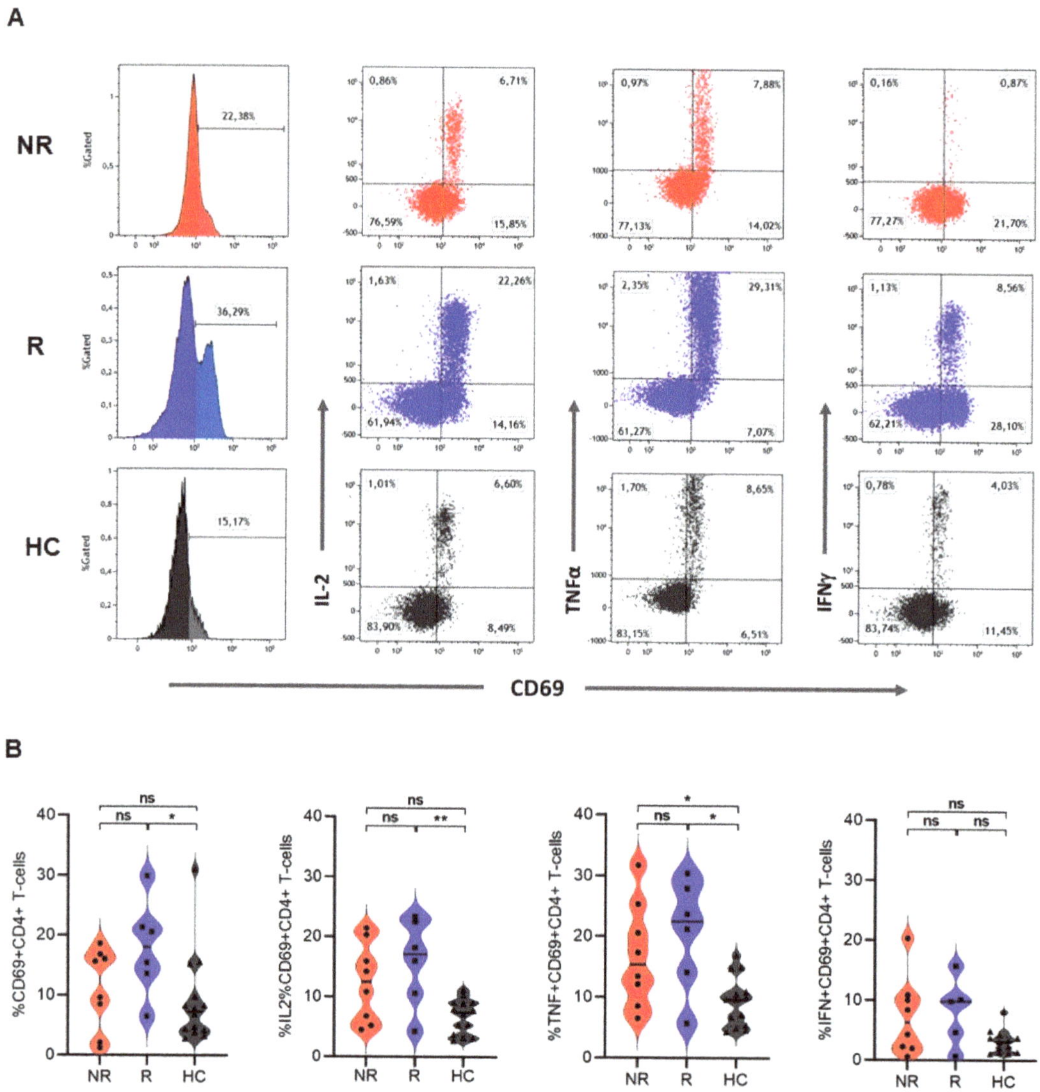

Figure 3. Activation and intracellular cytokine expression in CD4⁺ T cells after stimulation with SEB. (**A**) Representative flow cytometry plots from SEB-stimulated CD4⁺ T cells showing CD69 expression alone (histograms) or in CD69 expression combination with expression of intracellular cytokines IL-2, TNFα, and IFNγ (dot-plots) in a non-responder (NR; upper plots), a responder (R; middle plots), and an age- and sex-matched healthy control (HC; lower plots). Values in each gate represent the percentage of positive cells. (**B**) Frequencies of CD69, IL-2, TNFα, and IFNγ expression within responding CD4⁺ T cells from NR (n = 8), R (n = 6), and HC (n = 14) after in vitro stimulation with SEB. Horizontal lines represent the median percentage. p-values were calculated using the Mann–Whitney U test. *: p <0.05. **: p <0.001. ns: non-significant.

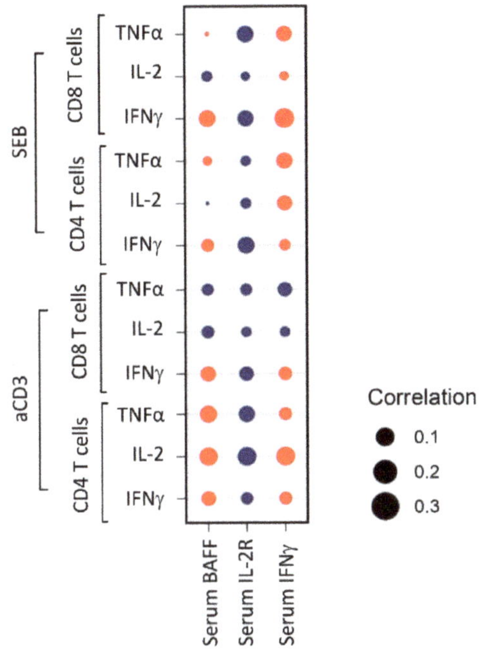

Figure 4. Correlation between baseline intracellular and serum cytokine expression after stimulation with SEB and anti-CD3. A red circle indicates a positive correlation, a blue circle a negative correlation. aCD3: anti CD3. BAFF: B-cell-activating factor belonging to the TNF family. IFNγ: interferon gamma. IL-2: interleukin 2. SEB: Staphylococcal enterotoxin B. TNFα: tumor necrosis factor alpha.

4. Discussion

In this study, we provide evidence that the potency of T cell activation at baseline, as assessed through stimulated intracellular cytokine production, may serve as a non-invasive biomarker to discriminate responders from non-responders to ICI. Specifically, our findings indicate increased expression of cytokines, particularly IL-2 expression by $CD8^+$ T cells following ex vivo stimulation at baseline in patients who turned out to respond to ICI (in combination with SBRT to the primary tumor).

To the best of our knowledge, this is the first study in NSCLC exploring peripheral T cell activation potency as a predictor of response to ICI. T cell activation was assessed by measurement of intracellular cytokines (IL-2, IFNγ, TNFα) following stimulation of peripheral blood T cells with PHA, a tetanus and diphtheria cocktail, PPD, anti-CD3, or SEB. Stimulation of T cells using PHA involves several cell surface receptors, most notably CD3 and CD2 and is, as such, an unselective and strong mitogenic activator of T cells that partly bypasses co-stimulation [22,23]. The strong nature of stimulation using PHA masks possible subtle differences in the activation potency of T cells between groups. In line with this, we did not observe any differences between responders, non-responders, and healthy controls in the expression of cytokines by T cells upon stimulation with PHA. In contrast, the cocktail and PPD are weak antigenic recall stimuli, only activating a subpopulation of T cells, i.e., those that were previously exposed to these stimuli. Using these weak recall stimuli, no differences were noted between the three groups analyzed here, possibly reflecting a too-weak or not-sensitive-enough readout. Using a more intermediate stimulus, such as SEB (or anti-CD3), functional differences in the activation potency of T cells derived from the different groups might be revealed. Both SEB and anti-CD3 stimulate T cells specifically via the T cell receptor (TcR)/CD3 complex, where anti-CD3 does so with higher affinity compared to SEB. As such, SEB potentially represents the most optimal mode of T cell activation to reveal intrinsic differences in the T cell activation potency between the different groups studied here. Indeed, most notable and significant differences in the responses were observed upon SEB stimulation, especially in the $CD3^+CD8^+$ T cell compartment.

$CD8^+$ T cells play an important role in the immune response to cancer and the number of $CD8^+$ T cells, and their capability of infiltration into the tumor (environment) is associated with prolonged survival. Activated $CD8^+$ T cells are capable of killing cancer cells, which is associated with the release of cytokines [24]. Effective activation of (cytotoxic) $CD8^+$ T cells requires signaling via the TcR/CD3 complex in combination with one or more costimulatory signals. Using ICI, both anti-CTLA4 and anti-PD(L)1 enhance the activation of the T cell by blocking an inhibitory signal (the interaction between CTLA-4 and CD80 and PD-1 with PD-L1, respectively) allowing unrestrained T cell (co)stimulation [25,26].

In our study, the $CD8^+$ T cells of responders were significantly more prone to activation at baseline compared to non-responders and healthy controls and expressed more intracellular cytokines, especially IL-2, after ex vivo stimulation with SEB. This suggests that the $CD8^+$ T cells of responders have an increased potential to become activated via the TcR/CD3 complex even before starting ICI treatment than non-responders. This increased potential to be activated might reflect the observed response to ICI treatment and as such, may represent a potential baseline biomarker for ICI response. A similar difference, although non-significant, was seen in the activation potency of $CD4^+$ T cells of responders and non-responders. Alternatively, the poorer responsiveness of T cells to ex vivo stimulation in non-responders may be an indication of an increased presence or activity of regulatory T cells, which has been described as a resistance mechanism to ICI treatment [16]. Alternatively, it may be an indication of exhausted $CD8^+$ T cells, another resistance mechanism for immunotherapy to fail [27].

Absolute lymphocyte count, including different subtypes, was not significantly different between responders and non-responders. We compared the intracellular cytokine production of IL-2, IFNγ, and TNFα with the serum values of these cytokines. There was no correlation between the serum cytokines and T cell activation potency, nor with intracellular cytokine production upon stimulation of T cells at baseline. This is in line with the melanoma data and ICI of the research by Pedersen et al., where baseline serum derived cytokines including TNFα, IFNγ and IL-6 were not associated with PFS [28]. Cytokines released by T cells after stimulation in vivo are likely either directly used or bound by surrounding cells. Therefore, circulating serum cytokines levels may not be representative for T cell activation potency at baseline. Apparently, T cells prone to activation by ICI are thus not reflected in serum cytokines or lymphocyte subsets, and this specific characteristic needs to be measured with ex vivo stimulation in combination with intracellular cytokine expression.

Our study included patients with an indication for $\geq$second-line ICI. Unfortunately, we were not able to include more patients in this study. Due to advancements in knowledge, ICI became standard first-line treatment in patients with advanced NSCLC without a driver mutation during the period of our study. We realize that because of this, our sample size is limited. Nevertheless, this is an exploratory parameter and for the generalizability of our result, this should be repeated in an independent and larger cohort. We think that the use of SBRT in our patient cohort had little influence on the result because of the small treatment volumes. Radiation-induced lymphopenia is associated with multiple courses, multiple irradiated sites, and higher dose (>50 Gy) and is mainly seen when large vessels and the heart are part of the irradiated field [29,30]. Therefore, we suggest studying T cell activation in patients receiving immune checkpoint inhibition in a first-line setting. Our healthy controls were laboratory employees working on the day of blood withdrawal of the participating patients and were not age- and sex-matched with the patients. As such, we cannot rule out that differences between patients and healthy controls observed are due to differences in sex and or age between the groups included.

5. Conclusions

The potency to activate $CD8^+$ T cells at baseline, as determined through ex vivo activation and intra-cellular cytokine production, discriminates responders from non-responders to immune checkpoint inhibition combined with stereotactic radiotherapy in our patient cohort with non-small-cell lung cancer. Further research—for instance, in first-line ICI treatment in a larger cohort—is needed to confirm this potential biomarker.

Supplementary Materials: The following supporting information can be downloaded at: https://www.mdpi.com/article/10.3390/cancers16142592/s1, Figure S1: Representative gating strategy of quantifying cytokine-expressing $CD4^+$ and $CD8^+$ T cells using fluorescent cell barcoding (FCSB) technique; Figure S2: Overall survival; Table S1: Baseline characteristics; Table S2: Correlation between lymphocyte count and survival; Table S3: Median percentages of responding ($CD69^+$) $CD4^+$ and $CD8^+$ T—cells and cytokine production after stimulation; Table S4: Median percentages of responding ($CD69^+$) $CD4^+$ and $CD8^+$ T—cells and cytokine production after stimulation.

Author Contributions: H.K., M.B.M.-W., W.H.A., B.-J.K. and T.J.N.H. wrote the manuscript. W.H.A., B.-J.K. and T.J.N.H. designed the study protocol. H.K., M.B.M.-W., L.B.M.H.-K., B.I.H., R.W., J.F.U., H.J.M.G., A.J.v.d.W. and T.J.N.H. were responsible for data collection. H.K., M.B.M.-W., W.H.A., M.J.v.d.L. and J.B. performed the analyses. H.K., M.B.M.-W. and W.H.A. made the figures and tables. H.K., M.B.M.-W., W.H.A., A.R., H.A.M.K., B.-J.K. and T.J.N.H. interpreted the data. T.J.N.H. was responsible for funding acquisition. All authors contributed to the article and approved the submitted version. All authors have read and agreed to the published version of the manuscript.

Funding: This study was sponsored by a research grant from AstraZeneca, ESR 16-12461/D4191C00075. AstraZeneca had no role in the design of the study, data collection, data analysis, data interpretation, or the writing of the manuscript.

Institutional Review Board Statement: This study was performed in accordance with ethical principles that have their origin in the Declaration of Helsinki and are consistent with ICH-Good Clinical Practice, and applicable regulatory requirements concerning subject data protection. The study protocol was approved by the local medical ethic committee on 16 November 2017 (ICTRP: NL-OMON44296; EudraCT: 2017–002797-39).

Informed Consent Statement: All patients gave their written informed consent prior to the start of any study related procedures.

Data Availability Statement: The data generated in this study are available upon reasonable request from the corresponding author.

Acknowledgments: We would like to thank our patients and healthy controls for their contribution and Klaas Bakker for his help in making the funds for the translational research available.

Conflicts of Interest: This study received funding from AstraZeneca. The funder was not involved in the study design, collection, analysis, interpretation of data, the writing of this article, or the decision to submit it for publication. All authors declare no other competing interests.

References

1. Reck, M.; Rodríguez-Abreu, D.; Robinson, A.G.; Hui, R.; Csőszi, T.; Fülöp, A.; Gottfried, M.; Peled, N.; Tafreshi, A.; Cuffe, S.; et al. Pembrolizumab versus Chemotherapy for PD-L1–Positive Non–Small-Cell Lung Cancer. *N. Engl. J. Med.* **2016**, *375*, 1823–1833. [CrossRef] [PubMed]
2. Langer, C.J.; Gadgeel, S.M.; Borghaei, H.; Papadimitrakopoulou, V.A.; Patnaik, A.; Powell, S.F.; Gentzler, R.D.; Martins, R.G.; Stevenson, J.P.; Jalal, S.I.; et al. Carboplatin and pemetrexed with or without pembrolizumab for advanced, non-squamous non-small-cell lung cancer: A randomised, phase 2 cohort of the open-label KEYNOTE-021 study. *Lancet Oncol.* **2016**, *17*, 1497–1508. [CrossRef] [PubMed]
3. Gandhi, L.; Rodríguez-Abreu, D.; Gadgeel, S.; Esteban, E.; Felip, E.; De Angelis, F.; Domine, M.; Clingan, P.; Hochmair, M.J.; Powell, S.F.; et al. Pembrolizumab plus Chemotherapy in Metastatic Non–Small-Cell Lung Cancer. *N. Engl. J. Med.* **2018**, *378*, 2078–2092. [CrossRef] [PubMed]
4. Mok, T.S.K.; Wu, Y.L.; Kudaba, I.; Kowalski, D.M.; Cho, B.C.; Turna, H.Z.; Castro, G., Jr.; Srimuninnimit, V.; Laktionov, K.K.; Bondarenko, I.; et al. Pembrolizumab versus chemotherapy for previously untreated, PD-L1-expressing, locally advanced or metastatic non-small-cell lung cancer (KEYNOTE-042): A randomised, open-label, controlled, phase 3 trial. *Lancet* **2019**, *393*, 1819–1830. [CrossRef] [PubMed]

5. Cottrell, T.R.; Thompson, E.D.; Forde, P.M.; Stein, J.E.; Duffield, A.S.; Anagnostou, V.; Rekhtman, N.; Anders, R.A.; Cuda, J.D.; Illei, P.B.; et al. Pathologic features of response to neoadjuvant anti-PD-1 in resected non-small-cell lung carcinoma: A proposal for quantitative immune-related pathologic response criteria (irPRC). *Ann. Oncol.* **2018**, *29*, 1853–1860. [CrossRef] [PubMed]
6. Trujillo, J.A.; Sweis, R.F.; Bao, R.; Luke, J.J. Combination Therapy Selection. *Cancer Immunol. Res.* **2019**, *6*, 990–1000. [CrossRef] [PubMed]
7. Galon, J.; Bruni, D. Approaches to treat immune hot, altered and cold tumours with combination immunotherapies. *Nat. Rev. Drug Discov.* **2019**, *18*, 197–218. [CrossRef] [PubMed]
8. Goc, J.; Fridman, W.H.; Sautès-Fridman, C.; Dieu-Nosjean, M.C. Characteristics of tertiary lymphoid structures in primary cancers. *Oncoimmunology* **2013**, *2*, e26836. [CrossRef] [PubMed]
9. Tamminga, M.; Hiltermann, T.J.N.; Schuuring, E.; Timens, W.; Fehrmann, R.S.N.; Groen, H.J.M. Immune microenvironment composition in non-small cell lung cancer and its association with survival. *Clin. Transl. Immunol.* **2020**, *9*, e1142. [CrossRef]
10. Petitprez, F.; Meylan, M.; de Reyniès, A.; Sautès-Fridman, C.; Fridman, W.H. The Tumor Microenvironment in the Response to Immune Checkpoint Blockade Therapies. *Front. Immunol.* **2020**, *11*, 784. [CrossRef]
11. Prelaj, A.; Tay, R.; Ferrara, R.; Chaput, N.; Besse, B.; Califano, R. Predictive biomarkers of response for immune checkpoint inhibitors in non–small-cell lung cancer. *Eur. J. Cancer* **2019**, *106*, 144–159. [CrossRef] [PubMed]
12. Zhang, B.; Liu, Y.; Zhou, S.; Jiang, H.; Zhu, K.; Wang, R. Predictive effect of PD-L1 expression for immune checkpoint inhibitor (PD-1/PD-L1 inhibitors) treatment for non-small cell lung cancer: A meta-analysis. *Int. Immunopharmacol.* **2020**, *80*, 106214. [CrossRef]
13. Russo, A.; Russano, M.; Franchina, T.; Migliorino, M.R.; Aprile, G.; Mansueto, G.; Berruti, A.; Falcone, A.; Aieta, M.; Gelibter, A.; et al. Neutrophil-to-Lymphocyte Ratio (NLR), Platelet-to-Lymphocyte Ratio (PLR), and Outcomes with Nivolumab in Pretreated Non-Small Cell Lung Cancer (NSCLC): A Large Retrospective Multicenter Study. *Adv. Ther.* **2020**, *37*, 1145–1155. [CrossRef] [PubMed]
14. Diem, S.; Schmid, S.; Krapf, M.; Flatz, L.; Born, D.; Jochum, W.; Templeton, A.J.; Früh, M. Neutrophil-to-Lymphocyte ratio (NLR) and Platelet-to-Lymphocyte ratio (PLR) as prognostic markers in patients with non-small cell lung cancer (NSCLC) treated with nivolumab. *Lung Cancer* **2017**, *111*, 176–181. [CrossRef]
15. Kang, D.H.; Park, C.K.; Chung, C.; Oh, I.J.; Kim, Y.C.; Park, D.; Kim, J.; Kwon, G.C.; Kwon, I.; Sun, P.; et al. Baseline serum interleukin-6 levels predict the response of patients with advanced non-small cell lung cancer to pd-1/pd-l1 inhibitors. *Immune Netw.* **2020**, *20*, e27. [CrossRef]
16. Kagamu, H.; Kitano, S.; Yamaguchi, O.; Yoshimura, K.; Horimoto, K.; Kitazawa, M.; Fukui, K.; Shiono, A.; Mouri, A.; Nishihara, F.; et al. CD4+ T-cell immunity in the peripheral blood correlates with response to Anti-PD-1 therapy. *Cancer Immunol. Res.* **2020**, *8*, 334–344. [CrossRef]
17. Li, P.; Qin, P.; Fu, X.; Zhang, G.; Yan, X.; Zhang, M.; Zhang, X.; Yang, J.; Wang, H.; Ma, Z. Associations between peripheral blood lymphocyte subsets and clinical outcomes in patients with lung cancer treated with immune checkpoint inhibitor. *Ann. Palliat. Med.* **2021**, *10*, 3039–3049. [CrossRef] [PubMed]
18. Jiang, T.; Zhou, C.; Ren, S. Role of IL-2 in cancer immunotherapy. *Oncoimmunology* **2016**, *5*, e1163462. [CrossRef]
19. Higgs, B.W.; Morehouse, C.A.; Streicher, K.; Brohawn, P.Z.; Pilataxi, F.; Gupta, A.; Ranade, K. Interferon gamma messenger RNA Signature in tumor biopsies predicts outcomes in patients with non–small cell lung carcinoma or urothelial cancer treated with durvalumab. *Clin. Cancer Res.* **2018**, *24*, 3857–3866. [CrossRef]
20. Kievit, H.; Muntinghe-Wagenaar, M.B.; Hijmering-Kappelle, L.B.M.; Hiddinga, B.I.; Ubbels, J.F.; Wijsman, R.; Slingers, G.; de Vries, R.; Groen, H.J.M.; Kerstjens, H.A.M.; et al. Safety and tolerability of stereotactic radiotherapy combined with durvalumab with or without tremelimumab in advanced non-small cell lung cancer, the phase I SICI trial. *Lung Cancer* **2023**, *178*, 96–102. [CrossRef]
21. Stam, J.; Abdulahad, W.; Huitema, M.G.; Roozendaal, C.; Limburg, P.C.; van Stuijvenberg, M.; Schölvinck, E.H. Fluorescent cell barcoding as a tool to assess the age-related development of intracellular cytokine production in small amounts of blood from infants. *PLoS ONE* **2011**, *6*, e25690. [CrossRef]
22. Elshari, Z.S.; Nepesov, S.; Tahrali, I.; Kiykim, A.; Camcioglu, Y.; Deniz, G.; Kucuksezer, U.C. Comparison of mitogen-induced proliferation in child and adult healthy groups by flow cytometry revealed similarities. *Immunol. Res.* **2023**, *71*, 51–59. [CrossRef] [PubMed]
23. Kashef, S.; Moghtaderi, M.; Hatami, H.R.; Kalani, M.; Alyasin, S.; Nabavizadeh, H.; Farjadian, S. Evaluation of T Cell Proliferation Using CFSE Dilution Assay: A Comparison between Stimulation with PHA and Anti-CD3/Anti-CD28 Coated Beads. *Iran. J. Allergy Asthma Immunol.* **2022**, *21*, 458–466. [CrossRef] [PubMed]
24. Durgeau, A.; Virk, Y.; Corgnac, S.; Mami-Chouaib, F. Recent advances in targeting CD8 T-cell immunity for more effective cancer immunotherapy. *Front. Immunol.* **2018**, *9*, 14. [CrossRef]
25. Martínez-Lostao, L.; Anel, A.; Pardo, J. How Do Cytotoxic Lymphocytes Kill Cancer Cells? *Clin. Cancer Res.* **2015**, *21*, 5047–5056. [CrossRef] [PubMed]
26. de Mello, R.A.; Veloso, A.F.; Catarina, P.E.; Nadine, S.; Antoniou, G. Potential role of immunotherapy in advanced non-small-cell lung cancer. *Onco Targets Ther.* **2017**, *10*, 21–30. [CrossRef] [PubMed]
27. Chow, A.; Perica, K.; Klebanoff, C.A.; Wolchok, J.D. Clinical implications of T cell exhaustion for cancer immunotherapy. *Nat. Rev. Clin. Oncol.* **2022**, *19*, 775–790. [CrossRef] [PubMed]

28. Pedersen, J.G.; Sokac, M.; Sørensen, B.S.; Luczak, A.A.; Aggerholm-Pedersen, N.; Birkbak, N.J.; Øllegaard, T.H.; Jakobsen, M.R. Increased Soluble PD-1 Predicts Response to Nivolumab plus Ipilimumab in Melanoma. *Cancers* **2022**, *14*, 3342. [CrossRef]
29. Upadhyay, R.; Venkatesulu, B.P.; Giridhar, P.; Kim, B.K.; Sharma, A.; Elghazawy, H.; Dhanireddy, B.; Elumalai, T.; Mallick, S.; Harkenrider, M. Risk and impact of radiation related lymphopenia in lung cancer: A systematic review and meta-analysis. *Radiother. Oncol.* **2021**, *157*, 225–233. [CrossRef]
30. Cho, Y.; Park, S.; Byun, H.K.; Lee, C.G.; Cho, J.; Hong, M.H.; Kim, H.R.; Cho, B.C.; Kim, S.; Park, J.; et al. Impact of Treatment-Related Lymphopenia on Immunotherapy for Advanced Non-Small Cell Lung Cancer. *Int. J. Radiat. Oncol. Biol. Phys.* **2019**, *105*, 1065–1073. [CrossRef]

cancers

MDPI

Article

Peripheral Blood TCRβ Repertoire, IL15, IL2 and Soluble Ligands for NKG2D Activating Receptor Predict Efficacy of Immune Checkpoint Inhibitors in Lung Cancer

Andrea Sesma [1,2,*,†], Julian Pardo [2,3,4], Dolores Isla [1,2], Eva M. Gálvez [3,5], Marta Gascón-Ruiz [1,2], Luis Martínez-Lostao [2,6,7,8,9], Alba Moratiel [1,2], J. Ramón Paño-Pardo [2,3,10], Elisa Quílez [1,2], Irene Torres-Ramón [1,2], Alfonso Yubero [1,2] María Zapata-García [1,2], María Pilar Domingo [9], Patricia Esteban [2], Rebeca Sanz Pamplona [2], Rodrigo Lastra [1,2,†] and Ariel Ramírez-Labrada [2,3,*,†]

1 Medical Oncology Department, University Hospital Lozano Blesa, 50009 Zaragoza, Spain; disla@salud.aragon.es (D.I.); mgasconr@salud.aragon.es (M.G.-R.); amoratielp@salud.aragon.es (A.M.); equilezb@salud.aragon.es (E.Q.); irentorresr@salud.aragon.es (I.T.-R.); ayuberoe@salud.aragon.es (A.Y.); mzapatag@salud.aragon.es (M.Z.-G.); rlastrad@salud.aragon.es (R.L.)
2 Aragon Health Research Institute (IIS Aragón), 50009 Zaragoza, Spain; pardojim@unizar.es (J.P.); lmartinezlos@salud.aragon.es (L.M.-L.); jrpanno@salud.aragon.es (J.R.P.-P.); pesteban@iisaragon.es (P.E.); rebecasanz@iconcologia.net (R.S.P.)
3 CIBER de Enfermedades Infecciosas (CIBERINFEC), 28029 Madrid, Spain; eva@icb.csic.es
4 Microbiology, Radiology, Pediatry and Public Health Department Medicine, University of Zaragoza, 50009 Zaragoza, Spain
5 Instituto de Carboquímica (ICB-CSIC), Miguel Luesma 4, 50018 Zaragoza, Spain
6 Department of Microbiology, Pediatrics, Radiology and Public Health, University of Zaragoza, 50009 Zaragoza, Spain
7 Aragon Nanoscience Institute, 50018 Zaragoza, Spain
8 Aragon Materials Science Institute, 50009 Zaragoza, Spain
9 Immunology Department, University Hospital Lozano Blesa, 50009 Zaragoza, Spain; mpdomingo@icb.csic.es
10 Infectious Disease Department, University Hospital Lozano Blesa, 50009 Zaragoza, Spain
* Correspondence: asesmag@salud.aragon.es (A.S.); aramirez@iisaragon.es (A.R.-L.)
† These authors contributed equally to this work.

Simple Summary: The development of immune checkpoint inhibitors has revolutionized the treatment of lung cancer by becoming the standard therapy for advanced non-small cell lung cancer that lacks specific genetic mutations. However, not all patients respond equally, underscoring the need for biomarkers to predict treatment response. To address this, a study was conducted with 55 lung cancer patients treated with immune checkpoint inhibitors to investigate whether biomarkers like TCRβ diversity and certain cytokines linked to T cell activity could predict the response to immunotherapy. While higher TCRβ clonality and specific cytokine levels appeared to be associated with improved survival rates, the findings were not statistically significant. Specifically, higher levels of IL-2 and IL-15 were linked to shorter overall survival, with high IL-15 levels increasing the risk of death threefold in multivariable analysis. Although further research with larger sample sizes is needed for confirmation, these results offer promising insights into potential markers for predicting responses to immune checkpoint inhibitors.

Abstract: The development of immune checkpoint inhibitors (ICIs) has changed the therapeutic paradigm of lung cancer (LC), becoming the standard of treatment for previously untreated advanced non-small cell lung cancer (NSCLC) without actionable mutations. It has allowed the achievement of durable responses and resulted in significant survival benefits. However, not all patients respond; hence, molecular biomarkers are needed to help us predict which patients will respond. With this objective, a prospective observational study was designed, including a cohort of 55 patients with NSCLC who received ICIs. We studied whether biomarkers such as TCRβ and specific cytokines involved in the regulation of T cell activity were related to the immunotherapy response. In the survival analysis, it was found that patients with higher TCRβ clonality, lower TCRβ evenness, higher TCRβ Shannon diversity and lower TCRβ convergence had higher overall survival (OS) and progression-free survival (PFS). However, no statistically significant association was observed. Regarding cytokines, those patients with higher levels of IL-2 and IL-15 presented statistically significantly shorter OS and PFS, respectively. In fact, in the multivariable analysis, the high IL-15 level increased the risk of death by three times. Although the sample size was small and more studies are needed to confirm our results, our study reveals promising markers of responses to ICIs.

Keywords: lung cancer; immune checkpoint inhibitors; predictive biomarker; T cell receptor repertoire; cytokines

1. Introduction

Lung cancer (LC) has traditionally been considered a poorly immunogenic tumor. Currently, we know that it is one of the tumors with the highest mutational burden (MBT), with a median of approximately 10 mutations per megabase [1]. The development of immune checkpoint inhibitors (ICIs) has spurred a change in the therapeutic paradigm of LC, becoming the standard of treatment for previously untreated advanced non-small cell lung cancer (NSCLC) without actionable mutations, achieving durable responses and resulting in significant survival benefits.

In LC, an immunoresistance mechanism adopted by tumor cells is the expression of immuno-inhibitory molecules in the tumor microenvironment such as PD-L1 and CTLA-4 [2]. Therefore, using monoclonal antibodies directed against these molecules to block their action and re-establish T cell-mediated antitumor immunity has become an effective therapeutic option [3,4].

Despite the current success of ICIs, not all patients will benefit from this therapeutic strategy. Up to 50% develop resistance to treatment, a complex and dynamic mechanism in which alterations in the antigenic processing and presentation machinery, epigenetic modifications, alterations in signaling pathways (MAPK, PI3K, WNT) and the modulation of the tumor microenvironment towards a tolerogenic state can arise [5,6]. Due to this heterogeneity in responses, the need arises to identify predictive biomarkers of responses to ICIs that allow us to select which patients will benefit and which will not.

A tumor's molecular and phenotypic characteristics are altered and modified throughout the disease and depending on the treatment, so it would be ideal to find biomarkers that reflect the changes in tumor characteristics and help us identify which tumors will respond to treatment with immunotherapy [7]. In this context, biomarkers, such as TMB and PD-L1 expression, have emerged, albeit with inconclusive results. In the search to find markers of response, the determination of the T cell receptor repertoire (TCRβ) in peripheral blood has emerged as a new predictive biomarker of the response to IT [8–11].

TCR is the antigen-specific receptor essential for the specific immune response located on the cell surface of helper and cytotoxic T lymphocytes [11]. TCR is a complex formed by two variant chains linked by disulfide bonds forming a heterodimer (αβ or γδ) that gives it the unique specificity for the antigen. Each chain has a variable (V) and a constant (C) immunoglobulin-type domain. The complementarity-determining regions (CDRs) are part of the variable chains in the TCR and are crucial for the diversity of antigen specificity generated by lymphocytes. For each chain, there are three CDRs; CDR3s have the highest variability and are encoded by the combination between V(D)J regions and are the primary sites of antigen contact. The CDR3 region of the β-chain accounts for most of the variation [11].

Statistically derived descriptive indexes have emerged to estimate the repertoire diversity and homology of TCR. TCR richness refers to the number of clonotypes that comprise the repertoire defined by the number of unique CDR3 TCRβ sequences. Clonal diversity is determined by the Shannon index, which identifies the proportions of the repertoire containing an expanded clone. Evenness is known as Shannon's normalized diversity. It measures clonotype size similarity and ranges from 0 (the sample contains clonotypes of non-equivalent sizes as occurs in clonal expansion) to 1 (the sample is composed of clonotypes that are at the same frequency). TCR convergence refers to the frequency of identical clonotypes in amino acid sequences but different in nucleotide sequences due to codon degeneration.

TCR richness and convergence could be predictive markers of responses to ICIs. A high T cell richness and convergence level before initiating treatment with ICIs in NSCLC patients has been described as a predictive marker of response to ICIs [12]. It has also been observed that an increase in TCRβ richness and convergence during treatment with ICIs is associated with better outcomes [13].

Tumors produce a wide variety of neoantigens presented by MHC molecules and recognized by specific TCRs via TCR/peptide/MHC interactions. These T cell clones with specificity for each respective neoantigen can be reactivated again with ICIs, thereby augmenting the host antitumor immune response. Therefore, determining the TCR repertoire could allow a more precise approach and could be a more selective and effective biomarker than TMB and PD-L1 [14,15].

In contrast to TMB, TCR convergence is able to detect T cell responses to any antigen, including neoantigens arising not only from nonsynonymous mutations but also aberrant posttranslational modifications, ectopic gene expression, splicing defects, self-antigens and virus-derived antigens [16–18].

Determining the TCR repertoire during treatment may be a helpful tool to establish the evolution and prognosis of patients with LC. The combination of these features together with other established biomarkers, such as PD-L1 expression, would improve response prediction [17]. Even so, both TMB and TCR present certain limitations in their determination, such as the lack of standardization in the cut-off points used, and the cost of the sequencing techniques employed.

Thus, our study aims to prospectively analyze a cohort of patients with LC receiving treatment with ICIs to determine whether the TCRβ repertoire and certain soluble factors in peripheral blood involved in the regulation of T cell antitumor function such as IL-2, IL-4, IL-10, IL-12, IL-15, MICA, MICB, ULBP1, ULBP2, ULBP4, IFN-γ and CXCL10 could predict the efficacy of treatment with ICIs.

2. Materials and Methods

2.1. Population

A prospective observational study was performed on a 55-patient cohort with locally advanced and metastatic non-small cell lung cancer (NSCLC) (stage III and IV) who received treatment with PD1/PD-L1/CTLA-4 ICIs at Hospital Clínico Universitario Lozano Blesa, a tertiary hospital in Zaragoza (Spain) from April 2019 to October 2020.

Patients with histology other than NSCLC, those with contraindications for treatment with ICIs such as autoimmune diseases, synchronous tumors of other origins, patients with immunodeficiencies or at grade 2 or higher in the Eastern Cooperative Oncology Group (ECOG) scale were excluded from the study.

Patients previously treated with immunotherapy were excluded from the study. A total of 23 of 55 patients included had previously received chemotherapy, 14 patients had received treatment of localized disease (chemotherapy + radiotherapy) and 18 patients were treatment naïve.

The demographic and clinicopathologic characteristics of the patients were collected from anonymized medical records and by direct interview with the patient. The functional status of the patients was assessed using the ECOG scale. The responsible physician carried out treatment indications. Response to treatment was evaluated according to the criteria for response evaluation in solid tumors (RECIST) carried out by radiodiagnosis. The prognostic indicator variables to predict the benefit to ICIs were progression-free survival time (PFS) and overall survival (OS), defined as the time elapsed from the start of treatment to the date of disease progression or death, and to death, respectively. For those patients who did not experience progression or death, the last recorded assessment was considered the end of the study (1 October 2021).

The treatment of personal data corresponded to the Biobank of the Aragon Health System (integrated into the Spanish National Biobanks Network (PT20/00112)) after informed consent had been given and signed by the patients included in the study. The Clinical Research Ethics Committee of Aragón (CEICA) approved and evaluated the study with code (CI PI19/052). All research was performed in accordance with relevant guidelines/regulations and with the Declaration of Helsinki.

2.2. Sample Type and Processing

Plasma from patients was collected at baseline before starting any treatment with immunotherapy alone or immunotherapy combined with chemotherapy. Patient samples were incorporated as a final destination in the Biobank of the Aragon Health System. The analyses were performed at the Instituto de Investigación Sanitaria de Aragón (IIS Aragón) with the technological infrastructure available at the Service of Functional Genomics (SAI, Unizar/IACS) at the Biomedical Research Center of Aragon (CIBA).

2.2.1. Sample Processing

The 55 peripheral blood samples were collected in tubes containing ethylenediaminetetraacetic acid (EDTA) to prevent coagulation and centrifuged at room temperature for 10 min at 2600 revolutions per minute (rpm) to separate the cells from the plasma. Subsequently, peripheral blood mononuclear cells (PBMCs) were isolated using a density gradient medium (Ficoll-Paque Plus) and incubated in RNAlater® solution (Invitrogen, Carlsbad, CA, USA) overnight at 4 °C and then stored at −80 °C until processing.

A kit (MagMAX mirVana Total RNA Isolation Kit (Thermo Fisher Scientific, Waltham, MA, USA)) was used for RNA extraction from the lymphocytes isolated. The quantification, quality and measurement of the integrity of the extracted RNA were evaluated using the Qubit fluorometer, and the integrity of the extracted RNA was evaluated using the TapeStation 2200 bioanalyzer. Complementary DNA (cDNA) synthesis was carried out using the SuperScript (IV) VILO cDNA Synthesis kit (Thermo Fisher Scientific).

2.2.2. Sequencing and Analysis of the TCRβ Repertoire

For sequencing, we used the Oncomine TCR Beta-SR Assay system (Thermo Fisher Scientific) targeting the CDR3 region of the TCR B-chain responsible for antigen recognition, allowing us to identify a clonotype of T lymphocytes with the same TCR (same VDJ rearrangement). For each clonotype, the nucleotide sequence of the CDR3 region (CDR3NT), the corresponding amino acid sequence (CDR3AA), and the V (variable) and J (junction) segments that compose it were reported. This platform thus allows the identification of rare or abundant clones and enables TCR convergence profiling that can measure tumor immunogenicity. The productive sequences obtained from each library were used to determine the indices that characterize the repertoire: richness, Shannon diversity, evenness and convergence.

2.2.3. Determination of Cytokines in Peripheral Blood

Luminex® Discovery Assay (R&D Systems a bio-techne brand) was run according to the manufacturer's instructions in plasma, using a human premixed multi-analyte kit cytokine panel. The next soluble factors were included (IL-2, IL-4, IL-10, IL-12, IL-15, MICA, MICB, ULBP1, ULBP2, ULBP4, IFN-γ and CXCL10). Briefly, supernatants were mixed with beads coated with capture antibodies, incubated, washed and incubated with biotin-labeled detection antibodies, followed by a final incubation with streptavidin-PE. Assay plates were measured using a Luminex 200 instrument (ThermoFisher, catalog no. APX10031). Data acquisition and analysis were performed using 57 xPONENT software. The standard curve for each analyte had a five-parameter R2 value > 0.95 with or without minor fitting using xPONENT software. These determinations were carried out in collaboration with the National Centre of Oncological Investigation (CNIO) in Madrid.

2.3. Statistical Analysis

The population included in the study was 50 patients plus an additional 10% (5 patients) to account for possible losses, resulting in a total of 55 patients. Qualitative variables were expressed as percentages, and quantitative variables as median and standard deviation. The Kolmogorov–Smirnov test was used to test the normal distribution of a variable. The T-Student U or Mann–Whitney U test was used for independent samples according to whether the variable followed the normal distribution or not, respectively. Spearman's correlation analysis was used to examine the association between two quantitative variables that did not follow the normal distribution (and Pearson's correlation analysis was used if they did). For the survival analysis (OS and PFS), the nonparametric Kaplan–Meier estimator and the Mantel–Cox test were used to determine statistical significance in the comparative analysis. The Cox proportional hazard model was used on the variables detected as significant with Kaplan–Meier to determine the Hazard ratio (HR) and 95% confidence interval (95%CI).

All statistical analyses were performed using Statistical Package for the Social Sciences (SPSS) software version 24.0. Two-sided p-values < 0.05 were considered statistically significant.

3. Results

3.1. Descriptive Analysis

Patient and Tumor Disease Characteristics

The cohort of patients in our study was the same as that included in another recently published study, in which other different variables (the frequency of peripheral blood T and NK cell subsets) were analyzed [18]. A total of 55 patients were included in the study with a mean age of 65.02 years, 70.9% were male, and 98.2% were Caucasian. The vast majority (96.4%) were smokers or ex-smokers, with an Eastern Cooperative Oncology Group (ECOG) of 0 (65.5%) and no concomitant chronic infections (90.9%). Of the patients, 60% had lung adenocarcinoma, and 40% had squamous cell lung cancer. A total of 70.9% of the patients were classified with stage IV lung cancer, and only 29.1% had stage III.

PDL1 determination was performed in 47 patient tumor samples, 34% had high PDL1 expression ≥50%, 44.7% had PDL1 expression from 1 to 49% and 21.3% had an expression of PDL1 < 1%. A total of 63.6% had a baseline blood lactate dehydrogenase (LDH) level (≤214 U/L). An intermediate prognostic index LIPI score calculated by LDH and derived neutrophil/lymphocyte ratio (derived neutrophil/lymphocyte ratio ≥ 3 or LDH ≥ ULN) was present in 49.1% or poor (neutrophil/lymphocyte ratio ≥ 3 + LDH ≥ ULN) in another 49.1%.

The main indication for treatment with ICIs was being palliative in successive lines (41.8%) followed by being palliative first line (32.75%) and locally advanced (25.5%). Pembrolizumab (38.2%) was the ICI most frequently administered followed by Atezolizumab (32.7%), Durvalumab (25.5%) and Nivolumab (3.6%).

Regarding response to ICIs, 18.2% [10] experienced complete response (CR), 23.6% [13] experienced partial response (PR) and 21.8% [12] experienced disease stabilization (DS). The average time to response to ICIs was 2.74 months (95% CI 1.85–3.63), with a response duration of 8.06 months (95% CI 4.54–11.58). Despite the initial response, 56.4% (31 patients) of patients experienced disease progression.

A total of 45.5% [19] of patients presented immune-mediated toxicity: cutaneous (10.9%) and pneumonitis (10.9%) followed by endocrine (9.1%), musculoskeletal (9.1%), renal (7.3%), hepatic (5.5%) and colitis (3.6%). Immune-mediated adverse events occurred early, within the first 3 months, for endocrine, neurologic, skin and cardiovascular toxicities and late (>3 months) for the rest. Most toxicities were grade 1 and 2, and there was only one case of liver toxicity that was grade 4 (Table 1).

Table 1. Overview of the patient cohort comprised in the study that includes 55 LC patients. Demographic parameters such as sex, age, gender, race, smoking habits, PD-L1, LDH, LIPI score, histology, tumor stage, treatment indication, ICI, ICI response, death, immune-mediated toxicity and type of immune-mediated toxicity are included.

Variable	Total: 55 Patients (N)	(%)
Sex		
Males	39	70.9
Female	16	29.1
Age		
Mean age: 65.02		
<5	47	85
≥75	8	15
ECOG		
ECOG 0	36	65.5
ECOG 1	19	34.5
IMC		
<30 kg/m^2	46	83.6
≥30 kg/m^2	9	16.4
Race		
Caucasian	54	98.2
Others	1	1.8
Smoking Habit		
Never smoker	2	3.60
Former smoker/Current smoker	53	96.40
PD-L1		
<1%	10	18.20
1–49%	21	38.20
≥50%	16	29.10
Unknown	8	14.50
LDH		
Normal (≤214 U/L)	35	63.60
High (>214 U/L)	20	36.40
LIPI Score		
Poor	27	49.10
Intermediate	27	49.10
Good	1	1.80
Histology		
Adenocarcinoma	22	40
Squamous	33	60
Tumor Stage		
III	16	29.1
IV	39	70.9
Treatment Indication		
Locally advanced	14	25.50
First line	18	32.70
Second line or more	23	41.80
ICI		
Durvalumab	14	25.50
Pembrolizumab	21	38.20
Atezolizumab	18	32.70
Nivolumab	2	3.60
ICI Response		
Complete response (CR)	10	18.80
Partial response (PR)	13	23.60
Stable disease (SD)	12	21.80
Progressive disease (PD)	15	27.30
Not evaluable (NE)	5	

Table 1. *Cont.*

Variable	Total: 55 Patients (N)	(%)
Death		
Yes	32	58.20
No	23	41.80
Immune-Mediated Toxicity		
Yes	25	45.50
No	30	54.50
Immune-Mediated Toxicity		
Skin	6	24
Pneumonitis	6	24
Endocrine	5	20
Musculoskeletal	5	20
Renal	4	16
Liver	3	12
Colitis	2	8

ECOG: Eastern Cooperative Oncology Group, LDH: lactate dehydrogenase, LIPI: Lung Immune Prognostic Index.

3.2. Survival Analysis

3.2.1. Clinical Pathological Features

The average PFS to ICI therapy was 6.42 months (95% CI 3.97–8.87). Thirty-two patients (58.2%) had died at the time of analysis, and the time to death had a mean of 8.38 months (95% CI 5.61–11.14). The OS median was 19 months. No statistically significant differences were found in OS according to age, sex, race, smoking, BMI, latent infections, histology, immunoreactive toxicity, degree of toxicity, or PDL1 expression level. Statistically significant differences were found in OS according to ECOG (ECOG0, ECOG1), disease stage, treatment indication, the type of ICIs, the administration of corticosteroids, LIPI score, and LDH levels (Table 2).

Regarding PFS, a statistically significant association was observed between PFS and immunorelated toxicity and tumor stage (Table 2).

Table 2. Analysis of the influence of each variable on patient survival (OS and PFS). Statistical analysis was conducted to assess the influence of each variable (sex, age, ECOG, BMI, race, smoking habit, PD-L1 expression, LDH, LIPI score, histology, tumor stage, treatment indication, ICI, ICI response, immune-mediated toxicity, type of immune-mediated toxicity) on patient survival (OS and PFS). Nonparametric Kaplan–Meier estimators and the Mantel–Cox test were utilized to determine statistical significance in the comparative analysis.

	Overall Survival	Progression-Free Survival
	IC (11.13–26.87)	IC (2.81–17.19)
	p value	*p* value
Sex Males Female	0.396	0.646
Age Mean age: 65.02 <75 ≥75	0.065	0.170
ECOG ECOG 0 ECOG 1	0.000	0.643
IMC <30 kg/m^2 ≥30 kg/m^2	0.695	0.889

Table 2. *Cont.*

	Overall Survival	Progression-Free Survival
	IC (11.13–26.87)	IC (2.81–17.19)
	p value	*p* value
Race Caucasian Others	0.196	0.113
Smoking Habit Never smoker Former smoker/Current smoker	0.165	0.528
PD-L1 <1% 1–49% ≥50% Unknown	0.194	0.389
LDH Normal (≤214 U/L) High (>214 U/L)	0.017	0.086
LIPI Score Poor Intermediate Good	0.000	0.005
Histology Adenocarcinoma Squamous	0.487	0.713
Tumor Stage III IV	0.000	0.034
Treatment Indication Locally advanced First line Second line or more	0.000	0.076
ICI Durvalumab Pembrolizumab Atezolizumab Nivolumab	0.000	0.354
ICI Response Complete response (CR) Partial response (PR) Stable disease (SD) Progressive disease (PD) Not evaluable (NE)	0.000	0.000
Immune-Mediated Toxicity Yes No	0.051	0.030
Immune-Mediated Toxicity Skin Pneumonitis Endocrine Musculoskeletal Renal Liver Colitis	0.588	0.697

ECOG: Eastern Cooperative Oncology Group, LDH: lactate dehydrogenase, LIPI: Lung Immune Prognostic Index.

3.2.2. TCRβ Repertoire

TCRβ Sequencing Data Analysis

TCR analysis was performed on cell populations from peripheral blood samples, before starting treatment with immunotherapy. TCRβ analysis and CDR3 sequencing were finally performed on 44 samples. A total of 11 samples could not be used due to insufficient quantity or low RNA quality. The mean RNA concentration was 51.8 ng/µL, and their integrity (RIN values) had a mean of 5.2. The finally sequenced libraries had an average depth of 878,956.64 reads with an average read length of 88 base pairs per library. From the total number of libraries, three were discarded due to low quality, and the rest had 50% or more productive reads, indicating that the sequencing of all samples was satisfactory and sequencing analysis could continue.

The productive sequences obtained from each library were used to determine the indices that characterize the TCRβ repertoire, as explained in the methodology. The mean of TCRβ evenness was 0.77 ± 0.14 (median: 0.81 ± 0.14), the Shannon diversity index was 10.7 ± 2.56 (median: 10.9 ± 2.56) and the TCRβ convergence pre-treatment was 0.01 ± 0.01 (median: 0.007 ± 0.01). The median of the different TCRβ variables was chosen to divide the cohort into two groups. The cohort was divided into two groups for each variable according to the median to analyze the impact of TCRβ repertoire characteristics on OS and PFS (Figure 1).

No statistically significant differences were observed in OS or PFS according to TCRβ clonality (Figure 1A) and TCRβ evenness (Figure 1B). However, there was a tendency to increase the OS and PFS with higher clonality and, therefore, a better ICI response. Regarding TCRβ Shannon diversity, those patients with a survival ≥24 months had higher levels of TCRβ Shannon diversity with a trend to statistical significance ($p = 0.089$) (Figure 1C). Finally, when analyzing TCRβ convergence, there were no differences in OS ($p = 0.096$) nor in PFS between the TCRβ convergence groups, although there was a trend that at a lower convergence, OS was higher (Figure 1D). When comparing TCRβ convergence between patients with PFS < or ≥ 12 months, statistically significant differences were observed with lower convergence in patients with PFS ≥ 12 months (Figure 1E).

3.2.3. Cytokines and Other Soluble Factors

Analysis of the Determination of Cytokines

The levels of 10 cytokines in 54 pre-treatment peripheral blood patient samples were analyzed (Table 3).

Table 3. Analysis of the determination of cytokines.

Cytokine	Median	Mean	Standard Deviation	Interquartile Range
MICA	119.5	128.0	49.5	44.9
MICB	117.4	126.7	42.4	49.6
ULBP1	21.0	42.7	96.7	16.1
ULBP2	157.5	159.5	22.9	27.9
CXCL10	16.4	19.0	11.0	11.1
IL10	2.8	6.7	9.2	9.9
ULBP4	0.0	27.6	81.4	20.6
IFNγ	25.5	26.5	8.0	5.1
IL4	0.0	18.0	23.8	37.3
IL2	26.1	26.7	3.5	2.9
IL15	6.7	7.0	3.8	2.8
IL12	0.0	137.7	412.2	153.9

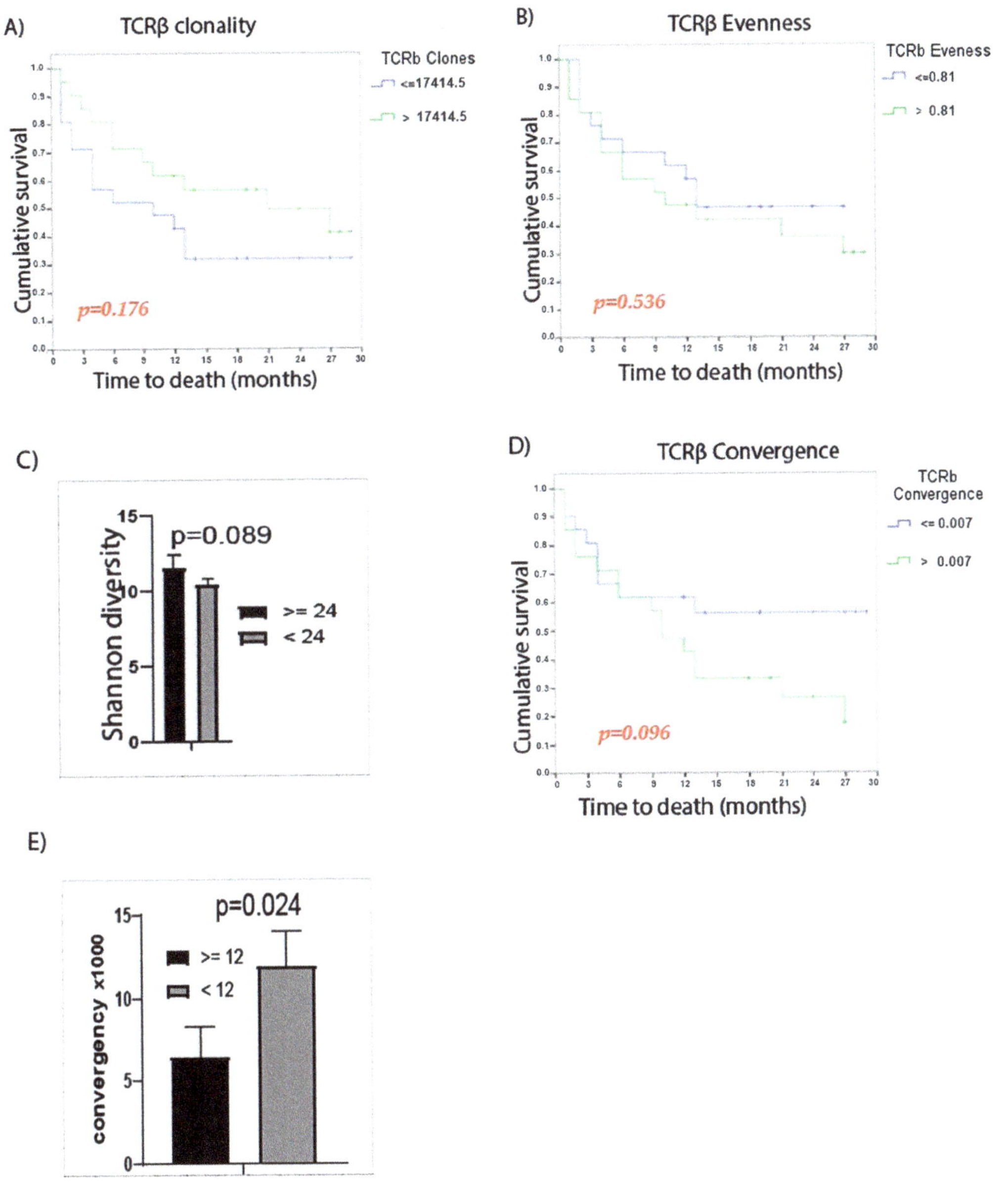

Figure 1. Characteristics of the TCRβ repertoire and its relationship with OS or PFS. (**A**) Kaplan–Meier curves for OS of patients with TCRβ clonality that was high vs. low ($p = 0.176$), (**B**) Kaplan–Meier curves for OS of patients with TCRβ evenness that was low vs. high ($p = 0.536$), (**C**) TCRβ Shannon diversity between patients with OS < or ≥24 months, (**D**) Kaplan–Meier curves for OS of patients with TCRβ convergence that was high vs. low ($p = 0.096$), (**E**) TCRβ convergence between patients with PFS < or ≥12 months. Mantel–Cox test was used to determine statistical significance.

An analysis was performed between ICI responders and non-responders, selecting only those cytokines or soluble factors that were statistically significant or close to significance such as MICB, CXCL10, IFNγ and IL15 (Table 4). Responders showed lower MICB values before initiating therapy, reaching statistical significance ($p = 0.037$) (Figure 2).

Table 4. Analysis of cytokines between ICI responders and non-responders.

Cytokines	Response to ICI	N	Mean	*p* Value
MICA	No	20	133.7	0.516
	Yes	34	124.6	
MICB	No	20	140.2	0.037
	Yes	34	118.8	
CXCL10	No	20	21.6	0.094
	Yes	34	17.5	
IFNγ	No	20	28.4	0.092
	Yes	34	25.4	
ULBP1	No	20	52.1	0.589
	Yes	34	37.2	
ULBP2	No	20	162.8	0.206
	Yes	34	157.5	
IL10	No	20	4.0	0.162
	Yes	34	5.6	
ULBP4	No	20	9.2	0.302
	Yes	34	7.1	
IL4	No	20	19.6	0.356
	Yes	34	17.0	
IL2	No	20	26.3	0.319
	Yes	34	26.0	
IL15	No	20	7.4	0.061
	Yes	34	6.2	
IL12	No	20	204.6	0.183
	Yes	34	98.4	

Patients with no disease progression also had significantly lower levels of MICB ($p = 0.049$), ULBP1 ($p = 0.041$) and IL2 ($p = 0.043$). Likewise, those who had not died at the end of the analysis, showed lower levels of MICB ($p = 0.029$) and IL-2 ($p = 0.04$), and those who survived 12 months or more had statistically significantly lower levels of MICB ($p = 0.024$) and ULBP1 ($p = 0.044$) (Figure 2).

A second analysis was carried out. The cohort was divided for each soluble factor using the median to analyze the impact of each one on OS and PFS. It was observed that those patients with levels of IL-2 > 26.1 presented shorter OS ($p = 0.037$) and shorter PFS ($p = 0.009$). Similarly, those patients with levels of IL-15 > 6.7 had shorter OS ($p = 0.033$) and PFS ($p = 0.050$). Significant differences were observed in PFS between groups according to ULBP1 levels ($p = 0.002$), with PFS being lower if the levels of ULBP1 > 21.0. Although not statistically significant ($p = 0.058$), it was observed that those patients with levels of IL-10 > 2.8 tended to have higher OS (Figure 3). Regarding the rest of the cytokines: IL-4, IL-12, MICA, MICB, ULBP2, ULBP4, IFN-γ and CXCL10 no statistically significant differences were observed between expression level and OS or PFS (Figure 3).

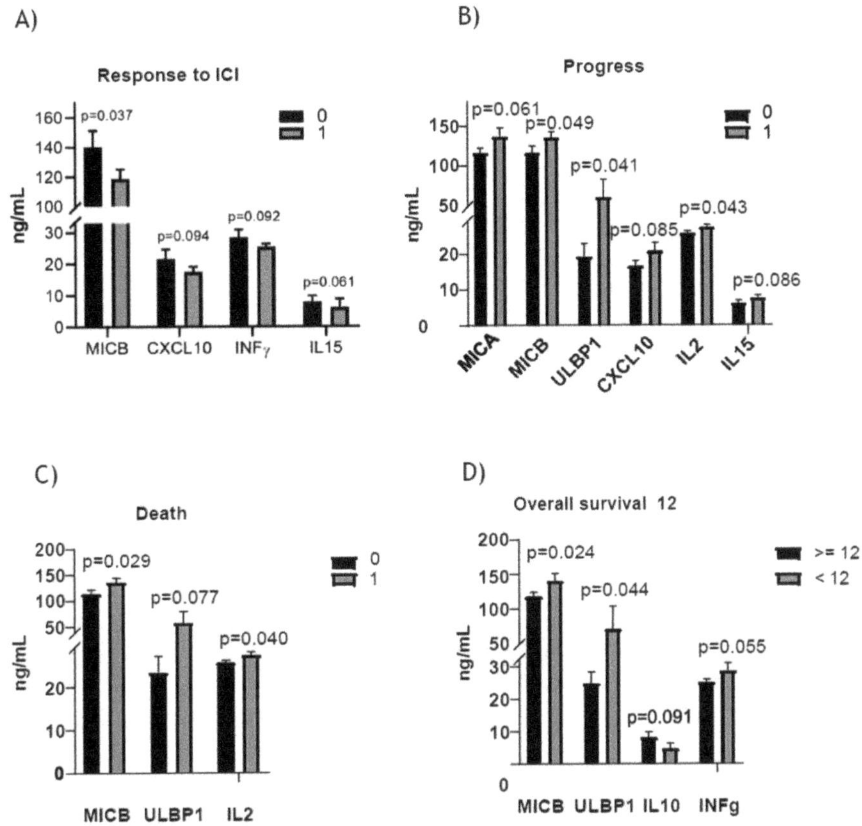

Figure 2. Analysis of expression of cytokines. (**A**) Levels of cytokines between ICI responders (=1) and non-responders (=0), (**B**) Levels of cytokines between patients with disease in progression (=1) and no progression (=0), (**C**) Levels of cytokines between dead patients (=1) and not dead patients (=0), and (**D**) Levels of cytokines in those with overall survival ≥12 months.

3.3. Multivariate Analysis

In order to develop an analysis for identifying independent predictive biomarkers, we performed a multivariate analysis. We studied the relationship between variables with *p*-values < 0.05 in the univariate analysis and OS. For this purpose, the Cox regression model was used. The variables included were ECOG, tumor stage, the indication for treatment, the type of ICI, the best response, LDH, IL-2 and IL-15. The variables of immune-related toxicity and IL-10 were also added, as they were close to being significant. The coefficient estimate, standard error, significance, hazard ratio and confidence interval are shown in Supplementary Table S1 as the hazard ratio (log HR) confidence interval for each variable (Supplementary Table S1).

The multivariate COX analyses (Table 5) showed the simultaneous analysis of more than one response variable according to the Cox analysis. Some variables appeared to be associated with OS. For the same reason, these variables validate our statistical model and the use of IL-15 as a biomarker to predict the ICI therapy response. As can be seen, patients with high IL-15 levels in plasma had a shorter OS, indicating that IL-15 is an independent prognostic factor. Elevated IL-15 levels in plasma increased the risk of death by three times (HR = 3, 95% CI: 1.368–6.578, $p < 0.006$).

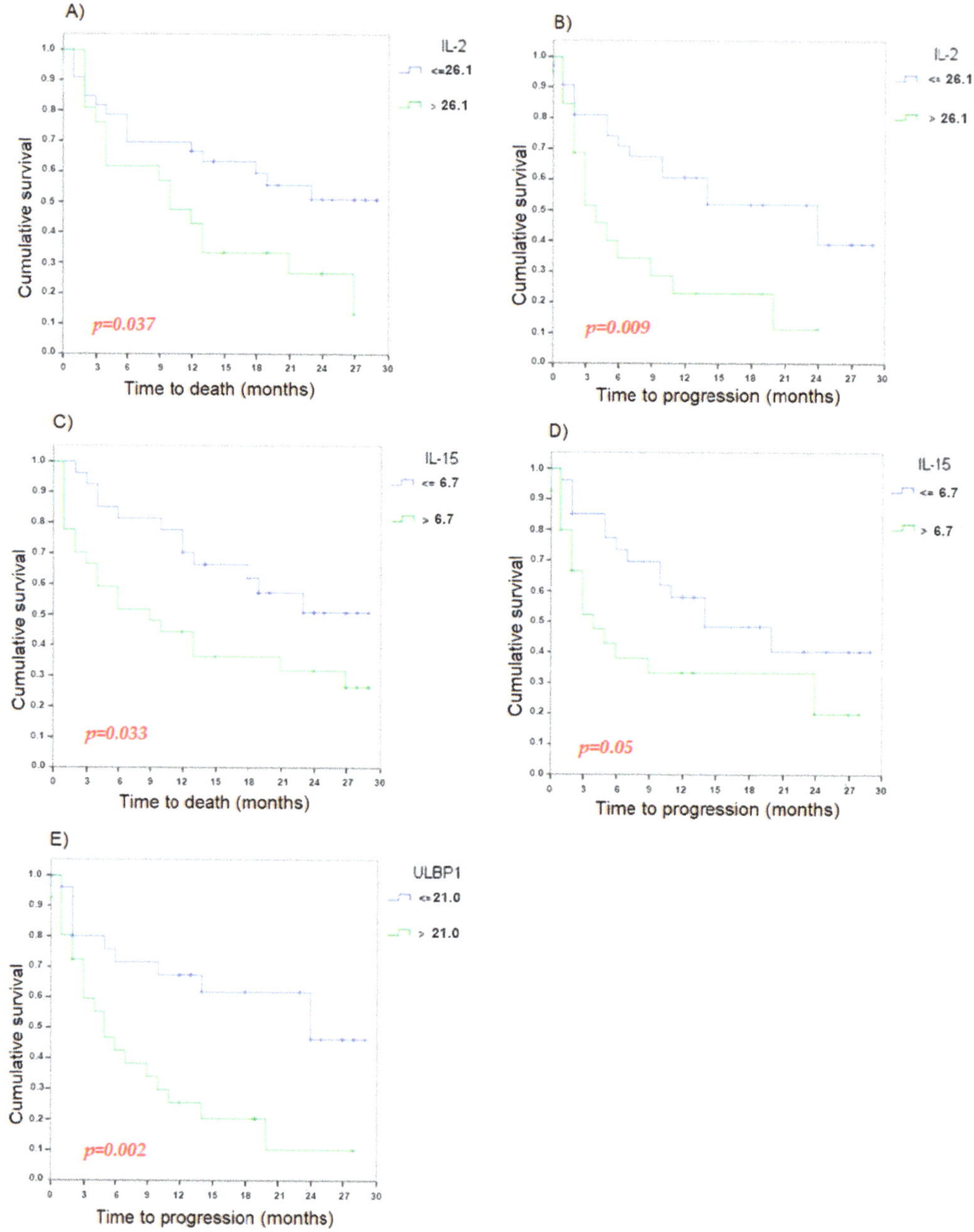

Figure 3. Characteristics of cytokines and their relationships with overall survival and progression-free survival. (**A**) Kaplan–Meier curves for OS of patients with levels of IL-2 >26.1 vs. levels of IL-2 ≤26.1 (p = 0.037), (**B**) Kaplan–Meier curves for PFS of patients with levels of IL-2 > 26.1 vs. levels of IL-2 ≤ 26.1 (p = 0.009), (**C**) Kaplan–Meier curves for OS of patients with levels of IL-15 > 6.7 vs. levels of IL-15 ≤ 6.7 (p = 0.033), (**D**) Kaplan–Meier curves for PFS of patients with levels of IL-15 > 6.7 vs. levels of IL-15 ≤ 6.7 (p = 0.050), and (**E**) Kaplan–Meier curves for PFS of patients with levels of ULBP1 >> 21.0 vs. levels of ULBP ≤ 21.0 (p = 0.002).

Table 5. Multivariate analysis.

Covariable	Coefficient Estimate (B_i)	SD Estimation	Sig.	Exp (B) (HR.)	IC (HR) 95%
TI Locally advanced	-	-	0.006	-	-
TI First line	2.363	0.901	0.009	10.627	1.816–62.181
TI Second line or more	2.708	0.852	0.001	14.998	2.823–79.679
ECOG [1]	0.915	0.393	0.020	2.496	1.155–5.391
Staging (IV)	2.295	1.081	0.034	9.929	1.193–82.625
IL-15 (>6.7)	1.098	0.401	0.006	3.000	1.368–6.578
Atezolizumab	-	-	0.008	-	-
Nivolumab	−1.615	1.082	0.135	0.199	0.024–1.657
Pembrolizumab	−1.457	0.444	0.001	0.233	0.098–0.557
Durvalumab	−0.790	1.277	0.536	0.454	0.037–5.540

In the same multivariate model, other variables showed a statistical correlation. Patients with an indication for palliative first-line treatment had a 10.63 times higher risk of death than those with locally advanced treatment. On the other hand, patients with palliative successive lines of treatment had a 15-fold increased risk of death. The ICI response variable showed great variability (possibly due to the scarcity of data within each category). Patients with PD increased the risk of death 53.256 times over CR patients. ECOG1 patients had 2.5 times the risk of death over ECOG0, in any time unit and adjusted for all other confounding variables. Those patients with stage IV tumor disease had a 10-fold higher risk of death than stage III patients.

In previous studies, all these variables have already been demonstrated to affect OS and are used in clinical practice. Hence, they validate our model and the use of IL-15 as a biomarker to predict the ICI therapy response. The following graphs show the survival functions estimated with the proposed model for the different levels of the influential categorical factors (Figure 4).

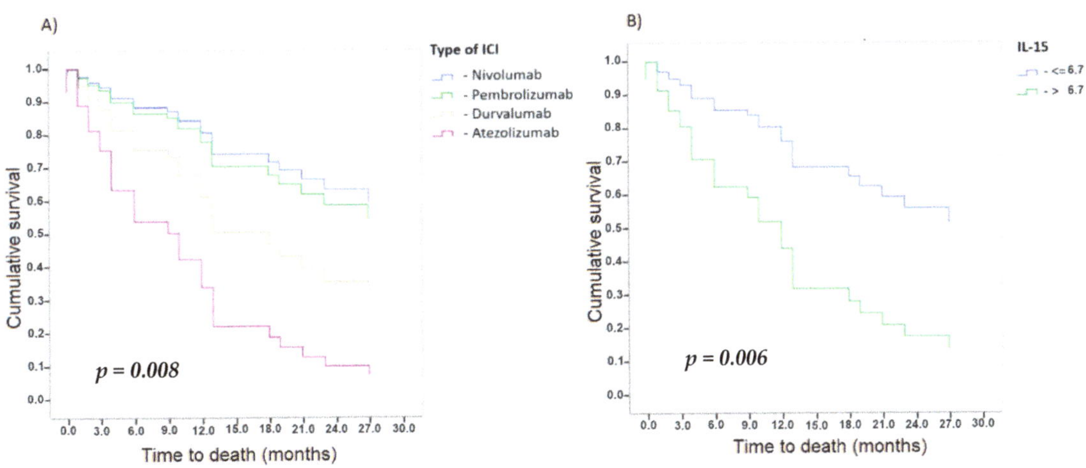

Figure 4. Kaplan–Meier curves of survival functions for the different levels of the influential categorical factors. (**A**) OS according to type of ICI, (**B**) OS according to IL-15.

4. Discussion

Although immunotherapy has revolutionized lung cancer management, we still need to identify markers to help us better select patients who will benefit from it, thus avoiding toxic treatments and a lack of therapeutic benefit in non-responders.

With this objective, we designed a study that included a cohort of patients with advanced and locally advanced lung cancer in which the characteristics of the TCRβ repertoire in mononuclear cells and certain soluble factors were analyzed in peripheral blood samples before the initiation of ICIs in monotherapy or in combination with chemotherapy. Although our study's sample size is limited, the cohort included is representative of the lung cancer population, which may help us draw certain conclusions in this regard.

There is evidence that the analysis of the TCRβ profile provides predictive information on responses to ICIs [8]. However, the evidence in this regard is quite variable, possibly due to small sample sizes, studies performed in different tumor pathologies, the type of immunotherapeutic agent administered and the sequencing methods performed [20]. One of the advantages of using TCRβ as a biomarker is that it can be determined in an easily accessible sample, such as peripheral blood, without resorting to tumor tissue, which is often scarce in lung cancer [21]. Evaluating TCR clonality in both compartments, blood and tissue, can provide valuable insights into the dynamics of the immune response in lung cancer patients. In lung cancer, it has been observed that a more clonal TCR repertoire in tumor tissue is associated with greater T cell infiltration and a better response to immunotherapy. However, the exact relationship between clonality in peripheral blood and tumor tissue may vary, as peripheral blood does not always reflect the clonal diversity present in the tumor microenvironment [19]. Even so, TCRβ has certain limitations regarding its determination, such as the lack of standardization in the cut-off points, the platforms and the cost of the sequencing techniques employed.

In our study, it was observed that there is a trend, although not significant, that the greater the richness or baseline number of different TCRβ clonotypes, the greater the clinical benefit, achieving an increase in OS. Previous studies have shown that treatment with ICIs can have a pharmacodynamic effect by increasing the number of unique TCR clonotypes in peripheral blood and, consequently, a greater possibility of recognizing tumor neoantigens, thus improving therapeutic response [14,19]. This has been proven by analyzing tumor tissues in responder patients with NSCLC receiving treatment with neoadjuvant ICIs, observing tumor tissue enriched with expanded clonotypes [14]. Thus, it has been observed that the pre-treatment presence of clones in the tumor shows clonal expansion in blood after treatment, correlating with better response and the clearance of circulating tumor DNA [20].

TCRβ diversity refers to the number of clonotypes present. It is to be expected that the greater the diversity of the TCRβ repertoire, the greater the probability that T lymphocytes will recognize tumor antigens and, therefore, have the best antitumor response. Han, J. et al. 2020 and Huang, A.C. et al. 2017 showed that the diversity of the PD-1+ CD8+ TCR population in blood could indicate that there is a greater proportion of exhausted T cells that can be reactivated with ICIs, leading to a more effective immune response in patients with non-small cell lung cancer [22,23]. However, the conclusions reached by previous studies regarding TCRβ diversity are pretty mixed. In fact, in our study, no significant differences in OS or PFS were observed according to the TCRβ Shannon diversity index. However, there was a trend towards better OS after treatment in those with greater TCRβ Shannon diversity. This can be explained by the fact that there are antigen-specific TCRs in the peripheral blood mononuclear cells that are non-tumorigenic and can dilute tumor-specific TCRs [19].

Therefore, our study confirms that a wide repertoire of TCR in blood with greater richness and diversity makes the recognition of tumor antigens more likely and that they are reactivated later with the action of ICIs, thus decreasing the immune escape of tumor cells.

In terms of the convergence analysis, it has been seen that patients with a greater response to ICIs have a greater pre-treatment TCR convergence, thus reinforcing the idea that T cells with convergent TCRs target tumor antigens [16]. An advantage of TCR convergence as a biomarker is that it is able to detect the T cell response to tumor neoantigens beyond those originating from nonsynonymous mutations (point mutations that alter the resulting protein sequences). The evidence in this regard is contradictory to what was observed in our study, in which no significant differences in OS or PFS were observed according to TCRβ convergence, and even a tendency was observed that the lower the TCRβ convergence, the higher the OS and PFS. TCR convergence is a process by which a tumor antigen determined by antigenic specificity leads to the expansion of T cells that share TCRs with antigen specificity. Furthermore, some studies suggest that successful immunotherapy was not reliant on the select expansion of specific T cell clones, but instead induced the relatively uniform expansion of most tumor-infiltrating T cells, enhancing effector capabilities [24]. However, different therapeutic approaches and distinct cancers could likely yield different results.

It has also been shown that the sequencing method or platform chosen for analysis can vary in determining TCRβ [16]. Thus, a study comparing TCRβ sequencing using the Illumina platform with the Oncomine assay was performed and differences were seen between them. Both were consistent in detecting TCR clonality and diversity, but Illumina resulted in a higher detection of convergent TCRs [16]. With prior knowledge of the different substitution error rates in the different sequencing platforms, the most appropriate platform can be chosen accordingly.

In our study, the Oncomine platform was used. Perhaps the sequencing platform employed with a lower detection rate than others, such as Illumina along with the paucity of peripheral blood samples that could be used for TCRβ determination, contributed to these results. Still, further understanding of the mechanisms involved and studies involving larger cohorts are required to consider TCR as a predictive biomarker of responses to ICIs.

It has been shown that several factors in peripheral blood prior to initiating treatment with ICIs, such as the number of activated CD4 memory T cells and a more clonally diverse TCR repertoire, are associated with the development of severe immune-mediated adverse effects and with a greater response to ICIs [25]. In addition, it has been observed that patients receiving treatment with ICIs experience changes in TCR clonality that may be related to the severity of the immune-mediated event and the timing of the event [25]. In fact, according to other studies, patients with NSCLC with greater increases in PD-1+ CD8+ TCR repertoire clonality intra and post-ICI presented greater PFS and OS, reflecting an expansion of a successful anti-tumoral clonotype. On the contrary, pre-treatment TCR repertoire diversity could be a treatment-agnostic prognostic factor [26].

These findings could be of great utility because the modification in the characteristics of the TCRβ repertoire during treatment could serve as a tool to predict and identify which patients are at a higher risk of developing them. In our cohort, it was seen that patients with higher toxicity had higher values of TCRβ convergence and as mentioned, this has been found to be associated with higher response to ICIs.

A proinflammatory gene expression profile in pre-treatment samples is associated with a superior pathologic response after treatment with chemotherapy and immunotherapy [27]. Thus, pre-treatment peripheral blood analysis of some cytokines was performed to study whether certain pre-treatment blood soluble factors could predict the response to or benefit of ICIs.

In our cohort, it was observed that higher pre-treatment levels of IL-2 and IL-15 were associated with more aggressive tumor behavior and worse outcomes: patients survived less and had lower PFS. This is consistent with evidence from other studies [28]. In patients with lung cancer, high concentrations of intratumoral IL-15 are associated with a worse prognosis [29]. It appears that intratumoral production and/or circulating sIL-15/IL-15Rα complexes contribute to developing a tumor microenvironment favorable for tumor progression and immune escape [30]. However, there are exceptions to the behavior described above since, in some solid tumors, the IL-15/IL-15R complex may also play an antineoplastic role [29,30]. Therefore, the role of intratumoral and circulating IL-15 is complex and depends on several factors, such as the type of IL-15 produced, the IL-15-Rα chain isoforms involved in sIL-15/IL-15Rα, the presence of functional IL-15 receptors on tumor cells, as well as their response to stromal and endogenous IL-15. Therefore, it is difficult to say whether it predicts aggressive behavior [30]. Although there is no data regarding levels of circulating IL-15 in relation with antiPD1/PDL1 efficacy, a previous study showed that low serum IL-15 levels correlates with better responses to antiCTLA4 treatment in melanoma [31].

Regarding IL-2, this cytokine seems to have a dual effect on the tumor immune microenvironment. On the one hand, ICIs could, by interacting with T cells, increase IL-2 secretion, enhancing the immune response, but recent studies show that IL-2 also induces immunosuppressive activity by promoting Treg proliferation and activation, which inhibits the antitumor response [32]. Therefore, further studies are needed to investigate and clarify the relationship between IL-2 and ICIs.

No statistically significant association was observed in the univariate analysis between IL-10 levels and OS. IL-10 is an immunosuppressive and anti-inflammatory cytokine that regulates the growth and differentiation of different cell types. It is well known that in cancer patients, higher levels of IL-10 in serum correlate inversely with oncologic prognosis [32,33]. Despite this, it has recently been observed that IL-10 may play a role in CD8+ T cell activation and proliferation in cancer and chronic inflammation [32]. In addition, IL-10 and PD-1 play immunosuppressive roles through very different pathways [33], and a dual blockade has synergistic antitumor action [34–36]. Their efficacy and safety as antitumor therapy has been proven in several studies [37]. Considering the heterogeneity of the findings regarding this cytokine, the small size of our sample, and the arbitrary value taken to consider high IL-10 expression (>2.8), more studies and data are needed to clarify the prognostic significance of IL-10 in the treatment of ICIs.

IFN-γ exerts a dual role. On the one hand, it is a potent inducer of the antitumor immune response, but it can also serve as a tumor escape mechanism. Its direction towards one or the other action will depend on tumor specificity, signal intensity and the tumor microenvironment [38]. In our work, we found no association between an IFN-γ expression signature and a response to ICIs; in fact, responders were observed to have lower levels of MICB, CXCL10 and IFN-γ. Several studies have shown that ICIs increase IFN-γ production, contributing to tumor clearance, and it has been demonstrated that resistance to IT could be due to defects in the IFN-γ signaling pathway [38,39]. The IFN-γ and PD-L1 gene signature combination has been associated with a greater therapeutic benefit to IT and could constitute a predictive biomarker of response to ICIs [40–43].

It has been shown in different studies that tumors with higher CXCL10 expression correlate with a better prognosis [44]. On the contrary, in our cohort, it was seen that responder patients had lower pre-treatment levels of CXCL10. This soluble factor is involved in T and NK cell mobilization. Thus, patients with low levels of this factor may have more of a CXCL10 increase after ICI treatment and, therefore, a better response. Deep research is needed to confirm this hypothesis.

The important role of NK cells and NKG2D ligands in cancer immunosurveillance suggests that their presence in serum could serve as a prognostic marker [45]. Their relationship with survival in cancer patients has been studied [46–49]. The ligands of NKG2D are MICA, MICB and six members of the ULBP family [50]. The release of soluble NKG2D ligands represents a form of tumor cell immune evasion strategy since these ligands deregulate NKG2D expression by decreasing NK cell function and T cell activation, and furthermore these soluble ligands compete in receptor binding with ligands expressed on the surface of tumor cells [45]. Therefore, it is to be expected that higher levels are associated with worse prognosis and disease progression, although its relation with ICI response has not been previously determined.

These findings are consistent with our work, in which patients with lower levels of ULBP1, ULBP2, MICB and MICA had better survival outcomes. Higher levels of ULBP1 were significantly associated with lower PFS. Patients with OS $\geq$ 12 months had statistically lower levels of MICB, ULBP1 and ULBP4. Although these ligands may play an essential role as predictors of the evolutionary course of cancer, the complex regulation of NKG2D ligands, their variation, and specificity depending on tumor type will have to be taken into account, and a better future understanding of the effects of these soluble factors on immune cells will be necessary [51].

The difference in cytokine serum level values between responders and non-responders could be appreciated as small. However, serum determination is an indicator of the different levels in the tumor microenvironment (TME), meaning the differences could be higher in the TME. Other publications report similar or minor differences between healthy and patient serum samples and between treated and non-treated patients with statistical differences and biological significance [52]. In our research, serum determinations were performed in all patients before starting treatment, so it is normal not to see large differences. However, these slight variations seem to help predict the response to ICIs. Of course, these findings need to be corroborated by subsequent studies using a bigger cohort of patients, but if so, they would be a valuable tool in clinical practice. Other publications in pediatric autoimmune hepatitis also detected minor differences between groups and showed how IL-2 levels predicted treatment response [53]. The relevance of our research just lies in the relatively few studies that have been published describing the relationship between cytokines and ICI responses.

The differences (or trends) in PFS and OS associated with TCR diversity and cytokines might be indicating a prognostic role of these biomarkers but there is not enough evidence to confirm the predictive role of the proposed biomarkers.

The cohort included in the study was heterogeneous, as it included patients with both localized and metastatic lung cancer, who therefore had different tumor burdens. Moreover, it is well established that the efficacy and possibly the underlying molecular mechanisms of immune checkpoint inhibitors (ICIs) in patients receiving immunotherapy after progression on platinum-based chemotherapy is different from untreated patients. To verify that the predictive value of cytokines and TCR diversity as markers of response to immunotherapy is not influenced by tumor burden among patients with localized and advanced lung cancer, we conducted a comparison using a t-test (Supplementary Materials Table S2). As shown, no significant differences were observed between the two groups, indicating that the results are not influenced by tumor burden. Although larger cohorts would be necessary to confirm these results, we suggest that the expression levels of certain cytokines and TCR diversity may have a predictive value for immunotherapy response.

In summary, we found in our study that the high clonality and diversity of the TCR repertoire and low levels of IL-2 and IL-15 are associated with a greater response to immunotherapy. This relationship between lower levels of interleukins and a greater response to immunotherapy treatment may seem paradoxical and contradictory, as both cytokines are known to induce the proliferation and activity of T and NK cells. This could be due to the fact that interleukins IL-2 and IL-15 play a role in the regulation and counter-regulation of the immune response. Very high levels of these cytokines can lead to an overactivation of the immune system, which may result in the induction of tolerance and immunosuppressive mechanisms [54]. On the other hand, the immune response is dynamic, and low levels of IL-2 and IL-15 may indicate a more balanced immune environment that favors an effective antitumor response. However, persistently high levels can result in a chronic inflammatory response, contributing to an immunosuppressive environment [55].

Additionally, it has been observed that ICIs can be more effective in the presence of not very high levels of IL-2 and IL-15, allowing T and NK cells to respond more effectively to the treatment.

Concerning markers predictive of immune toxicity during treatment, in our study, an association was observed between IL-15 and MICB expression and the development or not of immune-mediated toxicity; thus, those who did not present toxicity had higher levels of IL-15 and MICB and lower survival. Therefore, although more evidence is needed, it would be interesting to validate these findings in future studies and to observe the role of these soluble factors as predictors of immune toxicity and thus, of responses to ICIs.

5. Limitations

The small sample size is the main limitation of our study, which probably prevented us from reaching statistical significance for several of the variables studied. Another limitation is the heterogeneity of the cohort, related to the inclusion of patients with localized NSCLC and metastatic patients, which affects intrinsic characteristics of response to immunotherapy treatment.

TCRβ is a dynamic marker that can be modified throughout the disease course of a cancer patient being treated with IT. In the same way, ICIs can modify the basal characteristics of that TCRβ repertoire. In our study, the TCRβ analysis was only performed before the start of treatment. Therefore, although it may help us predict which patients may benefit most from treatment, it would have been interesting to analyze how the TCRβ repertoire is modified. Another limitation of our study is that we did not study the characteristics of the TCRβ repertoire according to the patient's demographic characteristics and smoking habits and that these models will require validation in larger independent cohorts of LC patients.

6. Conclusions

Characteristics of the TCR repertoire and cytokines such as IL-2 and IL-15 constitute promising molecular markers of response to immunotherapy treatment. In addition, IL-15 appears to be involved in immune-mediated toxicity. However, future studies are needed to consolidate our results in order to apply them in clinical practice.

Supplementary Materials: The following supporting information can be downloaded at: https://www.mdpi.com/article/10.3390/cancers16162798/s1, Supplementary Table S1. Individual coefficients of the variables included in the Cox regression. Supplementary Table S2. Comparison of Variables Between Localized and Advanced NSCLC Patients. To ensure that the observed differences in various variables are not attributable to tumor burden, patients with localized Non-Small Cell Lung Cancer (NSCLC) were compared to those with advanced NSCLC for each variable. As demonstrated, no statistically significant differences were observed between the two patient groups. Statistical analysis was performed using an independent samples *t*-test.

Author Contributions: Conceptualization and methodology, A.R.-L. and J.P.; investigation, A.R.-L., L.M.-L., M.P.D., P.E. and E.M.G.; data curation, A.S., D.I., M.G.-R., R.L., R.S.P., J.R.P.-P., M.Z.-G., E.Q., A.M., I.T.-R. and A.Y.; writing, A.S.; writing—review and editing, A.S., A.R.-L. and J.P. All authors have read and agreed to the published version of the manuscript.

Funding: Work in the JP lab is funded by FEDER (Fondo Europeo de Desarrollo Regional, Gobierno de Aragón, Group B29_20R), Grant PID2020-113963RBI00 by MICIN/AEI, CIBER—Consorcio Centro de Investigación Biomédica en Red—(CIBERINFEC, CB21/13/00087), Grant PTA2019-016739-I funded by MCIN/AEI/10.13039/501100011033 to PD, Instituto de Salud Carlos III, Aspanoa and Carrera de la mujer de Monzón. Grant PID2020-113963RBI00 by MICIN/AEI to EMGB. Contrato Ramón y Cajal RYC2022-036627-I (AR-L). Bristol Myers Squibb. Roche Farma S.A. Funders did not make any decisions about the study or the results to post.

Institutional Review Board Statement: The study was conducted in accordance with the Declaration of Helsinki, and approved by the Clinical Research Ethics Committee of Aragon (CEICA) with code (CI PI19/052) on 27 February 2019.

Informed Consent Statement: The treatment of personal data corresponded to the Biobank of the Aragon Health System (integrated into the Spanish National Biobanks Network (PT20/00112)). Written informed consent was given and signed by the patients included in the study. The Clinical Research Ethics Committee of Aragón (CEICA) approved and evaluated the study with code (CI PI19/052).

Data Availability Statement: The data presented in this study are available on request from the corresponding author.

Acknowledgments: To the Biobank of the Aragon Health System integrated in the Spanish National Biobanks Network (PT20/00112) and the patients included in the study for their collaboration.

Conflicts of Interest: The authors declare no conflicts of interest.

References

1. Antón, A. (Ed.) *Inmunología Tumoral e Inmunoterapia del Cáncer*; Amazing Books: Zaragoza, Spain, 2018.
2. Lim, S.W.; Ahn, M.J. Current status of immune checkpoint inhibitors in treatment of non-small cell lung cancer. *Korean J. Intern. Med.* **2019**, *34*, 50–59. [CrossRef]
3. Sánchez de Cos Escuín, J. Nueva inmunoterapia y cáncer de pulmón. *Arch. Bronconeumol.* **2017**, *53*, 682–687. [CrossRef] [PubMed]
4. Chen, D.S.; Mellman, I. Elements of cancer immunity and the cancer–immune set point. *Nature* **2017**, *541*, 321–330. [CrossRef]
5. Bai, J.; Gao, Z.; Li, X.; Dong, L.; Han, W.; Nie, J. Regulation of PD-1/PD-L1 pathway and resistance to PD-1/PD-L1 blockade. *Oncotarget* **2017**, *8*, 110693. [CrossRef]
6. Sharma, P.; Hu-Lieskovan, S.; Wargo, J.A.; Ribas, A. Primary, Adaptive, and Acquired Resistance to Cancer Immunotherapy. *Cell* **2017**, *168*, 707–723. [CrossRef] [PubMed]
7. Lesterhuis, W.J.; Bosco, A.; Millward, M.J.; Small, M.; Nowak, A.K.; Lake, R.A. Dynamic versus static biomarkers in cancer immune checkpoint blockade: Unravelling complexity. *Nat. Rev. Drug Discov.* **2017**, *16*, 264–272. [CrossRef]
8. Aversa, I.; Malanga, D.; Fiume, G.; Palmieri, C. Molecular T-cell repertoire analysis as source of prognostic and predictive biomarkers for checkpoint blockade immunotherapy. *Int. J. Mol. Sci.* **2020**, *21*, 2378. [CrossRef]
9. McNeel, D.G. TCR diversity—A universal cancer immunotherapy biomarker? *J. Immunother. Cancer* **2016**, *4*, 69. [CrossRef] [PubMed]
10. Cha, E.; Klinger, M.; Hou, Y.; Cummings, C.; Ribas, A.; Faham, M.; Fong, L. Improved survival with T cell clonotype stability after anti-CTLA-4 treatment in cancer patients. *Sci. Transl. Med.* **2014**, *6*, 238ra70. [CrossRef]
11. Lichtman, A. *Inmunología Celular y Molecular*, 6th ed.; Journal of Chemical Information and Modeling; Elsevier Saunders: Amsterdam, The Netherlands, 2015; Volume 53, pp. 1–532.
12. Quagliata, L.; Looney, T.; Storkus, W.; Taylor, J.; Topacio-Hall, D.; Lowman, G. T cell repertoire sequencing reveals dynamics of response to dendritic cell vaccine plus dasatinib for checkpoint blockade resistant metastatic melanoma. *Ann. Oncol.* **2019**, *30*, v496–v497. [CrossRef]
13. Dong, N.; Moreno-Manuel, A.; Calabuig-Fariñas, S.; Gallach, S.; Zhang, F.; Blasco, A.; Aparisi, F.; Meri-Abad, M.; Guijarro, R.; Sirera, R.; et al. Characterization of circulating t cell receptor repertoire provides information about clinical outcome after pd-1 blockade in advanced non-small cell lung cancer patients. *Cancers* **2021**, *13*, 2950. [CrossRef]
14. Robert, L.; Tsoi, J.; Wang, X.; Emerson, R.; Homet, B.; Chodon, T.; Mok, S.; Huang, R.R.; Cochran, A.J.; Comin-Anduix, B.; et al. CTLA4 blockade broadens the peripheral T-cell receptor repertoire. *Clin. Cancer Res.* **2014**, *20*, 2424–2432. [CrossRef]
15. Sesma, A.; Pardo, J.; Cruellas, M.; Gálvez, E.M.; Gascón, M.; Isla, D.; Martínez-Lostao, L.; Ocáriz, M.; Paño, J.R.; Quílez, E.; et al. From tumor mutational burden to blood T cell receptor: Looking for the best predictive biomarker in lung cancer treated with immunotherapy. *Cancers* **2020**, *12*, 2974. [CrossRef]
16. Looney, T.J.; Topacio-Hall, D.; Lowman, G.; Conroy, J.; Morrison, C.; Oh, D.; Fong, L.; Zhang, L. TCR Convergence in Individuals Treated with Immune Checkpoint Inhibition for Cancer. *Front. Immunol.* **2020**, *10*, 2985. [CrossRef]
17. Zhang, L.; Looney, T.; Lowman, G.; Oh, D.; Fong, L. Peripheral blood TCRB repertoire convergence and clonal expansion predict response to anti-CTLA-4 monotherapy for cancer. *Res. Sq.* 2024, *in press*.
18. Gascón-Ruiz, M.; Ramírez-Labrada, A.; Lastra, R.; Martínez-Lostao, L.; Paño-Pardo, J.R.; Sesma, A.; Zapata-García, M.; Moratiel, A.; Quílez, E.; Torres-Ramón, I.; et al. A Subset of PD-1-Expressing CD56bright NK Cells Identifies Patients with Good Response to Immune Checkpoint Inhibitors in Lung Cancer. *Cancers* **2023**, *15*, 329. [CrossRef]
19. Han, J.; Duan, J.; Bai, H.; Wang, Y.; Wan, R.; Wang, X.; Chen, S.; Tian, Y.; Wang, D.; Fei, K.; et al. TCR repertoire diversity of peripheral PD-1þCD8þ T cells predicts clinical outcomes after immunotherapy in patients with non–small cell lung cancer. *Cancer Immunol. Res.* **2020**, *8*, 146–154. [CrossRef]
20. Anagnostou, V.; Forde, P.M.; White, J.R.; Niknafs, N.; Hruban, C.; Naidoo, J.; Marrone, K.; Sivakumar, I.K.A.; Bruhm, D.C.; Rosner, S.; et al. Dynamics of Tumor and Immune Responses during Immune Checkpoint Blockade in Non-Small Cell Lung Cancer. *HHS Public Access* **2019**, *79*, 1214–1225. [CrossRef]
21. Binnewies, M.; Roberts, E.W.; Kersten, K.; Chan, V.; Fearon, D.F.; Merad, M.; Coussens, L.M.; Gabrilovich, D.I.; Ostrand-Rosenberg, S.; Hedrick, C.C.; et al. Understanding the tumor immune microenvironment (TIME) for effective therapy. *Nat. Med.* **2018**, *24*, 541–550. [CrossRef]
22. Zhang, J.; Ji, Z.; Caushi, J.X.; El Asmar, M.; Anagnostou, V.; Cottrell, T.R.; Chan, H.Y.; Suri, P.; Guo, H.; Merghoub, T.; et al. Compartmental analysis of T-Cell clonal dynamics as a function of pathologic response to neoadjuvant PD-1 blockade in resectable non-small cell lung cancer. *Clin. Cancer Res.* **2020**, *26*, 1327–1337. [CrossRef]
23. Huang, A.C.; Postow, M.A.; Orlowski, R.J.; Mick, R.; Bengsch, B.; Manne, S.; Xu, W.; Harmon, S.; Giles, J.R.; Wenz, B.; et al. T-cell invigoration to tumour burden ratio associated with anti-PD-1 response. *Nature* **2017**, *545*, 60–65. [CrossRef]
24. Puig-Saus, C.; Sennino, B.; Peng, S.; Wang, C.L.; Pan, Z.; Yuen, B.; Purandare, B.; An, D.; Quach, B.B.; Nguyen, D.; et al. Neoantigen-targeted CD8(+) T cell responses with PD-1 blockade therapy. *Nature* **2023**, *615*, 697–704. [CrossRef]
25. Kuehm, L.M.; Wolf, K.; Zahour, J.; DiPaolo, R.J.; Teague, R.M. Checkpoint blockade immunotherapy enhances the frequency and effector function of murine tumor-infiltrating T cells but does not alter TCRβ diversity. *Cancer Immunol. Immunother.* **2019**, *68*, 1095–1106. [CrossRef]

26. Lozano, A.X.; Chaudhuri, A.A.; Nene, A.; Bacchiocchi, A.; Earland, N.; Vesely, M.D.; Usmani, A.; Turner, B.E.; Steen, C.B.; Luca, B.A.; et al. T cell characteristics associated with toxicity to immune checkpoint blockade in patients with melanoma. *Nat. Med.* **2022**, *28*, 353–362. [CrossRef]
27. Balbach, M.L.; Axelrod, M.L.; Balko, J.M.; Bankhead, A.; Shaffer, T.; Lim, L.; Guo, J.; Hernandez, J.; Li, M.; Iams, W.T. Peripheral T-cell receptor repertoire dynamics in small cell lung cancer. *Transl. Lung Cancer Res.* **2023**, *12*, 257–265. [CrossRef]
28. Laza-Briviesca, R.; Cruz-Bermúdez, A.; Nadal, E.; Insa, A.; García-Campelo, M.d.R.; Huidobro, G.; Dómine, M.; Majem, M.; Rodríguez-Abreu, D.; Martínez-Martí, A.; et al. Blood biomarkers associated to complete pathological response on NSCLC patients treated with neoadjuvant chemoimmunotherapy included in NADIM clinical trial. *Clin. Transl. Med.* **2021**, *11*, e491. [CrossRef]
29. Badoual, C.; Bouchaud, G.; Agueznay, N.E.H.; Mortier, E.; Hans, S.; Gey, A.; Fernani, F.; Peyrard, S.; -Puig, P.L.; Bruneval, P.; et al. The soluble alpha chain of interleukin-15 receptor: A proinflammatory molecule associated with tumor progression in head and neck cancer. *Cancer Res.* **2008**, *68*, 3907–3914. [CrossRef]
30. Seike, M.; Yanaihara, N.; Bowman, E.D.; Zanetti, K.A.; Budhu, A.; Kumamoto, K.; Mechanic, L.E.; Matsumoto, S.; Yokota, J.; Shibata, T.; et al. Use of a cytokine gene expression signature in lung adenocarcinoma and the surrounding tissue as a prognostic classifier. *J. Natl. Cancer Inst.* **2007**, *99*, 1257–1269. [CrossRef]
31. Fiore, P.F.; Di Matteo, S.; Tumino, N.; Mariotti, F.R.; Pietra, G.; Ottonello, S.; Negrini, S.; Bottazzi, B.; Moretta, L.; Mortier, E.; et al. Interleukin-15 and cancer: Some solved and many unsolved questions. *J. Immunother. Cancer* **2020**, *8*, e001428. [CrossRef]
32. Mlecnik, B.; Bindea, G.; Angell, H.K.; Sasso, M.S.; Obenauf, A.C.; Fredriksen, T.; Lafontaine, L.; Bilocq, A.M.; Kirilovsky, A.; Tosolini, M.; et al. Functional network pipeline reveals genetic determinants associated with in situ lymphocyte proliferation and survival of cancer patients. *Sci. Transl. Med.* **2014**, *6*, 228ra37. [CrossRef]
33. Mao, X.-C.; Yang, C.-C.; Yang, Y.-F.; Yan, L.-J.; Ding, Z.-N.; Liu, H.; Yan, Y.-C.; Dong, Z.-R.; Wang, D.-X.; Li, T. Peripheral cytokine levels as novel predictors of survival in cancer patients treated with immune checkpoint inhibitors: A systematic review and meta-analysis. *Front. Immunol.* **2022**, *13*, 884592. [CrossRef]
34. Ni, G.; Zhang, L.; Yang, X.; Li, H.; Ma, B.; Walton, S.; Wu, X.; Yuan, J.; Wang, T.; Liu, X. Targeting interleukin-10 signalling for cancer immunotherapy, a promising and complicated task. *Hum. Vaccin. Immunother.* **2020**, *16*, 2328–2332. Available online: https://pubmed.ncbi.nlm.nih.gov/32159421 (accessed on 11 March 2020). [CrossRef]
35. Brooks, D.G.; Ha, S.-J.; Elsaesser, H.; Sharpe, A.H.; Freeman, G.J.; Oldstone, M.B.A. IL-10 and PD-L1 operate through distinct pathways to suppress T-cell activity during persistent viral infection. *Proc. Natl. Acad. Sci. USA* **2008**, *105*, 20428–20433. [CrossRef]
36. Lamichhane, P.; Karyampudi, L.; Shreeder, B.; Krempski, J.; Bahr, D.; Daum, J.; Kalli, K.R.; Goode, E.L.; Block, M.S.; Cannon, M.J.; et al. IL10 Release upon PD-1 Blockade Sustains Immunosuppression in Ovarian Cancer. *Cancer Res.* **2017**, *77*, 6667–6678. [CrossRef]
37. Sun, Z.; Fourcade, J.; Pagliano, O.; Chauvin, J.-M.; Sander, C.; Kirkwood, J.M.; Zarour, H.M. IL10 and PD-1 Cooperate to Limit the Activity of Tumor-Specific CD8+ T Cells. *Cancer Res.* **2015**, *75*, 1635–1644. [CrossRef]
38. Naing, A.; Papadopoulos, K.P.; Autio, K.A.; Ott, P.A.; Patel, M.R.; Wong, D.J.; Falchook, C.S.; Pant, S.; Whiteside, M.; Rasco, D.R.; et al. Safety, Antitumor Activity, and Immune Activation of Pegylated Recombinant Human Interleukin-10 (AM0010) in Patients with Advanced Solid Tumors. *J. Clin. Oncol.* **2016**, *34*, 3562–3569. [CrossRef]
39. Castro, F.; Cardoso, A.P.; Gonçalves, R.M.; Serre, K.; Oliveira, M.J. Interferon-gamma at the crossroads of tumor immune surveillance or evasion. *Front. Immunol.* **2018**, *9*, 847. [CrossRef]
40. Chen, H.; Liakou, C.I.; Kamat, A.; Pettaway, C.; Ward, J.F.; Tang, D.N.; Sun, J.; Jungbluth, A.A.; Troncoso, P.; Logothetis, C.; et al. Anti-CTLA-4 therapy results in higher CD4+ICOShi T cell frequency and IFN-gamma levels in both nonmalignant and malignant prostate tissues. *Proc. Natl. Acad. Sci. USA* **2009**, *106*, 2729–2734. [CrossRef]
41. Peng, W.; Liu, C.; Xu, C.; Lou, Y.; Chen, J.; Yang, Y.; Yagita, H.; Overwijk, W.W.; Lizée, G.; Radvanyi, L.; et al. PD-1 blockade enhances T-cell migration to tumors by elevating IFN-γ inducible chemokines. *Cancer Res.* **2012**, *72*, 5209–5218. [CrossRef]
42. Higgs, B.W.; Morehouse, C.A.; Streicher, K.; Brohawn, P.Z.; Pilataxi, F.; Gupta, A.; Ranade, K. Interferon Gamma Messenger RNA Signature in Tumor Biopsies Predicts Outcomes in Patients with Non-Small Cell Lung Carcinoma or Urothelial Cancer Treated with Durvalumab. *Clin. Cancer Res.* **2018**, *24*, 3857–3866. [CrossRef]
43. Karachaliou, N.; Gonzalez-Cao, M.; Crespo, G.; Drozdowskyj, A.; Aldeguer, E.; Gimenez-Capitan, A.; Teixido, C.; Molina-Vila, M.A.; Viteri, S.; de los Llanos Gil, M.; et al. Interferon gamma, an important marker of response to immune checkpoint blockade in non-small cell lung cancer and melanoma patients. *Ther. Adv. Med. Oncol.* **2018**, *10*, 1758834017749748. [CrossRef] [PubMed]
44. Toiyama, Y.; Fujikawa, H.; Kawamura, M.; Matsushita, K.; Saigusa, S.; Tanaka, K.; Inoue, Y.; Uchida, K.; Mohri, Y.; Kusunoki, M. Evaluation of CXCL10 as a novel serum marker for predicting liver metastasis and prognosis in colorectal cancer. *Int. J. Oncol.* **2012**, *40*, 560–566. [CrossRef]
45. Hilpert, J.; Grosse-Hovest, L.; Grünebach, F.; Buechele, C.; Nuebling, T.; Raum, T.; Steinle, A.; Salih, H.R. Comprehensive analysis of NKG2D ligand expression and release in leukemia: Implications for NKG2D-mediated NK cell responses. *J. Immunol.* **2012**, *189*, 1360–1371. [CrossRef]
46. Salih, H.R.; Goehlsdorf, D.; Steinle, A. Release of MICB Molecules by Tumor Cells: Mechanism and Soluble MICB in Sera of Cancer Patients. *Hum. Immunol.* **2006**, *67*, 188–195. Available online: https://www.sciencedirect.com/science/article/pii/S0198885906000292 (accessed on 26 June 2024). [CrossRef]

47. Waldhauer, I.; Steinle, A. Proteolytic release of soluble UL16-binding protein 2 from tumor cells. *Cancer Res.* **2006**, *66*, 2520–2526. [CrossRef]
48. Madjd, Z.; Spendlove, I.; Moss, R.; Bevin, S.; E Pinder, S.; Watson, N.F.S.; Ellis, I.; Durrant, L.G. Upregulation of MICA on high-grade invasive operable breast carcinoma. *Cancer Immun.* **2007**, *7*, 17.
49. Li, K.; Mandai, M.; Hamanishi, J.; Matsumura, N.; Suzuki, A.; Yagi, H.; Yamaguchi, K.; Baba, T.; Fujii, S.; Konishi, I. Clinical significance of the NKG2D ligands, MICA/B and ULBP2 in ovarian cancer: High expression of ULBP2 is an indicator of poor prognosis. *Cancer Immunol. Immunother.* **2009**, *58*, 641–652. [CrossRef] [PubMed]
50. Wensveen, F.M.; Jelenčić, V.; Polić, B. NKG2D: A master regulator of immune cell responsiveness. *Front. Immunol.* **2018**, *9*, 441. [CrossRef] [PubMed]
51. Le Bert, N.; Gasser, S. Advances in NKG2D ligand recognition and responses by NK cells. *Immunol. Cell Biol.* **2014**, *92*, 230–236. [CrossRef]
52. Van Tong, H.; Song, L.H.; Hoan, N.X.; Cuong, B.K.; Sy, B.T.; Son, H.A.; Quyet, D.; Binh, V.Q.; Kremsner, P.G.; Bock, C.T.; et al. Soluble MICB protein levels and platelet counts during hepatitis B virus infection and response to hepatocellular carcinoma treatment. *BMC Infect. Dis.* **2015**, *15*, 25. [CrossRef]
53. Diestelhorst, J.; Junge, N.; Jonigk, D.; Schlue, J.; Falk, C.S.; Manns, M.P.; Baumann, U.; Jaeckel, E.; Taubert, R. Baseline IL-2 and the AIH score can predict the response to standard therapy in paediatric autoimmune hepatitis. *Sci. Rep.* **2018**, *8*, 419. [CrossRef] [PubMed]
54. Boyman, O.; Sprent, J. The role of interleukin-2 during homeostasis and activation of the immune system. *Nat. Rev. Immunol.* **2012**, *12*, 180–190. [CrossRef] [PubMed]
55. Waldmann, T.A. The biology of interleukin-2 and interleukin-15: Implications for cancer therapy and vaccine design. *Nat. Rev. Immunol.* **2006**, *6*, 595–601. [CrossRef] [PubMed]

cancers

MDPI

Article

Predictive Value of Circulatory Total VEGF-A and VEGF-A Isoforms for the Efficacy of Anti-PD-1/PD-L1 Antibodies in Patients with Non-Small-Cell Lung Cancer

Tetsu Hirakawa [1], Kakuhiro Yamaguchi [1,*] Kunihiko Funaishi [1], Kiyofumi Shimoji [1], Shinjiro Sakamoto [1], Yasushi Horimasu [1], Takeshi Masuda [1], Taku Nakashima [1], Hiroshi Iwamoto [1], Hironobu Hamada [2], Shingo Yamada [3] and Noboru Hattori [1]

[1] Department of Molecular and Internal Medicine, Graduate School of Biomedical and Health Sciences, Hiroshima University, Hiroshima 734-8551, Japan; bgdwj032@yahoo.co.jp (T.H.); ta-masuda@hiroshima-u.ac.jp (T.M.); tnaka@hiroshima-u.ac.jp (T.N.); nhattori@hiroshima-u.ac.jp (N.H.)

[2] Department of Physical Analysis and Therapeutic Sciences, Graduate School of Biomedical and Health Sciences, Hiroshima University, Hiroshima 734-8551, Japan

[3] Shino-Test Corporation, Sagamihara 252-0331, Japan

* Correspondence: yamaguchikakuhiro@gmail.com; Tel.: +81-82-257-5196

Simple Summary: Vascular endothelial growth factor (VEGF)-A is known to play a crucial role in the tumor microenvironment. This study investigated the relationship between circulating total VEGF-A (tVEGF-A) and its isoforms with the therapeutic effects of anti-programmed cell death 1 (PD-1)/programmed cell death ligand 1 (PD-L1) antibody monotherapy in patients with non-small-cell lung cancer (NSCLC). Higher levels of tVEGF-A were associated with shorter progression-free survival (PFS) in anti-PD-1/PD-L1 antibody monotherapy only when measured in serum, not in plasma. Notably, higher levels of serum $VEGF_{121}$, an isoform of VEGF-A, were significantly associated with not only shorter PFS but also a lower objective response rate. Serum $VEGF_{121}$ levels could serve as a useful biomarker for predicting anti-PD-1/PD-L1 antibody monotherapy efficacy in patients with NSCLC.

Abstract: Background/Objectives: Vascular endothelial growth factor (VEGF)-A promotes an immunosuppressive tumor microenvironment, potentially affecting the efficacy of anti-programmed cell death 1 (PD-1)/programmed cell death ligand 1 (PD-L1) antibody therapy. $VEGF_{121}$ and $VEGF_{165}$, VEGF-A isoforms, promote and inhibit tumor growth, respectively. Additionally, VEGF-A levels differ depending on whether they are measured in serum or plasma. However, whether the serum or plasma levels of total VEGF-A (tVEGF-A) or its isoforms are the most suitable for predicting anti-PD-1/PD-L1 antibody therapy efficacy remains unclear. Methods: Eighty-six patients with non-small-cell lung cancer (NSCLC) who were treated with anti-PD-1/PD-L1 antibody monotherapy between December 2015 and December 2023 were retrospectively enrolled. The association between the serum and plasma levels of tVEGF-A and its isoforms ($VEGF_{121}$ and $VEGF_{165}$) and treatment outcomes was analyzed. Results: The median progression-free survival (PFS) was 2.9 months, and the objective response rate (ORR) was 23.3%. PFS was significantly shorter in patients with higher tVEGF-A serum levels ($\geq$484.2 pg/mL) than in those without (median PFS 2.1 vs. 3.7 months, $p = 0.004$). In contrast, plasma tVEGF-A levels could not be used to stratify PFS. Therefore, the serum levels of VEGF-A isoforms were measured. Patients with higher $VEGF_{121}$ serum levels ($\geq$523.5 pg/mL) showed both significantly shorter PFS (median PFS 2.3 vs. 3.3 months, $p = 0.022$) and a lower ORR (9.7% vs. 30.9%, $p = 0.033$) than those without. Multivariate Cox and logistic regression analyses showed that higher levels of serum $VEGF_{121}$ were significantly associated with shorter PFS and a lower ORR.

https://doi.org/10.3390/cancers17040572

Conclusions: Serum $VEGF_{121}$ levels may be useful in predicting anti-PD-1/PD-L1 antibody monotherapy efficacy.

Keywords: vascular endothelial growth factor-A; $VEGF_{121}$; non-small cell lung cancer; anti-PD-1/PD-L1 antibody; biomarker

1. Introduction

Non-small-cell lung cancer (NSCLC) has a poor prognosis compared to many other cancers [1]. The prognosis of advanced NSCLC has dramatically improved with the advent of nivolumab [2,3], pembrolizumab [4,5] (anti-programmed cell death 1 [PD-1] antibodies), and atezolizumab [6] (anti-programmed cell death ligand 1 [PD-L1] antibodies). Currently, the PD-L1 tumor proportion score (TPS) is a predictor of the response to anti-PD-1/PD-L1 antibody therapy and is used in clinical practice [7]. However, even in patients with advanced NSCLC harboring high PD-L1 expression, the response rate for anti-PD-1/PD-L1 antibody monotherapy as a first-line treatment is only 38.3–44.8% [4,6], and the accuracy of predictions based on PD-L1 expression is limited. Therefore, there is an urgent need to identify new biomarkers other than PD-L1 that can predict the efficacy of anti-PD-1/PD-L1 antibody therapy.

Vascular endothelial growth factor (VEGF)-A is a homodimer protein of 40–45 kDa that is secreted by various cells, including tumor cells, immune cells, and platelets [8–11]. VEGF-A binds to vascular endothelial growth factor receptor (VEGFR) and neuropilin (NRP) [12]. VEGF-A expression is regulated by hypoxia-inducible factor-1α and is induced under hypoxic conditions [13]. Secreted VEGF-A is involved in angiogenesis, tumor growth, and tumor metastasis [13,14]. VEGF-A is highly expressed in lung cancer tissues, and its overexpression is a poor prognostic factor [15]. Furthermore, VEGF-A increases the presence and function of myeloid-derived suppressor cells, regulatory T cells, and tumor-associated macrophages, which suppress anticancer immunity and inhibit cytotoxic T lymphocytes and dendritic cells [16]. Hence, VEGF-A promotes the development of an immunosuppressive tumor microenvironment, which may affect the therapeutic efficacy of anti-PD-1/PD-L1 antibodies.

The VEGF-A gene is located on chromosome 6p21.1 and consists of eight exons separated by seven introns [17]. The alternative splicing of VEGF-A mRNA from exons 5 to 8 produces different VEGF-A isoforms, such as $VEGF_{121}$, $VEGF_{165}$, $VEGF_{189}$, and $VEGF_{206}$ [18–20]. Of these, $VEGF_{121}$ and $VEGF_{165}$ are primarily secreted by tumor cells [21]. In a cancer mouse model with the overexpression of $VEGF_{121}$ or $VEGF_{165}$, $VEGF_{121}$ promotes tumor growth, whereas $VEGF_{165}$ suppresses it [22]. However, no studies have examined the association between the efficacy of anti-PD-1/PD-L1 antibody monotherapy and the levels of VEGF-A isoforms in the blood.

VEGF-A can be measured in both serum and plasma; however, because serum VEGF-A levels include VEGF-A pooled in the platelets, serum VEGF-A levels have been reported to be approximately 2–7 times higher than plasma levels [23,24]. Additionally, conflicting reports have demonstrated a relationship between the efficacy of anti-PD-1 antibody therapy in NSCLC and circulatory VEGF-A levels [25,26]. Shibaki et al. revealed that higher levels of VEGF-A in serum were associated with shorter survival [25], although Tiako et al. showed that there was no significant association between VEGF-A levels in plasma and efficacy in patients with NSCLC [26]. These data suggest that the usefulness of VEGF-A as a blood marker for predicting the efficacy of anti-PD-1/PD-L1 antibody therapy in patients with NSCLC depends on the sample type, such as serum or plasma.

Therefore, we investigated whether the association between the efficacy of anti-PD-1/PD-L1 antibody monotherapy and the circulatory levels of total VEGF-A (tVEGF-A) was dependent on sample types, such as serum and plasma, and compared the predictive value of tVEGF-A and its major isoforms, $VEGF_{121}$ and $VEGF_{165}$, for the efficacy of anti-PD-1/PD-L1 antibody monotherapy.

2. Materials and Methods

2.1. Study Population and Design

This study screened 137 patients with NSCLC treated with anti-PD-1/PD-L1 antibody monotherapy (nivolumab, pembrolizumab, or atezolizumab) at the Department of Respiratory Medicine, Hiroshima University Hospital, between December 2015 and December 2023 (Figure 1). Forty-two patients without serum and plasma samples were excluded. Because the administration of bevacizumab and ramucirumab has been reported to cause fluctuations in circulatory VEGF-A [27–29], eight patients with a history of bevacizumab or ramucirumab before anti-PD-1/PD-L1 antibody administration were also excluded. Moreover, one patient who developed radiation pneumonitis immediately before the initiation of anti-PD-1/PD-L1 antibody monotherapy was excluded because VEGF-A levels may fluctuate owing to the development of pneumonitis [30]. Ultimately, 86 patients with serum and plasma samples were included in this study. This study was performed in accordance with the principles of the Declaration of Helsinki and approved by the Ethics Committee of Hiroshima University Hospital (E2004-0326-23, approved 7 August 2024). Written informed consent was obtained from all the participants.

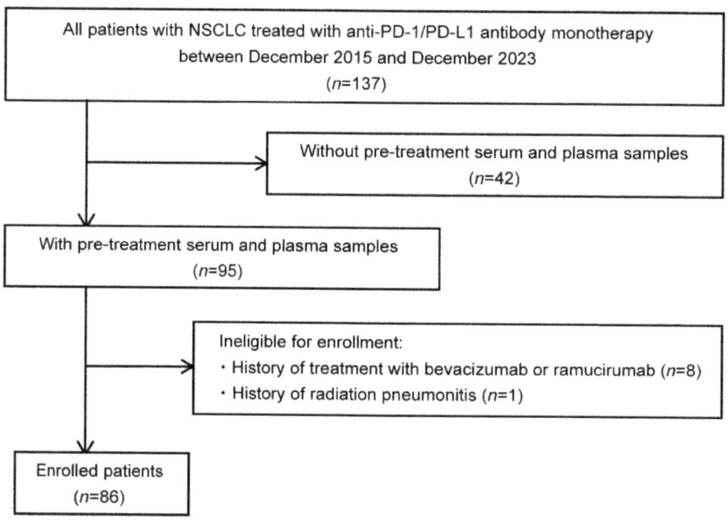

Figure 1. Flowchart of patient enrollment. This study included patients with non-small-cell lung cancer (NSCLC) treated with anti-programmed cell death 1(PD-1)/programmed cell death ligand 1(PD-L1) antibody monotherapy (nivolumab, pembrolizumab, or atezolizumab) at the Department of Respiratory Medicine, Hiroshima University Hospital, between December 2015 and December 2023, for whom serum and plasma samples were stored. After excluding eight patients who had a history of bevacizumab or ramucirumab prior to anti-PD-1/PD-L1 antibody administration and one patient who developed radiation pneumonitis just prior to the initiation of anti-PD-1/PD-L1 antibody monotherapy, 86 patients were finally included in the study. NSCLC, non-small-cell lung cancer; PD-1, programmed cell death 1; PD-L1, programmed cell death ligand 1.

2.2. Evaluations of the Objective Response Rate and Progression-Free Survival

Complete response (CR), partial response (PR), stable disease (SD), progressive disease (PD), and not evaluable (NE) were determined based on the Response Evaluation Criteria in Solid Tumors (RECIST) 1.1 [31]. The objective response rate (ORR) was defined as the proportion of patients who achieved CR or PR. Progression-free survival (PFS) was defined as the time from the start of each treatment until progression or death from any cause. Patients who failed to follow-up were censored on the date of their last known survival.

2.3. Measurement of tVEGF-A and Its Isoforms

Serum and plasma samples were collected prior to anti-PD-1/PD-L1 antibody administration and stored at −80 °C. Serum and plasma tVEGF-A levels were determined using an ELISA system developed by Shino-Test Corporation. Polystyrene microtiter plates were coated and incubated with 100 μL of anti-human VEGF-A polyclonal antibody (R&D Biosystems, Minneapolis, MN, USA) in PBS overnight at 4 °C. The plates were washed three times with PBS containing 0.05% Tween 20, and the remaining binding sites in the wells were blocked by incubating the plates for 2 h with 400 μL/well of PBS containing 0.5% casein. After the plates were washed, 100 μL of each dilution of the calibrator and samples (1:1 dilution in 0.2 mol/L Tris pH 8.5 and 0.15 mol/L sodium chloride containing 1% casein) was added to the wells. The plates were then incubated for 15 h at 25 °C. The plates were washed again and were incubated with 100 μL/well of peroxidase-conjugated anti-human VEGF-A monoclonal antibody (R&D Biosystems, Minneapolis, MN) for 2 h at 25 °C. After another washing step, chromogenic substrate 3,3′,5,5′-tetra-methylbenzidine (Dojindo Laboratories, Kumamoto, Japan) was added to each well. The reaction was terminated with sulfuric acid, and the absorbance at 450 nm was read using a microplate reader (Model 680, Bio-Rad, Irvine, CA, USA). $VEGF_{121}$ and $VEGF_{165}$ levels were measured using ELISA kits (Shino-Test, Kanagawa, Japan) [19].

2.4. Statistical Analysis

Values are expressed as a median (interquartile range [IQR]) unless stated otherwise. Differences among the groups were examined using the Fisher's exact, Wilcoxon signed-rank, and Mann–Whitney U tests. Spearman's rank correlation coefficient was calculated to evaluate the association between the levels of tVEGF-A and its isoforms. A receiver operating characteristic (ROC) curve analysis was performed to identify the optimal cut-off levels of tVEGF-A and its isoforms for predicting the objective response (CR or PR) to anti-PD-1/PD-L1 antibody monotherapy. The optimal cut-off level was determined by maximizing the sum of sensitivity plus specificity − 1. PFS was evaluated using a Kaplan–Meier analysis and the log-rank test. Median PFS intervals with a corresponding 95% confidence interval (CI) were calculated. Univariate and multivariate Cox proportional hazard models and logistic regression analyses were used to identify the independent predictors of PFS and the objective response for anti-PD-1/PD-L1 antibody monotherapy, respectively. Statistical significance was set at $p < 0.05$. All data analyses were performed using JMP statistical software version 17.0.0 (SAS Institute Inc., Cary, NC, USA).

3. Results

3.1. Patient Characteristics

The baseline characteristics of the patients are shown in Table 1. Of the 86 patients, the median age was 73 years (67–77), 60 (69.8%) were male, and 14 (16.3%) were NSCLC positive for driver oncogenes. PD-L1 TPS was ≥50% in 41 (47.7%), 1–49% in 18 (20.9%), <1% in 8 (9.3%), and unknown in 19 (22.1%). Anti-PD-1/PD-L1 antibody monotherapy was administered to 54 patients (62.8%) as a second- or later-line treatment.

Table 1. Baseline characteristics.

	All Patients (n = 86)	CR/PR (n = 20)	SD/PD/NE (n = 66)	p-Value
Age, years	73 (67–77)	70 (68–79)	73 (66–77)	0.705
Sex				0.163
Male, n (%)	60 (69.8)	11 (55.0)	49 (74.2)	
Female, n (%)	26 (30.2)	9 (45.0)	17 (25.8)	
Smoking history, pack-years †	50.0 (22.5–60.0)	56.3 (30.0–79.5)	49.8 (16.0–60.0)	0.242
BMI	21.3 (19.3–23.2)	21.2 (19.7–23.1)	21.3 (19.0–23.2)	0.759
PS				0.221
0–1, n (%)	68 (79.1)	18 (90.0)	50 (75.8)	
≥2, n (%)	18 (20.9)	2 (10.0)	16 (24.2)	
History of COPD				0.427
+, n (%)	31 (36.0)	9 (45.0)	22 (33.3)	
−, n (%)	55 (64.0)	11 (55.0)	44 (66.7)	
Previous thoracic RT				0.405
+, n (%)	26 (30.2)	4 (20.0)	22 (33.3)	
−, n (%)	60 (69.8)	16 (80.0)	44 (66.7)	
Stage				0.022 *
III, n (%)	8 (9.3)	5 (25.0)	3 (4.5)	
IV, n (%)	50 (58.1)	11 (55.0)	39 (59.1)	
Recurrence, n (%)	28 (32.6)	4 (20.0)	24 (36.4)	
Histological type				1.000
Squamous, n (%)	13 (15.1)	3 (15.0)	10 (15.2)	
Non-Squamous, n (%)	73 (84.9)	17 (85.0)	56 (84.8)	
Driver oncogene ‡				0.505
positive, n (%)	14 (16.3)	2 (10.0)	12 (18.2)	
negative, n (%)	72 (83.7)	18 (90.0)	54 (81.8)	
PD-L1 TPS				0.122
≥50%, n (%)	41 (47.7)	14 (70.0)	27 (40.9)	
1–49%, n (%)	18 (20.9)	3 (15.0)	15 (22.7)	
<1%, n (%)	8 (9.3)	0 (0.0)	8 (12.1)	
unknown, n (%)	19 (22.1)	3 (15.0)	16 (24.2)	
ICI treatment line				0.070
1st, n (%)	32 (37.2)	11 (55.0)	21 (31.8)	
2nd or later, n (%)	54 (62.8)	9 (45.0)	45 (68.2)	
ICI agent				0.009 **
Anti-PD-1 antibody, n (%)	69 (80.2)	20 (100.0)	49 (74.2)	
Anti-PD-L1 antibody, n (%)	17 (19.8)	0 (0.0)	17 (25.8)	

Data are presented as a median (interquartile range) unless stated otherwise. BMI, body mass index; COPD, chronic obstructive pulmonary disease; CR, complete response; ICI, immune checkpoint inhibitor; NE, not evaluable; PD, progressive disease; PD-1, programmed cell death 1; PD-L1, programmed cell death ligand 1; PR, partial response; PS, performance status; RT, radiotherapy; SD, stable disease; TPS, tumor proportion score. † There are missing data for one patient. ‡ Driver oncogenes included 12 patients with epidermal growth factor receptor gene mutations and 2 patients with mesenchymal–epithelial transition exon 14 skipping. * $p < 0.05$ and ** $p < 0.01$, comparison between CR/PR and SD/PD/NE using the Mann–Whitney U test or Fisher's exact test.

The median observation period was 10.6 months (4.6–30.6). At the data cut-off in July 2024, progression or death from any cause was observed in 78 patients (90.7%). The treatment responses to anti-PD-1/PD-L1 antibody monotherapy in 86 patients were classified as CR in 4 (4.7%), PR in 16 (18.6%), SD in 17 (19.8%), PD in 36 (41.9%), and NE in 13 (15.1%). The ORR and median PFS of anti-PD-1/PD-L1 antibody monotherapy were 23.3% and 2.9 months (95% CI: 2.1–3.4), respectively.

3.2. Prediction of the Therapeutic Effect of Anti-PD-1/PD-L1 Antibody Monotherapy by Serum and Plasma tVEGF-A

The serum and plasma levels of tVEGF-A were measured. Serum tVEGF-A levels were significantly higher than plasma levels (452.9 pg/mL [252.3–704.7] vs. 49.4 pg/mL [0.0–131.6], $p < 0.001$) (Figure 2a, Supplementary Table S1). Serum and plasma tVEGF-A levels were positively correlated ($\rho = 0.502$, $p < 0.001$) (Supplementary Figure S1a). The ROC curve analysis revealed that the optimal cut-off levels for predicting the objective response to anti-PD-1/PD-L1 antibody monotherapy were 484.2 pg/mL for serum tVEGF-A (area under the curve [AUC] = 0.54 [95% CI: 0.40–0.68], specificity = 48.5%, sensitivity = 70.0%) and 137.1 pg/mL for plasma tVEGF-A (AUC = 0.54 [95% CI: 0.39–0.68], specificity = 81.8%, sensitivity = 35.0%) (Supplementary Figure S2a,b). There was no significant difference in the ORR between the groups stratified by serum and plasma tVEGF-A cut-off levels (Figure 3a,b). Conversely, the Kaplan–Meier analysis showed that PFS was significantly shorter in patients with higher levels of serum tVEGF-A than in those with lower levels (median PFS 2.1 months [95% CI: 1.2–3.3] vs. 3.7 months [95% CI: 2.1–5.4], $p = 0.004$), but there was no significant difference in PFS between patients stratified by the cut-off levels of plasma tVEGF-A (median PFS 2.3 months [95% CI: 0.7–4.0] vs. 2.9 months [95% CI: 2.1–4.4], $p = 0.611$) (Figure 4a,b). The univariate Cox proportional hazards model revealed that serum tVEGF-A levels, a history of chronic obstructive pulmonary disease (COPD), an immune checkpoint inhibitor (ICI) treatment line, and the ICI agent were significant predictors of PFS (Table 2). Furthermore, the multivariate Cox proportional hazards model (model 1) revealed that serum tVEGF-A levels ($\geq$484.2 pg/mL) were independent predictors of shorter PFS when adjusted for a history of COPD, ICI treatment line, and ICI agent (Table 2).

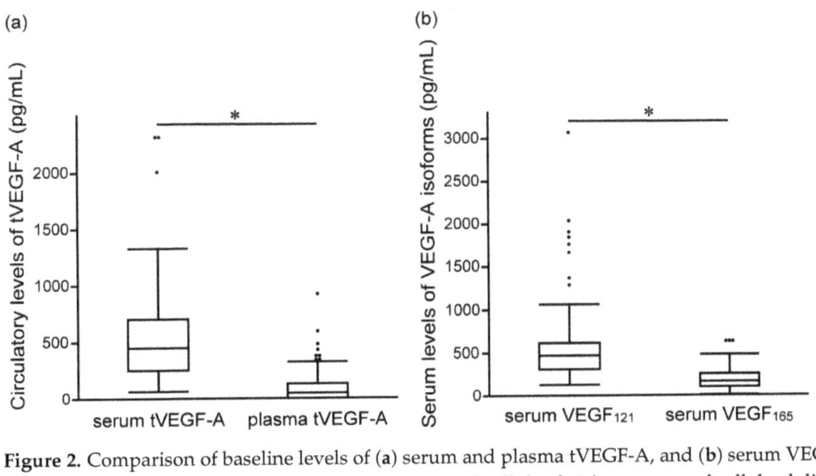

Figure 2. Comparison of baseline levels of (**a**) serum and plasma tVEGF-A, and (**b**) serum $VEGF_{121}$ and $VEGF_{165}$ before the initiation of anti-programmed cell death 1/programmed cell death ligand 1 antibody monotherapy. The serum levels of tVEGF-A are significantly higher than the plasma levels of tVEGF-A (452.9 pg/mL [interquartile range (IQR), 252.3–704.7] vs. 49.4 pg/mL [IQR, 0.0–131.6], $p < 0.001$) (**a**). The serum levels of $VEGF_{121}$ are significantly higher than the serum levels of $VEGF_{165}$ (466.4 pg/mL [IQR, 309.3–611.9] vs. 169.4 pg/mL [IQR, 98.8–251.8], $p < 0.001$) (**b**). The boxes represent the 25th to 75th percentiles; the solid lines within the boxes show the median values; the whiskers represent the 10th and 90th percentiles; the dots represent outliers. IQR, interquartile range; tVEGF, total vascular endothelial growth factor. * $p < 0.001$, using the Wilcoxon signed-rank test.

Table 2. Univariate and multivariate Cox proportional hazards model for predicting progression-free survival in patients with non-small-cell lung cancer treated with anti-programmed cell death 1(PD-1)/programmed cell death ligand 1(PD-L1) antibody monotherapy.

Variables	Univariate Analysis			Multivariate Analysis (Model 1)			Multivariate Analysis (Model 2)		
	HR	95% CI	*p*-Value	HR	95% CI	*p*-Value	HR	95% CI	*p*-Value
Age, ≥75	0.788	0.488–1.273	0.330						
Sex, male	1.587	0.927–2.717	0.092						
Smoking history, pack-years	0.996	0.989–1.003	0.236						
BMI	0.993	0.930–1.058	0.829						
PS, ≥2	1.231	0.705–2.149	0.466						
History of COPD	0.555	0.334–0.922	0.023 *	0.821	0.478–1.410	0.475	0.800	0.462–1.385	0.425
Previous thoracic RT	0.987	0.596–1.632	0.958						
Histological type, squamous	0.805	0.423–1.534	0.510						
Driver oncogene, positive	1.521	0.831–2.784	0.174						
PD-L1 TPS, ≥50%	0.722	0.450–1.159	0.178						
ICI treatment line, 1st	0.577	0.351–0.948	0.030 *	0.580	0.330–1.017	0.057	0.696	0.404–1.198	0.191
ICI agent, anti-PD-1 antibody	0.303	0.169–0.545	<0.001 ***	0.317	0.166–0.606	<0.001 ***	0.317	0.164–0.613	<0.001 ***
Serum tVEGF-A, ≥484.2 pg/mL	1.952	1.220–3.124	0.005 **	2.511	1.496–4.212	<0.001 ***			
Plasma tVEGF-A, ≥137.1 pg/mL	1.155	0.661–2.015	0.613						
Serum $VEGF_{121}$, ≥523.5 pg/mL	1.731	1.074–2.790	0.024 *				1.967	1.167–3.314	0.011 *
Serum $VEGF_{165}$, ≥165.0 pg/mL	1.194	0.749–1.904	0.457						

Two multivariate Cox proportional hazards models were analyzed because a positive correlation was observed between the serum levels of tVEGF-A and $VEGF_{121}$. BMI, body mass index; CI, confidence interval; COPD, chronic obstructive pulmonary disease; HR, hazard ratio; ICI, immune checkpoint inhibitor; PD-1, programmed cell death 1; PD-L1, programmed cell death ligand 1; PS, performance status; RT, radiotherapy; TPS, tumor proportion score; tVEGF, total vascular endothelial growth factor. * $p < 0.05$, ** $p < 0.01$ and *** $p < 0.001$, Cox proportional hazards models.

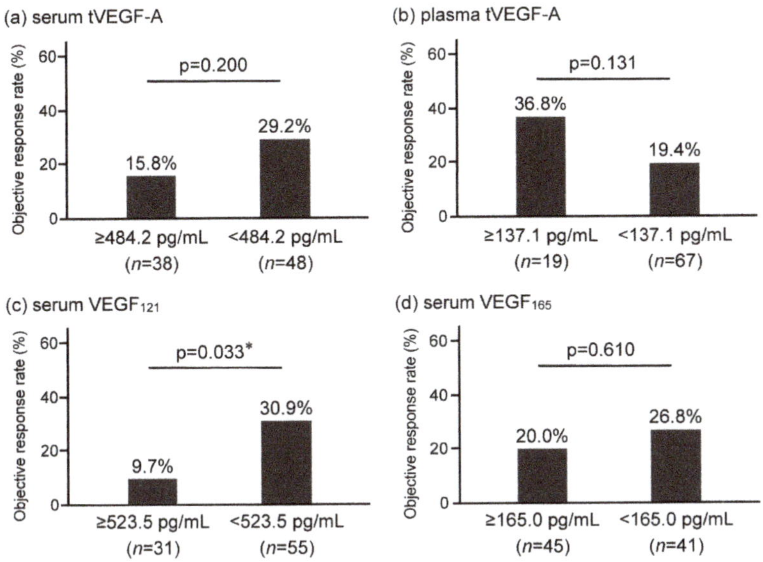

Figure 3. Comparison of the objective response rate (ORR) of anti-programmed cell death 1/programmed cell death ligand 1 antibody monotherapy in non-small-cell lung cancer stratified by baseline levels of (**a**) serum tVEGF-A, (**b**) plasma tVEGF-A, (**c**) serum $VEGF_{121}$, and (**d**) serum $VEGF_{165}$. The ORR is not significantly different for serum tVEGF-A (**a**) and plasma tVEGF-A (**b**). In contrast, the ORR is significantly lower in the high serum $VEGF_{121}$ group (9.7% vs. 30.9, $p = 0.033$) (**c**) but not significantly different in serum $VEGF_{165}$ (20.0% vs. 26.8%, $p = 0.610$) (**d**). ORR, objective response rate; tVEGF, total vascular endothelial growth factor. * $p < 0.05$, using the Fisher's exact test.

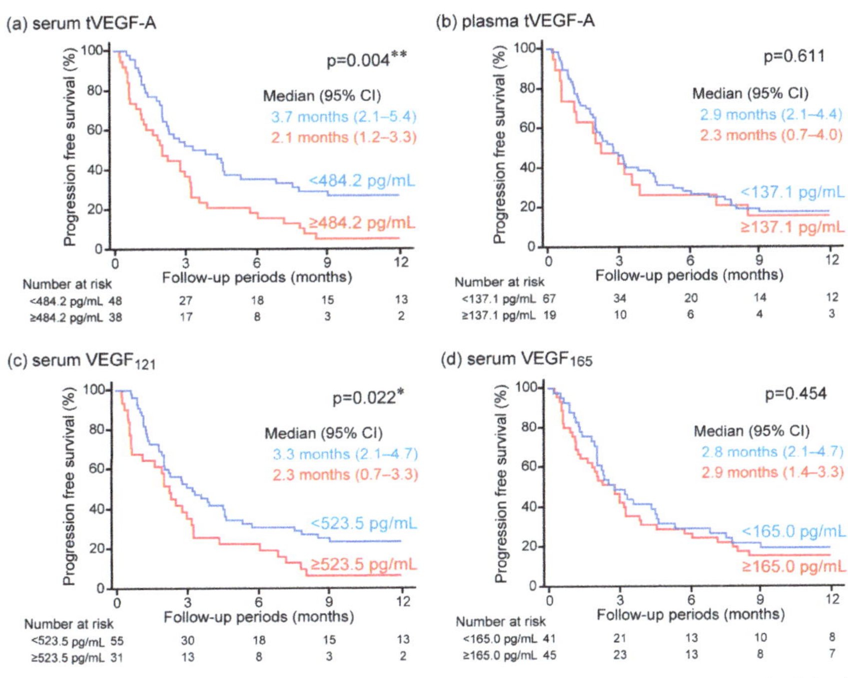

Figure 4. Kaplan–Meier analysis for progression-free survival (PFS) in anti-programmed cell death 1/programmed cell death ligand 1 antibody monotherapy in non-small-cell lung cancer stratified by

baseline (**a**) serum tVEGF-A, (**b**) plasma tVEGF-A, (**c**) serum $VEGF_{121}$, and (**d**) serum $VEGF_{165}$. Patients with higher levels of serum tVEGF-A showed significantly shorter PFS than those with lower levels (**a**), but no significant difference in PFS was observed when stratified by plasma tVEGF-A (**b**). Also, patients with higher levels of serum $VEGF_{121}$ showed significantly shorter PFS than those with lower levels (**c**), but there was no significant difference in PFS when evaluated by serum $VEGF_{165}$ (**d**). CI, confidence interval; PFS, progression-free survival; tVEGF, total vascular endothelial growth factor. * $p < 0.05$ and ** $p < 0.01$ using the log-rank test.

3.3. Prediction of the Therapeutic Effect of Anti-PD-1/PD-L1 Antibody Monotherapy by Serum VEGF-A Isoforms

This study additionally measured $VEGF_{121}$ and $VEGF_{165}$ levels using serum samples, as only the serum levels of tVEGF-A, not the plasma levels, were used to stratify PFS. The serum levels of $VEGF_{121}$ were significantly higher than the serum levels of $VEGF_{165}$ (466.4 pg/mL [309.3–611.9] vs. 169.4 pg/mL [98.8–251.8], $p < 0.001$) (Figure 2b, Supplementary Table S1). The serum levels of $VEGF_{121}$ and $VEGF_{165}$ were positively correlated with the serum levels of tVEGF-A ($\rho = 0.607$, $p < 0.001$ and $\rho = 0.865$, $p < 0.001$, respectively) (Supplementary Figure S1b,c). The ROC curve analysis revealed that the optimal cut-off levels for predicting the objective response to anti-PD-1/PD-L1 antibody monotherapy were 523.5 pg/mL for serum $VEGF_{121}$ (AUC = 0.61 [95% CI: 0.47–0.73], specificity = 42.4%, sensitivity = 85.0%) and 165.0 pg/mL for serum $VEGF_{165}$ (AUC = 0.50 [95% CI: 0.36–0.65], specificity = 54.6%, sensitivity = 55.0%) (Supplementary Figure S2c,d).

The ORR was significantly lower in patients with higher levels of serum $VEGF_{121}$ (≥523.5 pg/mL) than in those without (9.7% vs. 30.9%, $p = 0.033$), although there was no significant difference in the ORR between patients with and without $VEGF_{165}$ serum levels higher than 165.0 pg/mL (20.0% vs. 26.8%, $p = 0.610$) (Figure 3c,d). Univariate and multivariate logistic regression analyses revealed that, among circulatory tVEGF-A and its isoforms, only higher levels of serum $VEGF_{121}$ were independently and significantly associated with a failure to achieve the objective response (Table 3).

Table 3. Univariate and multivariate logistic regression analyses for predicting the objective response in patients with non-small-cell lung cancer treated with anti-programmed cell death 1(PD-1)/programmed cell death ligand 1(PD-L1) antibody monotherapy.

Variables	Univariate Analysis			Multivariate Analysis		
	OR	95% CI	*p*-Value	OR	95% CI	*p*-Value
Age, ≥75	1.342	0.479–3.698	0.570			
Sex, male	0.424	0.149–1.215	0.109			
Smoking history, pack-years	1.005	0.992–1.019	0.432			
BMI	1.006	0.867–1.165	0.932			
PS, ≥2	0.347	0.052–1.387	0.145			
History of COPD	1.636	0.581–4.549	0.346			
Previous thoracic RT	0.500	0.131–1.559	0.242			
Histological type, squamous	0.988	0.205–3.679	0.987			
Driver oncogene, positive	0.500	0.073–2.068	0.364			
PD-L1 TPS, ≥50%	3.370	1.193–10.551	0.021 *	2.645	0.864–8.845	0.089
ICI treatment line, 1st	2.619	0.947–7.451	0.064			
ICI agent, anti-PD-1 antibody	Not available †	Not available †	0.001 **	Not available †	Not available †	0.005 **
Serum tVEGF-A, ≥484.2 pg/mL	0.455	0.146–1.284	0.139			

Table 3. *Cont.*

Variables	Univariate Analysis			Multivariate Analysis		
	OR	95% CI	*p*-Value	OR	95% CI	*p*-Value
Plasma tVEGF-A, ≥137.1 pg/mL	2.423	0.774–7.337	0.126			
Serum $VEGF_{121}$, ≥523.5 pg/mL	0.239	0.052–0.799	0.019 *	0.231	0.049–0.819	0.022 *
Serum $VEGF_{165}$, ≥165.0 pg/mL	0.682	0.244–1.862	0.454			

BMI, body mass index; CI, confidence interval; COPD, chronic obstructive pulmonary disease; ICI, immune checkpoint inhibitor; OR, odds ratio; PD-1, programmed cell death 1; PD-L1, programmed cell death ligand 1; PS, performance status; RT, radiotherapy; TPS, tumor proportion score; tVEGF, total vascular endothelial growth factor. † OR and 95% CI could not be calculated because only one group responded to anti-PD-1/PD-L1 antibody monotherapy. * $p < 0.05$ and ** $p < 0.01$, logistic regression analysis.

The Kaplan–Meier analysis also showed a significantly shorter PFS in the group with higher levels of serum $VEGF_{121}$ than in the group without (median PFS 2.3 months [95% CI: 0.7–3.3] vs. 3.3 months [95% CI: 2.1–4.7] months, $p = 0.022$), but not in the group with and without higher levels of serum $VEGF_{165}$ (median PFS 2.9 months [95% CI: 1.4–3.3] vs. 2.8 months [95% CI: 2.1–4.7] months, $p = 0.454$) (Figure 4c,d). The univariate Cox proportional hazards model revealed that serum $VEGF_{121}$ levels were significant predictors of PFS (Table 2). A positive correlation was observed between the serum levels of tVEGF-A and $VEGF_{121}$; therefore, the association between PFS and serum $VEGF_{121}$ was analyzed in a multivariate Cox proportional hazards model not including serum tVEGF-A. The multivariate Cox proportional hazards model (model 2) revealed that higher levels of serum $VEGF_{121}$ were an independent predictor of shorter PFS when adjusted for a history of COPD, ICI treatment line, and ICI agent (Table 2).

4. Discussion

In this study, serum and plasma tVEGF-A levels were examined to predict the efficacy of anti-PD-1/PD-L1 antibody monotherapy in patients with NSCLC. Higher levels of tVEGF-A in serum, but not in plasma, were significantly associated with a shorter PFS. Furthermore, among the serum levels of tVEGF-A and its isoforms, higher levels of serum $VEGF_{121}$ were useful for predicting a lower ORR and shorter PFS in patients treated with anti-PD-1/PD-L1 antibody monotherapy.

This study demonstrated that serum samples were suitable for measuring tVEGF-A levels to stratify the PFS of patients receiving anti-PD-1/PD-L1 antibody monotherapy. Consistent with our results, high serum tVEGF-A levels have been reported to be associated with shorter PFS in patients with NSCLC who are elderly or have poor PS [25]; however, these associations have not been shown in other studies using plasma samples [26]. This discrepancy in the association between the efficacy of anti-PD-1/PD-L1 antibodies and serum or plasma tVEGF-A levels is potentially caused by platelet-derived VEGF-A in the serum. First, in circulation, most VEGF-A is pooled in the alpha granules of platelets, and VEGF-A is released from platelets, particularly when platelets are activated by several factors, including blood coagulation [32,33]. When serum samples are obtained, VEGF-A is released from platelets owing to blood coagulation in the serum collection tubes [33]. Accordingly, serum VEGF-A levels have been reported to be approximately two to seven times higher than plasma VEGF-A levels [23,24]. It has also been shown that the VEGF-A content of platelets increases with tumor progression, and much of the serum VEGF-A in patients with cancer is thought to be derived from VEGF-A pooled in platelets [10,11]. Secondly, VEGF-A released from activated platelets by tumor cells plays a role in promoting tumor progression and metastasis [34,35]. Additionally, VEGF-A promotes the development of an immunosuppressive tumor microenvironment associated with resistance to anti-PD-1/PD-L1 antibody therapy [16]. These data suggest that the measurement of VEGF-A levels pooled in platelets is needed to predict the efficacy of anti-PD-1/PD-L1 antibody therapy, and therefore, serum tVEGF-A could be a predictive biomarker for efficacy by reflecting the amount of VEGF-A in platelets in this study.

This study also showed that higher levels of serum $VEGF_{121}$ were significantly and independently associated with a lower ORR and shorter PFS in patients treated with anti-PD-1/PD-L1 antibody monotherapy, although high levels of serum tVEGF-A were associated with shorter PFS but not the ORR. Additionally, $VEGF_{165}$ serum levels could not be used to stratify PFS or the ORR. $VEGF_{121}$ promotes tumor angiogenesis and increases vascular permeability around the tumor [22,36–38]. Moreover, $VEGF_{121}$ promotes lymphangiogenesis in the sentinel lymph nodes of NSCLC cells [39]. Furthermore, in various types of cancer, including lung cancer, high expression of $VEGF_{121}$ evaluated using tumor samples has been shown to be associated with poor prognosis [40–42]. In contrast, $VEGF_{165}$ inhibits tumor growth by normalizing tumor blood vessels and reducing the hypoxic state of the tumors. $VEGF_{165}$ has an NRP-binding domain and can bind to NRP1, although $VEGF_{121}$ cannot accelerate NRP1 signaling [43]. The interaction between $VEGF_{165}$ and NRP1 recruits NRP1-expressing monocytes (NEMs) to newly formed blood vessels [43]. NEMs produce molecules that contribute to the stabilization of tumor blood vessels (such as transforming growth factor-β, platelet-derived growth factor-B, and stromal cell-derived factor-1) and chemokines with anti-tumor activity (C-C motif chemokine ligand [CCL]2, CCL4, CCL5, C-X-C motif chemokine ligand [CXCL]9, CXCL10, etc.), thereby inhibiting tumor growth [44]. Therefore, the use of $VEGF_{121}$, which is involved in tumor growth more specifically than tVEGF-A, including $VEGF_{165}$, could stratify both the PFS and ORR of anti-PD-1/PD-L1 antibody treatment.

This study has several limitations. First, this was a retrospective study with a limited sample size. Second, the study included patients who received anti-PD-1/PD-L1 antibody monotherapy from the second-line treatment onwards. Currently, ICI alone or in combination with chemotherapy is administered as the first-line therapy for the treatment of advanced-stage NSCLC in most cases. To further investigate the usefulness of $VEGF_{121}$, it is necessary to confirm the results of this study in a prospective cohort of patients with NSCLC who received anti-PD-1/PD-L1 antibody therapy as the first-line treatment. Additionally, to overcome ICI resistance in patients with higher levels of serum $VEGF_{121}$, our future perspective is that the efficacy of the combination therapy with molecular targeted therapies for VEGF-A or VEGFR needs to be evaluated. Third, the AUC value of serum $VEGF_{121}$ was 0.61, which is not sufficiently high. To enhance its predictive accuracy, a combination with other predictive markers may be necessary. Fourth, VEGF-A levels may have fluctuated due to the influence of the fasting state and time of day [45,46], as the conditions for obtaining blood samples were not unified with considering these factors in this study.

5. Conclusions

In patients with NSCLC who received anti-PD-1/PD-L1 antibody monotherapy, high serum tVEGF-A levels, but not plasma levels, were significantly associated with a shorter PFS. Furthermore, as the focus was on the VEGF-A isoforms $VEGF_{121}$ and $VEGF_{165}$, only $VEGF_{121}$ could be used as a predictive marker for the efficacy of anti-PD-1/PD-L1 antibody monotherapy, and high serum $VEGF_{121}$ levels were associated with a low ORR and shorter PFS. Therefore, VEGF-A levels, particularly $VEGF_{121}$, measured by serum levels, could be a predictive biomarker for the efficacy of anti-PD-1/PD-L1 antibody monotherapy.

Supplementary Materials: The following supporting information can be downloaded at: https://www.mdpi.com/article/10.3390/cancers17040572/s1. Figure S1: Association between circulatory levels of tVEGF-A and VEGF-A isoforms; Figure S2: Receiver operating characteristic (ROC) curve analysis for predicting the complete response or partial response to anti-programmed cell death 1/programmed cell death ligand 1 antibody monotherapy; Table S1: Baseline levels of circulatory tVEGF-A and VEGF-A isoforms before the initiation of anti-programmed cell death 1/programmed cell death ligand 1 antibody monotherapy.

Author Contributions: Conceptualization, K.Y.; methodology, K.Y.; validation, T.H. and K.Y.; formal analysis, T.H. and K.Y.; investigation, T.H.; resources, T.H., K.Y., K.F., K.S., S.S., Y.H., T.M., T.N., H.I., H.H., S.Y. and N.H.; data curation, T.H.; writing—original draft preparation, T.H.; writing—review and editing, K.Y., K.F., K.S., S.S., Y.H., T.M., T.N., H.I., H.H., S.Y. and N.H.; visualization, T.H.; supervision, H.I., H.H., and N.H.; project administration, N.H. All authors have read and agreed to the published version of the manuscript.

Funding: This research received no external funding.

Institutional Review Board Statement: The study was conducted in accordance with the Declaration of Helsinki and approved by the Ethics Committee of Hiroshima University Hospital (protocol code E2004-0326-23, approved 7 August 2024).

Informed Consent Statement: Informed consent was obtained from all subjects involved in the study.

Data Availability Statement: The raw data supporting the conclusions of this article will be made available by the authors on request.

Acknowledgments: We would like to thank Editage (https://www.editage.jp/, accessed on 9 January 2025) for the English language editing. Measurements of the levels of tVEGF-A, $VEGF_{121}$, and $VEGF_{165}$ were performed by Shino-Test Corporation.

Conflicts of Interest: Kakuhiro Yamaguchi reports personal fees from Chugai Pharmaceutical and Ono Pharmaceutical and research funding from Shino-Test Corporation. Takeshi Masuda reports personal fees from Daiichi-Sankyo, Taiho Pharmaceutical, Boehringer Ingelheim, Kyowa Kirin, Eli Lilly, Ono Pharmaceutical, Otsuka Pharmaceutical, Chugai Pharmaceutical, and AstraZeneca. Noboru Hattori reports personal fees from AstraZeneca, Boehringer Ingelheim, and Chugai Pharmaceutical and research funding from Ono Pharmaceutical, Chugai Pharmaceutical, and Shino-Test Corporation. All other authors declare no conflicts of interest.

Abbreviations

The following abbreviations are used in this manuscript:

AUC	area under the curve
CCL	C-C motif chemokine ligand
CI	confidence interval
COPD	chronic obstructive pulmonary disease
CR	complete response
CXCL	C-X-C motif chemokine ligand
ELISA	enzyme-linked immunosorbent assay
ICI	immune checkpoint inhibitor
IQR	interquartile range
NSCLC	non-small-cell lung cancer
NE	not evaluable
NRP	neuropilin
ORR	objective response rate
PD-1	programmed cell death 1
PD-L1	programmed cell death ligand 1
PFS	progression-free survival
PR	partial response

PD	progressive disease
RECIST	Response Evaluation Criteria in Solid Tumors
ROC	receiver operating characteristic
SD	stable disease
tVEGF	total vascular endothelial growth factor
TPS	tumor proportion score
VEGF	vascular endothelial growth factor
VEGFR	vascular endothelial growth factor receptor

References

1. Siegel, R.L.; Miller, K.D.; Jemal, A. Cancer statistics, 2020. *CA Cancer J. Clin.* **2020**, *70*, 7–30. [CrossRef] [PubMed]
2. Brahmer, J.; Reckamp, K.L.; Baas, P.; Crinò, L.; Eberhardt, W.E.E.; Poddubskaya, E.; Antonia, S.; Pluzanski, A.; Vokes, E.E.; Holgado, E.; et al. Nivolumab versus Docetaxel in Advanced Squamous-Cell Non-Small-Cell Lung Cancer. *N. Engl. J. Med.* **2015**, *373*, 123–135. [CrossRef]
3. Borghaei, H.; Paz-Ares, L.; Horn, L.; Spigel, D.R.; Steins, M.; Ready, N.E.; Chow, L.Q.; Vokes, E.E.; Felip, E.; Holgado, E.; et al. Nivolumab versus Docetaxel in Advanced Nonsquamous Non-Small-Cell Lung Cancer. *N. Engl. J. Med.* **2015**, *373*, 1627–1639. [CrossRef]
4. Reck, M.; Rodríguez-Abreu, D.; Robinson, A.G.; Hui, R.; Csőszi, T.; Fülöp, A.; Gottfried, M.; Peled, N.; Tafreshi, A.; Cuffe, S.; et al. Pembrolizumab versus Chemotherapy for PD-L1-Positive Non-Small-Cell Lung Cancer. *N. Engl. J. Med.* **2016**, *375*, 1823–1833. [CrossRef] [PubMed]
5. Mok, T.S.K.; Wu, Y.-L.; Kudaba, I.; Kowalski, D.M.; Cho, B.C.; Turna, H.Z.; Castro, G., Jr.; Srimuninnimit, V.; Laktionov, K.K.; Bondarenko, I.; et al. Pembrolizumab versus chemotherapy for previously untreated, PD-L1-expressing, locally advanced or metastatic non-small-cell lung cancer (KEYNOTE-042): A randomised, open-label, controlled, phase 3 trial. *Lancet* **2019**, *393*, 1819–1830. [CrossRef]
6. Herbst, R.S.; Giaccone, G.; de Marinis, F.; Reinmuth, N.; Vergnenegre, A.; Barrios, C.H.; Morise, M.; Felip, E.; Andric, Z.; Geater, S.; et al. Atezolizumab for First-Line Treatment of PD-L1-Selected Patients with NSCLC. *N. Engl. J. Med.* **2020**, *383*, 1328–1339. [CrossRef] [PubMed]
7. Lantuejoul, S.; Damotte, D.; Hofman, V.; Adam, J. Programmed death ligand 1 immunohistochemistry in non-small cell lung carcinoma. *J. Thorac. Dis.* **2019**, *11*, S89–S101. [CrossRef]
8. Nagy, J.A.; Dvorak, A.M.; Dvorak, H.F. VEGF-A and the induction of pathological angiogenesis. *Annu. Rev. Pathol.* **2007**, *2*, 251–275. [CrossRef]
9. Mabeta, P.; Steenkamp, V. The VEGF/VEGFR Axis Revisited: Implications for Cancer Therapy. *Int. J. Mol. Sci.* **2022**, *23*, 15585. [CrossRef]
10. Salgado, R.; Benoy, I.; Bogers, J.; Weytjens, R.; Vermeulen, P.; Dirix, L.; Van Marck, E. Platelets and vascular endothelial growth factor (VEGF): A morphological and functional study. *Angiogenesis* **2001**, *4*, 37–43. [CrossRef] [PubMed]
11. George, M.L.; Eccles, S.A.; Tutton, M.G.; Abulafi, A.M.; Swift, R.I. Correlation of plasma and serum vascular endothelial growth factor levels with platelet count in colorectal cancer: Clinical evidence of platelet scavenging? *Clin. Cancer Res. Off. J. Am. Assoc. Cancer Res.* **2000**, *6*, 3147–3152.
12. Silva-Hucha, S.; Pastor, A.M.; Morcuende, S. Neuroprotective Effect of Vascular Endothelial Growth Factor on Motoneurons of the Oculomotor System. *Int. J. Mol. Sci.* **2021**, *22*, 814. [CrossRef] [PubMed]
13. Lee, S.H.; Jeong, D.; Han, Y.S.; Baek, M.J. Pivotal role of vascular endothelial growth factor pathway in tumor angiogenesis. *Ann. Surg. Treat. Res.* **2015**, *89*, 1–8. [CrossRef] [PubMed]
14. Hanahan, D.; Weinberg, R.A. Hallmarks of cancer: The next generation. *Cell* **2011**, *144*, 646–674. [CrossRef]
15. Zhan, P.; Wang, J.; Lv, X.J.; Wang, Q.; Qiu, L.X.; Lin, X.Q.; Yu, L.K.; Song, Y. Prognostic value of vascular endothelial growth factor expression in patients with lung cancer: A systematic review with meta-analysis. *J. Thorac. Oncol. Off. Publ. Int. Assoc. Study Lung Cancer* **2009**, *4*, 1094–1103. [CrossRef] [PubMed]
16. Fukumura, D.; Kloepper, J.; Amoozgar, Z.; Duda, D.G.; Jain, R.K. Enhancing cancer immunotherapy using antiangiogenics: Opportunities and challenges. *Nat. Rev. Clin. Oncol.* **2018**, *15*, 325–340. [CrossRef] [PubMed]
17. Al Kawas, H.; Saaid, I.; Jank, P.; Westhoff, C.C.; Denkert, C.; Pross, T.; Weiler, K.B.S.; Karsten, M.M. How VEGF-A and its splice variants affect breast cancer development—Clinical implications. *Cell. Oncol.* **2022**, *45*, 227–239. [CrossRef] [PubMed]
18. White, A.L.; Bix, G.J. VEGFA Isoforms as Pro-Angiogenic Therapeutics for Cerebrovascular Diseases. *Biomolecules* **2023**, *13*, 702. [CrossRef]

19. Yamakuchi, M.; Okawa, M.; Takenouchi, K.; Bibek, A.; Yamada, S.; Inoue, K.; Higurashi, K.; Tabaru, A.; Tanoue, K.; Oyama, Y.; et al. VEGF-A165 is the predominant VEGF-A isoform in platelets, while VEGF-A121 is abundant in serum and plasma from healthy individuals. *PLoS ONE* **2023**, *18*, e0284131. [CrossRef]
20. Ferrara, N. Binding to the extracellular matrix and proteolytic processing: Two key mechanisms regulating vascular endothelial growth factor action. *Mol. Biol. Cell* **2010**, *21*, 687–690. [CrossRef]
21. Cheung, N.; Wong, M.P.; Yuen, S.T.; Leung, S.Y.; Chung, L.P. Tissue-specific expression pattern of vascular endothelial growth factor isoforms in the malignant transformation of lung and colon. *Hum. Pathol.* **1998**, *29*, 910–914. [CrossRef]
22. Kazemi, M.; Carrer, A.; Moimas, S.; Zandonà, L.; Bussani, R.; Casagranda, B.; Palmisano, S.; Prelazzi, P.; Giacca, M.; Zentilin, L.; et al. VEGF121 and VEGF165 differentially promote vessel maturation and tumor growth in mice and humans. *Cancer Gene Ther.* **2016**, *23*, 125–132. [CrossRef]
23. Hashiguchi, T.; Arimura, K.; Matsumuro, K.; Otsuka, R.; Watanabe, O.; Jonosono, M.; Maruyama, Y.; Maruyama, I.; Osame, M. Highly concentrated vascular endothelial growth factor in platelets in Crow-Fukase syndrome. *Muscle Nerve* **2000**, *23*, 1051–1056. [CrossRef]
24. Cappelletto, E.; Fasiolo, L.T.; Salizzato, V.; Piccin, L.; Fabozzi, A.; Contato, A.; Bianco, P.D.; Pasello, G.; Chiarion-Sileni, V.; Gion, M.; et al. Cytokine and soluble programmed death-ligand 1 levels in serum and plasma of cancer patients treated with immunotherapy: Preanalytical and analytical considerations. *Int. J. Biol. Markers* **2024**, *39*, 9–22. [CrossRef] [PubMed]
25. Shibaki, R.; Murakami, S.; Shinno, Y.; Matsumoto, Y.; Yoshida, T.; Goto, Y.; Kanda, S.; Horinouchi, H.; Fujiwara, Y.; Yamamoto, N.; et al. Predictive value of serum VEGF levels for elderly patients or for patients with poor performance status receiving anti-PD-1 antibody therapy for advanced non-small-cell lung cancer. *Cancer Immunol. Immunother.* **2020**, *69*, 1229–1236. [CrossRef] [PubMed]
26. Tiako Meyo, M.; Jouinot, A.; Giroux-Leprieur, E.; Fabre, E.; Wislez, M.; Alifano, M.; Leroy, K.; Boudou-Rouquette, P.; Tlemsani, C.; Khoudour, N.; et al. Predictive Value of Soluble PD-1, PD-L1, VEGFA, CD40 Ligand and CD44 for Nivolumab Therapy in Advanced Non-Small Cell Lung Cancer: A Case-Control Study. *Cancers* **2020**, *12*, 473. [CrossRef]
27. Ninomiya, H.; Ozeki, M.; Matsuzawa, Y.; Nozawa, A.; Yasue, S.; Kubota, K.; Endo, S.; Asano, T.; Taguchi, K.; Ohe, N.; et al. A pediatric case of anaplastic astrocytoma with a gliomatosis cerebri; the growth pattern and changes in serum VEGF-121 levels after bevacizumab treatment. *J. Clin. Neurosci.* **2020**, *81*, 431–433. [CrossRef]
28. Segerström, L.; Fuchs, D.; Bäckman, U.; Holmquist, K.; Christofferson, R.; Azarbayjani, F. The anti-VEGF antibody bevacizumab potently reduces the growth rate of high-risk neuroblastoma xenografts. *Pediatr. Res.* **2006**, *60*, 576–581. [CrossRef] [PubMed]
29. Spratlin, J.L.; Cohen, R.B.; Eadens, M.; Gore, L.; Camidge, D.R.; Diab, S.; Leong, S.; O'Bryant, C.; Chow, L.Q.; Serkova, N.J.; et al. Phase I pharmacologic and biologic study of ramucirumab (IMC-1121B), a fully human immunoglobulin G1 monoclonal antibody targeting the vascular endothelial growth factor receptor-2. *J. Clin. Oncol. Off. J. Am. Soc. Clin. Oncol.* **2010**, *28*, 780–787. [CrossRef] [PubMed]
30. Barratt, S.L.; Flower, V.A.; Pauling, J.D.; Millar, A.B. VEGF (Vascular Endothelial Growth Factor) and Fibrotic Lung Disease. *Int. J. Mol. Sci.* **2018**, *19*, 1269. [CrossRef]
31. Eisenhauer, E.A.; Therasse, P.; Bogaerts, J.; Schwartz, L.H.; Sargent, D.; Ford, R.; Dancey, J.; Arbuck, S.; Gwyther, S.; Mooney, M.; et al. New response evaluation criteria in solid tumours: Revised RECIST guideline (version 1.1). *Eur. J. Cancer* **2009**, *45*, 228–247. [CrossRef]
32. Wartiovaara, U.; Salven, P.; Mikkola, H.; Lassila, R.; Kaukonen, J.; Joukov, V.; Orpana, A.; Ristimäki, A.; Heikinheimo, M.; Joensuu, H.; et al. Peripheral blood platelets express VEGF-C and VEGF which are released during platelet activation. *Thromb. Haemost.* **1998**, *80*, 171–175. [CrossRef]
33. Banks, R.E.; Forbes, M.A.; Kinsey, S.E.; Stanley, A.; Ingham, E.; Walters, C.; Selby, P.J. Release of the angiogenic cytokine vascular endothelial growth factor (VEGF) from platelets: Significance for VEGF measurements and cancer biology. *Br. J. Cancer* **1998**, *77*, 956–964. [CrossRef] [PubMed]
34. Amirkhosravi, A.; Amaya, M.; Siddiqui, F.; Biggerstaff, J.P.; Meyer, T.V.; Francis, J.L. Blockade of GpIIb/IIIa inhibits the release of vascular endothelial growth factor (VEGF) from tumor cell-activated platelets and experimental metastasis. *Platelets* **1999**, *10*, 285–292. [CrossRef]
35. Zhou, L.; Zhang, Z.; Tian, Y.; Li, Z.; Liu, Z.; Zhu, S. The critical role of platelet in cancer progression and metastasis. *Eur. J. Med. Res.* **2023**, *28*, 385. [CrossRef] [PubMed]
36. Küsters, B.; de Waal, R.M.; Wesseling, P.; Verrijp, K.; Maass, C.; Heerschap, A.; Barentsz, J.O.; Sweep, F.; Ruiter, D.J.; Leenders, W.P. Differential effects of vascular endothelial growth factor A isoforms in a mouse brain metastasis model of human melanoma. *Cancer Res.* **2003**, *63*, 5408–5413.
37. Ferrara, N.; Gerber, H.P.; LeCouter, J. The biology of VEGF and its receptors. *Nat. Med.* **2003**, *9*, 669–676. [CrossRef] [PubMed]
38. Manenti, L.; Riccardi, E.; Marchini, S.; Naumova, E.; Floriani, I.; Garofalo, A.; Dossi, R.; Marrazzo, E.; Ribatti, D.; Scanziani, E.; et al. Circulating plasma vascular endothelial growth factor in mice bearing human ovarian carcinoma xenograft correlates with tumor progression and response to therapy. *Mol. Cancer Ther.* **2005**, *4*, 715–725. [CrossRef] [PubMed]

39. Kawai, H.; Minamiya, Y.; Ito, M.; Saito, H.; Ogawa, J. VEGF121 promotes lymphangiogenesis in the sentinel lymph nodes of non-small cell lung carcinoma patients. *Lung Cancer* **2008**, *59*, 41–47. [CrossRef] [PubMed]
40. Ohta, Y.; Watanabe, Y.; Murakami, S.; Oda, M.; Hayashi, Y.; Nonomura, A.; Endo, Y.; Sasaki, T. Vascular endothelial growth factor and lymph node metastasis in primary lung cancer. *Br. J. Cancer* **1997**, *76*, 1041–1045. [CrossRef]
41. Ljungberg, B.; Jacobsen, J.; Häggström-Rudolfssson, S.; Rasmuson, T.; Lindh, G.; Grankvist, K. Tumour vascular endothelial growth factor (VEGF) mRNA in relation to serum VEGF protein levels and tumour progression in human renal cell carcinoma. *Urol. Res.* **2003**, *31*, 335–340. [CrossRef] [PubMed]
42. Broséus, J.; Mourah, S.; Ramstein, G.; Bernard, S.; Mounier, N.; Cuccuini, W.; Gaulard, P.; Gisselbrecht, C.; Brière, J.; Houlgatte, R.; et al. VEGF(121), is predictor for survival in activated B-cell-like diffuse large B-cell lymphoma and is related to an immune response gene signature conserved in cancers. *Oncotarget* **2017**, *8*, 90808–90824. [CrossRef] [PubMed]
43. Zacchigna, S.; Pattarini, L.; Zentilin, L.; Moimas, S.; Carrer, A.; Sinigaglia, M.; Arsic, N.; Tafuro, S.; Sinagra, G.; Giacca, M. Bone marrow cells recruited through the neuropilin-1 receptor promote arterial formation at the sites of adult neoangiogenesis in mice. *J. Clin. Investig.* **2008**, *118*, 2062–2075. [CrossRef] [PubMed]
44. Carrer, A.; Moimas, S.; Zacchigna, S.; Pattarini, L.; Zentilin, L.; Ruozi, G.; Mano, M.; Sinigaglia, M.; Maione, F.; Serini, G.; et al. Neuropilin-1 identifies a subset of bone marrow Gr1- monocytes that can induce tumor vessel normalization and inhibit tumor growth. *Cancer Res.* **2012**, *72*, 6371–6381. [CrossRef]
45. Galeati, G.; Spinaci, M.; Govoni, N.; Zannoni, A.; Fantinati, P.; Seren, E.; Tamanini, C. Stimulatory effects of fasting on vascular endothelial growth factor (VEGF) production by growing pig ovarian follicles. *Reproduction* **2003**, *126*, 647–652. [CrossRef] [PubMed]
46. Hanefeld, M.; Engelmann, K.; Appelt, D.; Sandner, D.; Weigmann, I.; Ganz, X.; Pistrosch, F.; Köhler, C.; Gasparic, A.; Birkenfeld, A.L. Intra-individual variability and circadian rhythm of vascular endothelial growth factors in subjects with normal glucose tolerance and type 2 diabetes. *PLoS ONE* **2017**, *12*, e0184234. [CrossRef]

cancers

MDPI

Article

Immunological Network Signature of Naïve Non-Oncogene-Addicted Non-Small Cell Lung Cancer Patients Treated with Anti-PD1 Therapy: A Pilot Study

Pasquale Sibilio [1], Ilaria Grazia Zizzari [2,*], Alain Gelibter [3], Marco Siringo [3], Lucrezia Tuosto [2], Angelica Pace [2], Angela Asquino [2], Flavio Valentino [2], Arianna Sabatini [3], Manuela Petti [1], Filippo Bellati [4], Daniele Santini [3], Marianna Nuti [2], Lorenzo Farina [1], Aurelia Rughetti [2] and Chiara Napoletano [2]

[1] Department of Computer, Control and Management Engineering, Sapienza University of Rome, 00161 Rome, Italy; pasquale.sibilio@uniroma1.it (P.S.); manuela.petti@uniroma1.it (M.P.); lorenzo.farina@uniroma1.it (L.F.)
[2] Laboratory of Tumor Immunology and Cell Therapies, Department of Experimental Medicine, Sapienza University of Rome, 00161 Rome, Italy; lucrezia.tuosto@uniroma1.it (L.T.); angelica.pace@uniroma1.it (A.P.); angela.asquino@uniroma1.it (A.A.); flavio.valentino@uniroma1.it (F.V.); marianna.nuti@uniroma1.it (M.N.); aurelia.rughetti@uniroma1.it (A.R.); chiara.napoletano@uniroma1.it (C.N.)
[3] Division of Oncology, Department of Radiological, Oncological and Pathological Science, Policlinico Umberto I, Sapienza University of Rome, 00161 Rome, Italy; alain.gelibter@uniroma1.it (A.G.); marco.siringo@uniroma1.it (M.S.); arianna.sabatini@uniroma1.it (A.S.); daniele.santini@uniroma1.it (D.S.)
[4] Department of Medical and Surgical Sciences and Translational Medicine, Sant'Andrea University Hospital, Sapienza University of Rome, Via di Grottarossa 1035, 00189 Rome, Italy; filippo.bellati@uniroma1.it
* Correspondence: ilaria.zizzari@uniroma1.it; Tel.: +39-0649973025

Simple Summary: This research proposes distinct immunological profiles associated with non-responding NSCLC patients who have poor survival outcomes and those with a more favorable prognosis and better performance status. An inflammatory signature characterizes the patients in the first group, while a network based on checkpoint molecules identifies NSCLC patients with better outcomes. Defining the connectivity among the molecules of each profile serves as an optimal starting point for developing combinatory targeted drugs that aim to optimize the therapeutic strategies for each patient and avoid unnecessary, toxic treatments.

Abstract: Background/Objectives: Non-small cell lung cancer (NSCLC) patients without gene driver mutations receive anti-PD1 treatments either as monotherapy or in combination with chemotherapy based on PD-L1 expression in tumor tissue. Anti-PD1 antibodies target various immune system components, perturbing the balance between immune cells and soluble factors. In this study, we identified the immune signatures of NSCLC patients associated with different clinical outcomes through network analysis. **Methods**: Twenty-seven metastatic NSCLC patients were assessed at baseline for the levels of circulating $CD137^+$ T cells (total, $CD4^+$, and $CD8^+$) via cytofluorimetry, along with 14 soluble checkpoints and 20 cytokines through Luminex analysis. Hierarchical clustering and connectivity heatmaps were executed, analyzing the response to therapy (R vs. NR), performance status (PS = 0 vs. PS > 0), and overall survival (OS < 3 months vs. OS > 3 months). **Results:** The clustering of immune checkpoints revealed three groups with a significant differential proportion of six checkpoints between patients with PS = 0 and PS > 0 ($p < 0.0001$). Furthermore, significant pairwise correlations among immune factors evaluated in R were compared to the lack of significant correlations among the same immune factors in NR patients and vice versa. These comparisons were conducted for patients with PS = 0 vs. PS > 0 and OS < 3 months vs. OS > 3 months. The results indicated that NR with PS > 0 and OS $\leq$ 3 months exhibited an inflammatory-specific signature compared to the contrasting clinical conditions characterized by a checkpoint molecule-based network ($p < 0.05$). **Conclusions:** Identifying various connectivity immune profiles linked to response to therapy, PS, and survival in NSCLC patients represents significant findings that can optimize therapeutic choices.

Keywords: NSCLC; anti-PD1 therapy; network analysis

1. Introduction

Targeting the PD1/PD-L1 pathways enhances the immune response against tumor cells, providing clinical benefits for cancer patients with advanced solid tumors, such as non-small cell lung cancer (NSCLC) [1]. Current therapies for metastatic, non-oncogene-addicted NSCLC patients are based on the tumor expression of PD-L1 (tPD-L1), which is defined by the tumor proportional score (TPS). Patients receive pembrolizumab (anti-PD1) [2] when TPS $\geq$ 50%, and either pembrolizumab plus chemotherapy [3,4] or a combination of nivolumab (anti-PD1) and ipilimumab (anti-CTLA4) with chemotherapy [5] when TPS <50%. The latter two treatments demonstrate comparable efficacy, with an objective response rate of 48.3%, a median progression-free survival (PFS) of 9 months, and a 4-year overall survival (OS) rate of 23.6% for pembrolizumab plus chemotherapy. In contrast, the nivolumab/ipilimumab plus chemotherapy combination shows an objective response rate of 38%, a median PFS of 6.7 months, and a 4-year OS rate of 22% [3,5]. Although tPD-L1 expression is validated and employed in clinical practice, it remains an inadequate biomarker with limited predictive value.

It is well known that anti-PD-1/PD-L1 treatments act on various components of the immune system, altering the balance among immune cells and soluble factors [6]. Identifying the connections between each factor of the immune system could help define specific signatures of patients who may benefit from immunotherapy and have a favorable prognosis, serving as valuable biomarkers for identifying patients with a defined clinical outcome.

Among immune cells, $CD137^+$ lymphocytes form a T-cell subset that significantly contributes to the anti-tumor immune response. Activated $CD8^+$ and $CD4^+$ T cells express high levels of CD137 (4–1BB) marker, which induces effector functions, division, and survival of T cells [7,8], enhances mitochondrial metabolism in T cells [9], and promotes DNA methylation of CD8 genes [10]. These cells are recognized as tumor-specific T cells [11]. We have also demonstrated their role as predictive and prognostic biomarkers in NSCLC and other solid tumors [12,13].

Similarly, soluble checkpoints and cytokines are other critical players in the overall anti-tumor response in cancer patients. Immune cells release checkpoint molecules as alternative splice variants via microvesicles or proteolytic cleavage [14,15]. These molecules maintain their functional activity in modulating the anti-tumor immune response. Furthermore, their concentrations change during therapy, affecting the overall response rate of cancer patients [13,16,17]. Soluble PD1 (sPD1) is the most studied among patients with NSCLC. sPD1 inhibits the interaction between PD1 and PD-L1, enhancing T-cell responses, increasing the release of IFNγ, and reducing the percentage of regulatory T cells (Tregs) [18]. A similar effect has been observed for sCD80, which reverses PD-L1 signaling by binding to PD-L1 [19,20]. sPD1 is positively correlated with response and survival in NSCLC [17,21]. In contrast, high levels of sPD-L1 and sPD-L2 are associated with shorter progression-free survival and resistance to immunotherapy [21–24]. Likewise, sBTLA and sCTLA4 are considered negative regulators of the immune response linked to poor prognosis. sBTLA inhibits T-cell activation upon binding to its ligand HVEM, which is also expressed by antigen-presenting cells [25]. sCTLA4 acts as an immunosuppressive factor by blocking CD28-B7.1 ligation, inducing the release of IDO (indoleamine 2,3-deoxygenase), a tryptophan catabolic enzyme, and the FoxO3 transcription factor that regulates inflammatory cytokine production [25–28]. Recently, it was demonstrated that NK cells expressing CTLA-4 exhibited reduced cytotoxic activity, produced lower amounts of IFNγ and TNF-α, and increased IL-10 release [29].

Among the soluble factors, cytokines support the immune response toward inflammation (IL-1, IL-4, IL-6, IL-8, IL-13, IL-17, and TNFα), immune suppression (IL-10, TGF-β, and IL-35), and immune activation (IL-2, IL-12, and IFN-γ). Several cytokines, such as IFN-γ, exhibit pleiotropic activity and can function as both immune activators and suppressors [30]. Moreover, these molecules may reprogram the metabolic pathways of tumor cells, promoting metastasis and cell proliferation [31]. In recent years, serum cytokines have emerged as potential biomarkers for predicting treatment outcomes. High levels of IL-6 and IL-10 are correlated with poor survival in NSCLC patients undergoing immunotherapy [32]. Several cytokines, including IL-5, IL-6, IL-8, IL-4, and IL-10, have been identified as potential prognostic factors in NSCLC patients receiving anti-PD-1 treatment in combination with chemotherapy [33].

This evidence highlights that many immune parameters act simultaneously in the response against tumors (influencing reciprocally). The contribution of each factor strongly depends on its interaction with the immune context and tumor microenvironment that characterize each patient.

This study employs network analysis to evaluate the immunological connections among activated cells (including T-cell subsets), cytokines, and soluble immune checkpoints. These relationships were correlated with various clinical parameters, such as response to therapy, performance status, and overall survival, to identify specific immune signatures that indicate which patients are more suitable for immunotherapy.

2. Materials and Methods

2.1. Patients' Characteristics

Twenty-seven patients diagnosed with metastatic non-oncogene-addicted NSCLC (stage IV) were enrolled at Policlinico Umberto I Hospital between 2022 and 2023. NGS analysis confirmed the mutational profile for each patient. These patients received immune checkpoint inhibitor (ICI) treatment following Italian guidelines. Patients with a TPS of $\geq$50% received pembrolizumab as monotherapy, while those with a TPS of <50% were treated with a combination of chemotherapy and ICIs (either pembrolizumab or nivolumab and ipilimumab), based on the physicians' discretion. The Inclusion and Exclusion Criteria for NSCLC patients are illustrated in Supplementary Figure S1.

PS describes the patient's level of functioning based on physical ability, daily activities, and self-care capabilities. PS = 0 indicates fully active patients with no restrictions on activities; PS = 1 characterizes patients who cannot perform strenuous activities but are able to carry out light housework and sedentary tasks; PS = 2 defines patients who can walk and manage self-care but are unable to work; PS = 3 describes patients confined to bed or a chair for more than 50% of waking hours and capable of limited self-care; PS = 4 defines patients who are completely disabled.

Each patient's response to treatment and overall survival (OS) were evaluated. Responder (R) patients displayed a complete, partial response, or stable disease according to iRECIST criteria, whereas non-responders (NR) exhibited progression, both evaluated after 6 months of therapy. OS corresponded to the duration between the date of treatment initiation and death.

This study was conducted in accordance with good clinical practice guidelines and the Declaration of Helsinki, and it was approved by the Ethics Committee of Policlinico Umberto I (Ethical Committee Protocol, RIF.CE: 4181).

2.2. PBMC and Serum Collection

Peripheral blood mononuclear cells (PBMCs) and serum samples derived from 27 NSCLC patients were isolated prior to the initiation of immunotherapy. Specifically, blood samples were collected using BD Vacutainer EDTA tubes for PBMC isolation and BD Vacutainer Plus Plastic Serum tubes (both from Becton Dickinson, Franklin Lakes, NJ, USA) for serum isolation. PBMCs were stratified on Ficoll–Hypaque (Lympholite-H) (Cedarlane, Burlington, ON, Canada) and centrifuged for 30 min at 1400 rpm. The PBMCs were then collected and washed three times at 1200 rpm with PBS without Ca^{2+} and Mg^{2+} (Sigma-Aldrich, St. Louis, MO, USA). Serum samples were isolated by centrifuging the serum tubes for 30 min at 1800 rpm. PBMCs and serum were cryopreserved until use.

2.3. Flow Cytometry

The evaluation of T-cell subsets was conducted using cytofluorimetry with a multiparametric analysis employing the following monoclonal antibodies (MoAbs)—anti-CD3-BV510 (HIT3a clone), CD8-APC-H7 (SK1 clone), and CD137 (4–1BB)-APC (4B4–1 clone)—all sourced from BD Biosciences, San Jose, CA, USA. Live cells were identified utilizing the Live/Dead cell exclusion (Beckman Coulter, Brea, CA, USA). The negative controls were established using fluorescence minus one (FMO) and autofluorescence. All samples were processed using the DxFLEX Flow Cytometer (Beckman Coulter) and analyzed via FlowJo software (version 10.8.8, Becton Dickinson). Gating strategies are reported in the Supplementary Figure S2.

2.4. Cytokine and Chemokine Evaluation

Soluble immune checkpoints and cytokines were measured using the Immuno-Oncology Checkpoint 14 Plex Human ProcartaPlex Panel and the Inflammation 20 Plex Human ProcartaPlex Panel (both from ThermoFisher Scientific, Waltham, MA, USA) following the manufacturer's instructions. The 14 immune checkpoints and 20 cytokines analyzed were the BTLA, GITR, HVEM, IDO, LAG-3, PD-1, PD-L1, PD-L2, TIM-3, CD28, CD80, CD137, CD27, and CD152 checkpoints, as well as the sE-Selectin, GM-CSF, ICAM/CD54, IFNα, IFNγ, IL1α, IL1β, IL4, IL6, IL8, IL10, IL12p70, IL13, IL17A/CTLA8, IP10/CXCL10, MCP1/CCL2, MIP1α/CCL3, MIP1β/CCL4, sP-Selectin, and TNFα cytokines. All these factors were evaluated using Luminex multiplex assays and analyzed with Bioplex Manager MP 6.2 software (Bio-Rad, Hercules, CA, USA). The instrument did not reveal the quantities of HVEM, PD-L1, GM-CSF, and MIP1α because their values were below the standard curves. The median values of these soluble factors are reported in Supplementary Tables S1 and S2.

2.5. Hierarchical Clustering of Circulant Molecules' Expression Profile

The expression profiles of soluble molecules (cytokines and checkpoints) were logistically transformed. The checkpoint dataset was preprocessed by removing patients with outlier expression profiles through hierarchical clustering and applying a height threshold to the cluster dendrogram. Two patients were excluded from the analysis of checkpoints (25 patients in total) as outliers, while all patients were examined for the cytokine clustering study.

Unsupervised hierarchical clustering was conducted in the R environment using Euclidean metrics like distance and Ward's method. The D2 clustering algorithm was applied through the heatmap function in R. The differential expression analysis of checkpoint expression between the checkpoint-induced clusters was evaluated using the linear mixed model from the limma R package [34]. The resulting p-values were adjusted with the false discovery rate (FDR) to control for the expected proportion of false positives among the rejected hypotheses.

2.6. Differential Correlation Analysis of Multiple Clinical Conditions

Although the clustering analysis provided insights into the immune state of patients with the worst outcomes, it is limited by a small sample size. We performed a differential correlation analysis to address this limitation and provide more information about the molecular network of the immune system involved in immune therapy outcomes. In this case, the molecular profiles of patients were not evaluated in isolation; rather, the focus of the analysis shifted to all possible associations between the immune molecules and cell pairs, as well as how they change between patients with opposing therapeutic outcomes. We utilized the Differential Gene Correlation Analysis (DGCA) R package [35]. The Pearson correlation coefficient was used to assess the linear relationship between all immune molecules and cell pairs, given that a sample size of approximately 30 is generally considered acceptable when the data meet the assumptions of normality and linearity [36]. DGCA transforms sample correlation coefficients into z-scores to stabilize the variance of r, thereby allowing reliable comparisons of correlation coefficients across different subgroups. Then, DGCA uses the differences in z-scores to assess the statistical significance of the differentially correlated gene pairs.

In our study, we set DGCA to perform a differential correlation analysis of the immune molecule/cell pairwise Pearson's correlation coefficients among multiple clinical conditions. The dataset, which included circulating checkpoint and membrane markers ($CD3^+CD137^+$, $CD8^+CD137^+$, $CD4^+CD137^+$ T cells), as well as cytokines, was z-normalized and used as input for DGCA. Consequently, we calculated the immune molecule/cell pairs that exhibited a differential correlation between the following comparisons: (1) non-responding vs. responding patient groups, (2) PS > 0 vs. PS = 0, and (3) OS < 12 vs. OS > 12. To identify immune molecule/cell pairs that were differentially correlated across multiple conditions, we focused on those with a difference in the absolute value of the z-scores > 1.64 (p-value < 0.05).

The DGCA could categorize differentially correlated immune molecule/cell pairs into nine possible categories across two conditions: +/+: positive correlation in both conditions A and B; +/0: positive correlation in condition A and no significant correlation in condition B; +/−: positive correlation in condition A and negative correlation in condition B; 0/+: no significant correlation in condition A and positive correlation in condition B; 0/0: no significant correlation in either condition; 0/−: no significant correlation in condition A and negative correlation in condition B; −/+: negative correlation in condition A and positive correlation in condition B; −/0: negative correlation in condition A and no significant correlation in condition B; −/−: negative correlation in both conditions A and B. Molecule pairs that maintain the same sign of correlation between different conditions, such as the +/+ and −/− classes, were not further investigated.

3. Results

3.1. Patients' Characteristics

Twenty-seven patients with metastatic non-oncogenic-addicted NSCLC were enrolled in this study, as detailed in Table 1. Most histotypes were adenocarcinoma (21), with 16 patients having a TPS < 50%, while 11 had a TPS $\geq$ 50%. Fifteen patients were classified as PS = 0 (55%), and twelve were rated as PS > 0. Twenty patients were current or former smokers (74%), while 26% stated they had never smoked. Patients with an OS < 3 months (6 patients, 22%) were categorized as early progressors, whereas 21 exhibited an OS > 3 months. The response to treatments was assessed after six months of immunotherapy, with

14 patients considered responders (R) (52%) and 13 non-responders (NR) (48%). Median values of PFS and OS are reported in Supplementary Figure S3

Table 1. Patients' characteristics.

Tot	N°27 (100%)
Sex	
Male	18 (67%)
Female	9 (33%)
AgeMedian range	48–84
≤75	22 (81)
>75	5 (19)
Histotype	
Squamous	6 (22)
Adenocarcinoma	21 (78)
TPS	
<50%	16 (59)
≥50%	11 (41)
EOCG Performance Status	
0	15 (55)
>0	12 (45)
Overall Survival	
≤3 months	6 (22)
>3 months	21 (78)
Smoking Status	
Current	13 (48)
Former	7 (26)
Non-smoker	7 (26)
Response to immunotherapy	
Yes	14 (52)
No	13 (48)

3.2. Non-Responding Patients with PS > 0 Showed an Immunosuppressive Soluble Checkpoint Signature

The unsupervised hierarchical clustering of the expression profiles of circulating checkpoints resulted in three clusters (Cluster 1: $n = 3$ (left column), Cluster 2: n = 15 (central column), Cluster 3: n = 7 (right column), Figure 1). Cluster 1 is characterized by a higher expression of six circulating checkpoints (i.e., sCD80, sBTLA, sGITR, sCD137, sPD1, and sPD-L2) compared to Clusters 2 and 3 (log2 Fold Change > 1.5, FDR < 0.05, limma modified *t*-test, Supplementary Table S1). Cluster 3 is remarkable for its significantly higher expression of sCD27 compared to Clusters 1 and 2 (log2 Fold Change > 1.5, FDR < 0.05, limma modified *t*-test, Supplementary Table S3).

Furthermore, we observe a statistically significant difference in the proportion of patients' performance status (PS) among the clusters through the application of Fisher's exact test (*p*-value < 0.05, Table 2). Indeed, Cluster 1 consists entirely of patients with PS > 0 who are non-responsive to therapy, whereas Cluster 2 includes 60% of responders (9/15) and 53% of patients (8/15) with PS = 0. Interestingly, 85% of non-responding patients (5/6) have a PS > 0. Cluster 3 primarily comprises patients who respond to immunotherapy (4/7, 57%) or have PS = 0 (6/7, 85.7%); notably, the only patient with PS > 0 in this cluster was also non-responsive. Moreover, no differences between the three clusters were detected when evaluating the differential proportions of the variables according to immune response and OS. All these results suggest that non-responding patients with a worse PS

(Cluster 1) exhibit a more immunosuppressive profile than those with a longer response to ICIs and/or a PS = 0, confirming that poorer clinical status corresponds to a failure of the immune response.

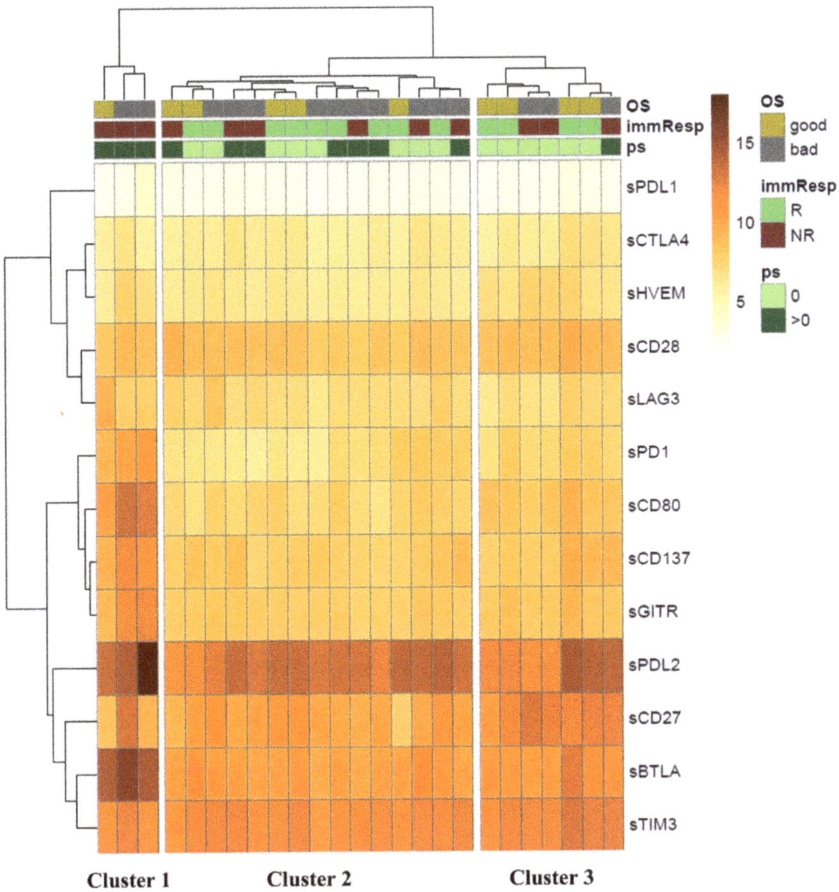

Figure 1. Unsupervised hierarchical clustering of the circulant checkpoints reveals three identified clusters: Cluster 1 corresponds to the left column, Cluster 2 to the central column, and Cluster 3 to the right column. The clustering includes 13 checkpoints analyzed across 25 patients (with two patients excluded due to outliers) and was conducted based on the following criteria: Overall survival (OS) categorized into good and poor (bad), with OS > 3 months indicated by a yellow square and ≤3 months by a gray square; response to treatment (immResp) after 6 months denoted as R for responders (green square) and NR for non-responders (red square); performance status (PS) = 0 indicated by a light green square and PS > 0 by a dark green square.

Although high levels of sPD1 and sCD80 are generally associated with better responses and improved survival, the increased expression of various immunosuppressive checkpoint molecules, such as sBTLA, sGITR, sCD137, and sPD-L2, shifts the immune balance toward a modified anti-tumor response. This pro-tumoral balance is further influenced by the release of multiple cytokines essential for maintaining the inflammatory environment in NR and PS > 0 patients. Indeed, unsupervised hierarchical clustering performed on the cytokines, chemokines, and adhesion molecules released into peripheral blood (Figure 2) identified two clusters. Cluster 1 (left column), consisting of 75% of NR patients with PS > 0, was primarily characterized by elevated levels of several molecules (log2 Fold Change > 1.5,

FDR < 0.05, limma modified *t*-test, Supplementary Table S4), including TNFα, IL1β, IL4, IL6, IL17, and CCL2, which are associated with an inflammatory network that supports tumor growth. Cluster 2 (right column) included 56% of responders and 61% of patients classified as PS = 0. Twenty-six percent of these patients (6/23) were simultaneously classified as NR and PS > 0, indicating that proinflammatory soluble factors are not the sole parameters influencing overall response rates and patient survival.

Table 2. The differential proportions of patients' overall survival, response to ICIs, and performance status evaluated in the soluble checkpoint inhibitor clustering.

Clinical Variables		Cluster 1	Cluster 2	Cluster 3	*p*-Value Fisher Test
Overall survival	Good	2	10	3	NS
	Bad	1	5	4	
Response to therapy	R	3	6	3	NS
	NR	0	9	4	
Performance Status	0	0	8	6	$p < 0.0005$
	>0	3	7	1	

NS: not statistically significant.

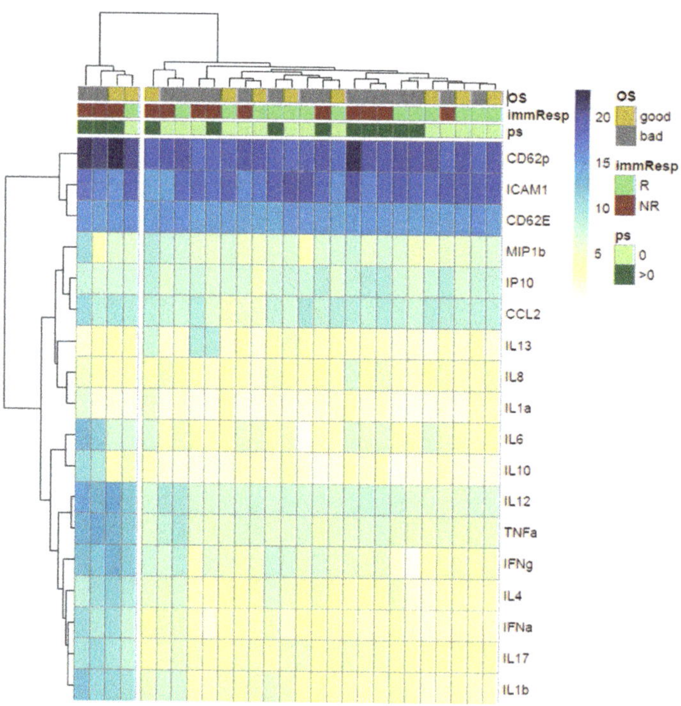

Figure 2. Unsupervised hierarchical clustering of the circulant cytokines. Two different clusters were identified: Cluster 1 corresponds to the left column, and Cluster 2 corresponds to the right column. The clusterization includes 18 cytokines, chemokines, and adhesion molecules analyzed in 27 patients and was performed according to 1. Overall survival (OS) divided for good and bad, corresponding to OS > 3 (yellow square) and ≤3 months (gray square), respectively; response to treatment (immResp) after 6 months (R: responders, green square; NR: non-responders, red square); performance status (PS) = 0 (light green square) and >0 (dark green square).

However, no significant differences were observed between the two clusters when analyzing the differential proportions of variables based on OS, response to therapy, and PS (Table 3).

Table 3. The differential proportions of patients' overall survival, response to ICIs, and performance status evaluated in the cytokine/chemokine clustering.

Clinical Variables		Cluster 1	Cluster 2	*p*-Value Fisher Test
Overall Survival	Good	15	2	NS
	Bad	8	2	
Response to therapy	R	10	3	NS
	NR	13	1	
Performance Status	0	3	1	NS
	>0	9	14	

NS: not statistically significant.

3.3. Non-Responding Patients Showed a Pronounced Inflammatory Network

The results of the DGCA analysis can be represented as a network, where the nodes are the immune soluble molecules (cytokines, chemokines, adhesion molecules, and checkpoints) and cellular subsets (CD137$^+$ T cells: total, CD4$^+$, and CD8$^+$). A link occurs when the molecules and cell subset pairs are differentially correlated between two conditions, according to DGCA. In the comparison between non-responding (NR) and responding (R) groups, we found 75 differentially correlated pairs of soluble molecules and cell subsets; 11 of these links were not represented because the correlation among the immune parameters showed a similar trend in both NR and R groups (NR/R: +/+ and −/−, corresponding to positive and negative correlation in both groups, respectively). Most of the differentially correlated immune molecule/cell pairs were included in the class +/0 (Figure 3a), where we observed 44 positive correlations (59%) involving diverse molecule species and lymphocyte subsets in NR, with no significant correlation in R. Similarly, other negative correlations were found to be significant in the NR group (six, accounting for 8% of the total links) compared to R, as observed in the class −/0 (Figure 3a). Conversely, 14 significant positive correlations (19%) were found in the R group compared to NR patients, as seen in class 0/+ (Figure 3a).

We highlighted the most connected nodes for each edge class (NR vs. R), specifically those with a class-specific degree above the 90th percentile of the entire class-specific degree distribution (Supplementary Table S5). This includes sBTLA, sCD80, IFNα, and IL12 for the +/0 class; sCD27 for the -/0 class; and sCTLA4, sCD28, and sLAG3 for the 0/+ class. In the non-responding group, there were limited negative connections, which in most cases involved a checkpoint molecule connected to one or more cytokines linked to the inflammatory pathway (IL6, IL4, IL13, and IL17) (Figure 3b). In the responding group, there were only positive connections, primarily among checkpoints, except for the interactions between IFNα and IL17-IL13. Interestingly, we found fewer connections among molecules in the responding group when comparing the network connectivity analysis between responders and non-responders. Furthermore, no common nodes were identified between the highlighted nodes within each edge class, indicating a distinct signature of network connectivity characterizing the responding and non-responding patients (Figure 3b).

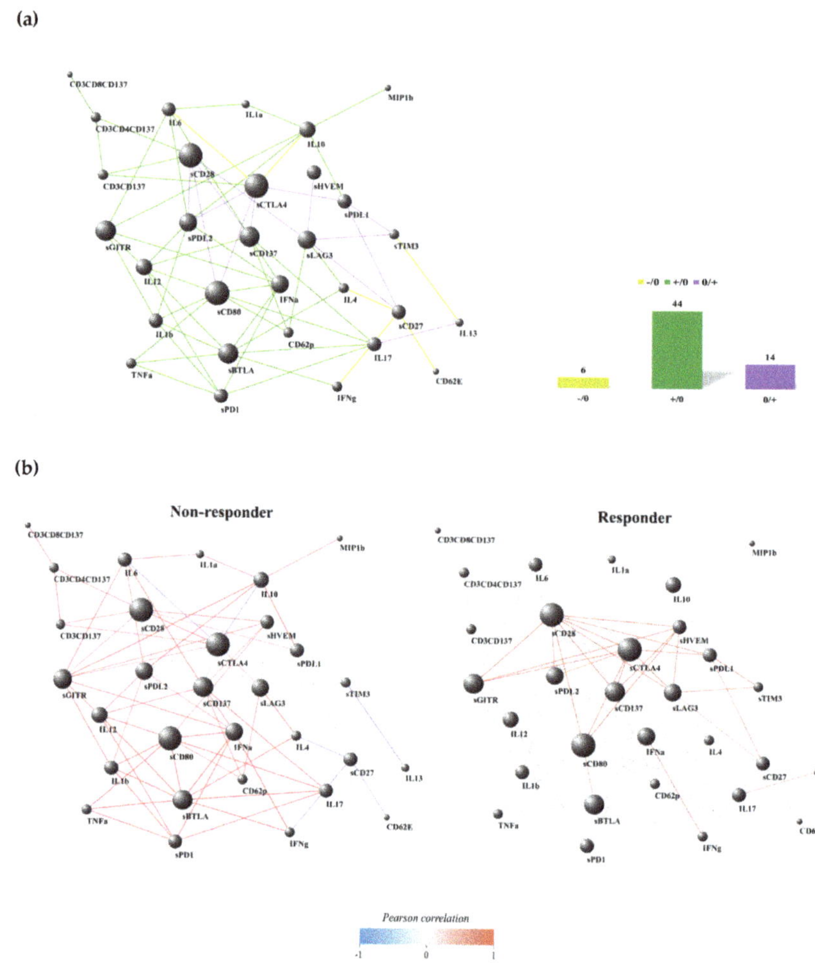

Figure 3. Connectivity network evaluation concerning the response to treatment in 27 naïve NSCLC patients (**a**) and dividing the analysis into non-responder and responder groups (**b**). In each network, nodes represent the immune factors (soluble cytokines/chemokines/adhesion and checkpoint molecules and CD137^{+} T-cell subsets); the link between two nodes is established when the absolute value of Spearman correlation between their expression levels is statistically significant (p-value ≤ 0.05). Node volume depends on the number of connections. The color of the connection depends on the class described. The yellow links and histogram identify the −/0 class that corresponds to the presence of negative correlations among the NR group, not present in the R group; the green connections and histogram discern the +/0 class that corresponds to the presence of positive correlations among the NR group, not present in the R group; the violet links and histogram are associated with the 0/+ class that corresponds to the presence of positive correlations among the R group, not present in the NR group.

3.4. PD-L2-IL6 and PD-L2-IL10 Connections Were Inversely Correlated in Patients with PS > 0 and PS = 0

The network analysis of performance status subgroups (PS > 0 vs. PS = 0) revealed 43 differentially correlated molecule/cellular pairs (PS > 0/PS = 0: +/0, −/0, 0/+, and +/−). Eight of these connections were excluded because a similar trend was observed between patients scored as PS > 0 and PS = 0 (+/+ and −/−). Interestingly, the two PS subgroups exhibited a similar differential correlation network between the R and NR groups (Figure 4) due to the significant overlap between the responding and PS score subgroups. Indeed, in the PS differential correlation network, most of the positive molecule pairs (24) were linked through the +/0 class (55%, Figure 4a), indicating a loss of correlation among different molecule species when transitioning from the worst state (PS > 0) to a better state (PS = 0) in the patients. A similar observation was made for the −/0 class (Figure 4a), where four negative links were noted exclusively in patients with a PS > 0 score (9%). In the 0/+ class, five significant positive correlations were identified with neighboring molecules in the PS = 0 state (12%) compared to the PS > 0 states (Figure 4a). Notably, we identified two molecule pairs (5%) that interacted through the +/− edge class (PD-L2-IL6 and PD-L2-IL10), altering the sign of their correlation when transitioning from PS > 0 to PS = 0 states (Figure 4a,b).

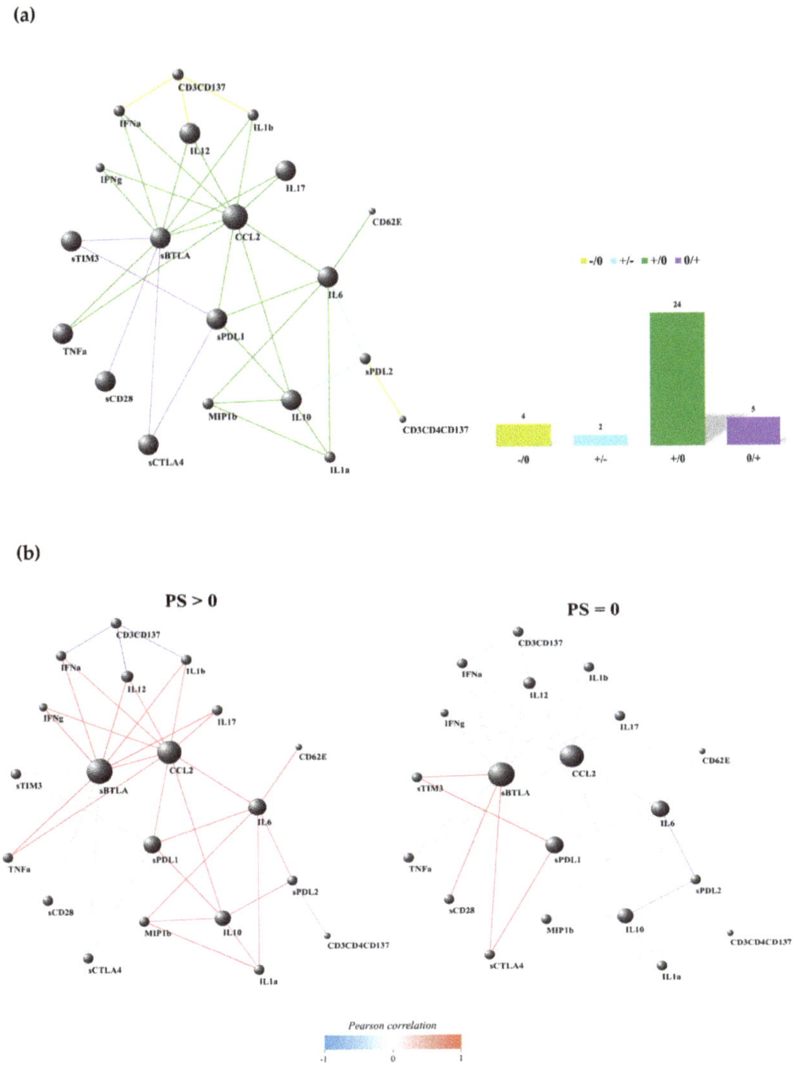

Figure 4. Connectivity network evaluation concerning performance status (PS) in the overall NSCLC population (**a**) and dividing the analysis into patients scored as PS > 0 and PS = 0 (**b**). In each network,

nodes represent the immune factors (soluble cytokines/chemokines/adhesion and checkpoint molecules and CD137$^+$ T-cell subsets); the link between two nodes is established when the absolute value of Spearman correlation between their expression levels is statistically significant (p-value ≤ 0.05). Node volume depends on the number of connections. The color of the connection depends on the class described. The yellow links and histogram identify the −/0 class that corresponds to the presence of negative correlations among patients with PS > 0, not present in patients with PS = 0; the sky blue links and histogram represent the +/− class that corresponds to the presence of positive and negative correlations between the PS > 0 and PS = 0 groups, respectively; the green connections and histogram represent the +/0 class that corresponds to the presence of positive correlations among the PS > 0 patients, not present in the PS = 0 group; the violet links and histogram are associated with the 0/+ class that corresponds to the presence of positive correlations among patients belonging to the PS = 0 group, not present in the patients with PS > 0.

We highlighted the most connected nodes within each edge class of the PS differential correlation network by selecting the nodes with a class-specific degree above the 90th percentile of the entire class-specific degree distribution (Supplementary Table S6). The highlighted molecules for the +/0 class were sBTLA and CCL2, while the highlighted node for the −/0 class was the cellular subset CD3$^+$CD137$^+$. The selected molecule for the +/− class was sPD-L2. Notably, sBTLA emerged as the circulating checkpoint with the highest degree in the 0/+ class, playing a significant role in the transition from a PS > 0 to a PS = 0 state. Furthermore, in the PS > 0 group, the only negative connection involved the cellular subset CD137$^+$ (total and CD4$^+$) T cells with cytokines (IFNα, IL1β, and IL12) and PD-L2, respectively. In the PS = 0 group, we mainly observed low positive connections and two negative associations between PD-L2 and the proinflammatory cytokines IL6 and IL10 (Figure 4b).

3.5. Most of the PD-L1 Connections Were Interrupted in Patients with a Favorable Prognosis

The analyses of DCGA were conducted by comparing patients with overall survival (OS) of less than 3 months to those with OS greater than 3 months. This timing identified patients with early progression (OS < 3 months). We found 52 differentially correlated molecule pairs (OS < 3 months/OS > 3 months: +/0, 0/−, and 0/+); 6 of these correlations were not included because the two OS groups exhibited a similar trend (+/+ and −/−). As previously described, most differentially correlated molecule pairs were found in the +/0 class (37 pairs, corresponding to 71%, Figure 5a), confirming the trend of certain molecule species, such as sPD-L1 and inflammatory cytokines like IL1β, TNFα, or IL17, being significantly correlated in the poor prognosis group (OS $\leq$ 3 months) and losing any significant correlation in the good prognosis group (OS > 3 months). In the 0/+ class, seven positive links were observed only in patients with OS > 3 months (13%, Figure 5a). Interestingly, we discovered two molecule pairs interacting in the 0/− class (4%, Figure 5a), represented by the nodes sTIM3-IL13 and sPD-L2-IL6.

We highlighted the most connected nodes for each edge class of the OS differential correlation network, i.e., nodes with a class-specific degree higher than the 90th percentile of the entire class-specific degree distribution (Figure 5b and Supplementary Table S7). This highlighted ICAM in the class 0/+ and sPD-L1, IL1a, sCD28, and sLAG3 for the class +/0. We only observed a positive connection for the patients in the poor survival group among the soluble factors. All positive connections were found in the group with OS > 3 months, except for sPD-L2-IL6 and IL13-TIM3, which were negatively correlated.

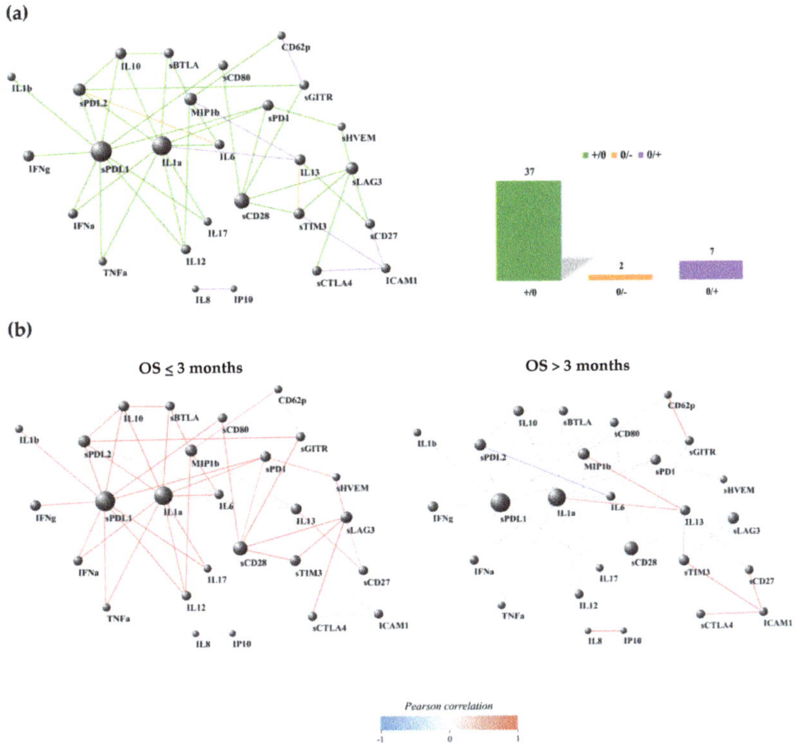

Figure 5. Connectivity network evaluation concerning Overall Survival (OS) in the entire NSCLC population (**a**) and dividing the analysis into patients with OS ≤ 3 months and OS > 3 months (**b**). In each network, nodes represent the immune factors (soluble cytokines/chemokines/adhesion and checkpoint molecules and CD137$^+$ T-cell subsets); the link between two nodes is established when the absolute value of Spearman correlation between their expression levels is statistically significant (p-value ≤ 0.05). Node volume depends on the number of connections. The color of the connection depends on the class described. The green connections and histogram identify the +/0 class that corresponds to the presence of positive correlations among patients with OS ≤ 3 months, not present in patients with OS > 3 months; the orange links and histogram correspond to the presence of negative correlations among patients with OS > 3 months, not present in patients with OS ≤ 3 months; the violet links and histogram are associated with the 0/+ class that corresponds to the presence of positive correlations among patients belonging to the OS > 3 months group, not present in the patients with OS ≤ 3 months.

4. Discussion

The immune network is a dynamic system that remodels the tumor microenvironment. It consists of various components (cellular and soluble factors) that continuously interact, shaping the immune response and, ultimately, the clinical outcome for cancer patients. Identifying these dynamic interactions is crucial for determining specific molecular profiles that can more accurately define the complex scenario characterizing each NSCLC patient. These profiles represent selected pathways of biomarkers for describing patients with similar clinical outcomes and potential multiple immune factors for combined therapies.

In this study, we evaluated the immune network of metastatic non-oncogene-addicted NSCLC patients before the beginning of immunotherapy to identify immunological connections associated with a better response, better performance status, and longer survival (see Supplementary Table S8).

We demonstrated that non-responding patients with PS > 0 exhibited elevated serum levels of multiple soluble activating immune checkpoints (sPD1 and sCD80) and suppressive immune checkpoints (sBTLA4, sCD137, sLAG3, and sPD-L2). Despite the beneficial effects of sPD1 and sCD80 [21,37], their activity was counteracted by the presence of various immunosuppressive factors that simultaneously inhibited T-cell activation [17,22,25,38] and contributed to treatment resistance. Among these parameters, sCD137 uniquely serves as an immunosuppressor in its soluble form while acting as an immune activator when located on the plasma membrane [8]. The other parameters induce the suppression of T-cell functions in their soluble form and associate with the plasma membrane through ligand binding [39].

Moreover, the analysis of the immune network highlighted that patients who benefited from immunotherapy had optimal performance status and longer survival, and exhibited a low number of connections compared to other conditions (NR, PS > 0, and OS < 3 months). These connections were primarily positive and mainly included checkpoint inhibitors rather than cytokines. As noted in another of our analyses on various solid tumors [40], this phenomenon was particularly evident when examining the group of responding patients, where we observed interactions solely among checkpoints, except for the correlations between IFNγ-IFNα and IL17-IL13. The IFNγ pathways shared the downstream IRF1 and STAT1 molecules that bound with the promoters of PD-L1 and CXCL10, enhancing the efficacy of anti-PD1/PD-L1 inhibitors and boosting T-cell infiltration in the tumor microenvironment [41]. Additionally, IFNα exerted immunomodulatory functions on the activities of dendritic cells (DCs) and lymphocytes [42]. Simultaneously, IFNγ operated through several mechanisms, including the inhibition of angiogenesis, the suppression of proliferation, and the induction of regulatory T-cell apoptosis [43]. The roles of IL17 and IL13 were more controversial. However, their influence became more significant during tumor progression, contributing to the spread of metastasis and tumor growth in lung cancer and inducing the shift of Th1-cytokine release toward Th2, respectively [44,45]. Furthermore, the responder group displayed a distinct profile of the most connected nodes compared to the non-responders. Indeed, sBTLA, sCD80, IFNα, and IL12 were the most interacted nodes among the non-responders, while sLAG3, sCTLA4, and sCD28 characterized the responding group. These results indicated that these two patient groups were distinguished by distinct networks of molecules and pathways that defined two specific soluble signatures capable of identifying different responsiveness profiles to therapy.

The immune network excluded the $CD137^+$ T-cell (total, $CD8^+$, and $CD4^+$) interactions in the responding group and in patients with PS = 0. These cellular subsets are identified as tumor-specific [11], and we demonstrated their role as predictive and prognostic factors in metastatic NSCLC patients [12] and in other solid tumors [46,47]. These data support the hypothesis that these cells independently exert their anti-tumor activity and are not positively associated with an inflammatory or immunosuppressive network. Conversely, in non-responders, $CD137^+$ T cells are positively linked to the inhibitory molecule sCD28, and in patients with PS > 0, $CD137^+$ T cells are negatively correlated with the cytokines IL1β, IL12, and IFNα. Surprisingly, in the analysis performed on OS, the nodes related to $CD137^+$ T cells were absent, confirming the role of these cells as independent prognostic factors.

The DGCA's performance, according to the PS, also indicated that the BTLA molecule was the most connected node in both the PS = 0 and PS > 0 groups. However, when we analyzed the BTLA-linked molecules in these two groups, we found that their connection profiles differed significantly. In patients with PS > 0, sBTLA was correlated with several cytokines, such as TNFα, IL1β, and IL17, contributing to a proinflammatory environment, along with sPD-L1 and sCCL2. These molecules negatively affect the cancer microenvironment [15,48]. Specifically, sPD-L1 inhibited T lymphocyte activation, while CCL2 recruited monocytes, dendritic cells (DCs), and other cells to the site of inflammation, thereby contributing to cancer pathogenesis [48]. Conversely, in patients with PS = 0, sBTLA exhibited fewer interactions and was connected only with checkpoint molecules. The connection with the sPD-L2 node was the only common link between patients with PS > 0 and PS = 0. Moreover, SPD-L2 was positively correlated with sIL6 and sIL10 cytokines in the PS > 0 group, while it was negatively associated with these cytokines in the PS = 0 group. The negative interaction between sPD-L2 and sIL6 was also observed in patients with longer overall survival (OS). These results align with the functions of these molecules. Patients with a poorer PS showed positive interactions among nodes that favored tumor growth, as they were closely correlated with the generation of a proinflammatory environment (IL6), a suppressive microenvironment (IL10), and resistance to immune checkpoints (PD-L2) [22].

The analysis of the immune network conducted based on survival revealed that the most connected node in the early progressors was sPD-L1. In the group with longer survival, all connections of sPD-L1 disappeared, except for the positive correlation with the checkpoint molecule sCD28. These two biomarkers have been proposed as negative predictive indicators of clinical response and prolonged survival in cancer patients with high levels of PD-L1 and low levels of sCD28, thereby confirming our data [49]. Furthermore, in patients with poor survival, sPD-L1 interacted with several proinflammatory and immunosuppressive cytokines, confirming that the immune connections in patients with the worst survival were also based on a proinflammatory network that supported tumor progression.

5. Conclusions

In conclusion, this pilot study analyzed the immunological network of several immune soluble factors and T-cell subsets that have been described as predictive and prognostic factors when evaluated individually, aiming to understand the complex immunological scenario of each NSCLC patient. We proposed several immune profiles to identify responding patients and those with longer survival, which could be utilized to optimize personalized therapeutic strategies. Despite the limited number of patients, it was evident that non-responding patients with poor clinical status and survival exhibited an inflammatory-specific signature, which was switched off in patients who responded to therapy and had improved performance status. Patients with a more favorable prognosis were characterized by a network based on checkpoint molecules, likely less conducive to promoting tumor growth. This study reported significant findings in the field of precision medicine, where identifying immune profiles and their connectivity represents a new challenge for further investigation.

Supplementary Materials: The following supporting information can be downloaded at https://www.mdpi.com/article/10.3390/cancers17060922/s1, Figure S1: Schematic representation of inclusion and exclusion criteria during NSCLC enrollment Figure S2: Cytofluorimetric analysis of CD137 T subsets; Figure S3: Kaplan Meier curves of Progression-Free Survival (PFS) and Overall Survival (OS) of NSCLC patients; Table S1: Median values and SEM of Cytokine and Chemokines/Adhesion molecules evaluated in 27 NSCLC patients; Table S2: Median values and SEM of checkpoint molecules evaluated in 27 NSCLC patients; Table S3: Differential expression analysis of circulating checkpoints induced clusters; Table S4: Differential expression analysis of circulating cytokines induced clusters; Table S5: Differential correlation analysis between NR vs R patients; Table S6: Differential correlation analysis between Performance Status (PS) > 0 vs PS = 0 patients; Table S7: Differential correlation analysis between Overal Survival (OS) < 3 months vs OS > 3 months; Table S8: Summury of the main results.

Author Contributions: P.S.: methodology, software, validation, formal analysis, data curation, writing—original draft, visualization; I.G.Z.: conceptualization, investigation, funding acquisition; A.G.: resources; M.S.: resources; L.T.: investigation; A.P.: investigation; A.A.: investigation; F.V.: investigation; A.S.: resources; M.P.: software, resources; F.B.: resources, funding acquisition; D.S.: resources; M.N.: resources, supervision; L.F.: methodology, software, validation, supervision; A.R.: conceptualization, funding acquisition; C.N.: conceptualization, validation, writing—original draft, visualization, supervision, funding acquisition; all authors: writing—review and editing. All authors have read and agreed to the published version of the manuscript.

Funding: This study was supported by the Sapienza University of Rome (CN: RM12117A7B767D0D, RP122181610A7266, and RM123188C3F3E706; IGZ: RM12117A85361029; AR: RM1221816BC0EAA; FB: RM120172B40A805A) and the "Dissecting naïve PD-1 + T-cells phenotype to generate new biological insights of immunotherapy response in advanced NSCLC" (P.I. IGZ) project funded by the European Union-NextGeneration EU within the PRIN 2022 program (D.D.104–02/02/2022, n° 2022K5RYF5 Ministero dell'Università e della Ricerca). This manuscript reflects only the authors' views and opinions, and the Ministry cannot be considered responsible for them.

Institutional Review Board Statement: This study was conducted in accordance with the Declaration of Helsinki, and approved by the Ethics Committee of Policlinico Umberto I (Ethical Committee Protocol, RIF.CE: 4181, 1 December 2016).

Informed Consent Statement: Informed consent was obtained from all subjects involved in this study.

Data Availability Statement: Data were generated by the authors and are included in the article.

Acknowledgments: Angela Asquino, Lucrezia Tuosto, and Flavio Valentino are fellows of the Ph.D. Course "Network Oncology and Precision Medicine", Department of Experimental Medicine, Sapienza University of Rome. Angela Asquino is supported by CN3 PNRR, and Lucrezia Tuosto and Flavio Valentino are supported by the C_PNRR–D.M. 118 founding programs.

Conflicts of Interest: The authors declare no potential conflicts of interest regarding this work. MS reports travel grants from Novartis grants and speaker's honoraria. AG received speaker's honoraria from and participated in the advisory board of AstraZeneca, MSD, Roche, BMS, Takeda.

Abbreviations

The following abbreviations are used in this manuscript:

NSCLC	Non-small cell lung cancer
PS	Performance status
OS	Overall survival
TPS	Tumor proportional score
IDO	Indoleamine 2,3-deoxygenase
FoxO3	Forkhead Box O3
EOCG	Eastern Cooperative Oncology Group
FDR	False discovery rate
DGCA	Different gene correlation

References

1. Liu, J.; Zhong, Y.; Peng, S.; Zhou, X.; Gan, X. Efficacy and safety of PD1/PD-L1 blockades versus docetaxel in patients with pretreated advanced non-small-cell lung cancer: A meta-analysis. *Onco Targets Ther.* **2018**, *11*, 8623–8632. [CrossRef] [PubMed]
2. Reck, M.; Rodríguez-Abreu, D.; Robinson, A.G.; Hui, R.; Csőszi, T.; Fülöp, A.; Gottfried, M.; Peled, N.; Tafreshi, A.; Cuffe, S.; et al. Pembrolizumab versus Chemotherapy for PD-L1-Positive Non-Small-Cell Lung Cancer. *N. Engl. J. Med.* **2016**, *375*, 1823–1833. [CrossRef] [PubMed]
3. Gandhi, L.; Rodríguez-Abreu, D.; Gadgeel, S.; Esteban, E.; Felip, E.; De Angelis, F.; Domine, M.; Clingan, P.; Hochmair, M.J.; Powell, S.F.; et al. Pembrolizumab plus Chemotherapy in Metastatic Non-Small-Cell Lung Cancer. *N. Engl. J. Med.* **2018**, *78*, 2078–2092. [CrossRef]

4. Paz-Ares, L.; Luft, A.; Vicente, D.; Tafreshi, A.; Gümüş, M.; Mazières, J.; Hermes, B.; Çay Şenler, F.; Csőszi, T.; Fülöp, A.; et al. Pembrolizumab plus Chemotherapy for Squamous Non-Small-Cell Lung Cancer. *N. Engl. J. Med.* **2018**, *379*, 2040–2051. [CrossRef]
5. Paz-Ares, L.; Ciuleanu, T.E.; Cobo, M.; Schenker, M.; Zurawski, B.; Menezes, J.; Richardet, E.; Bennouna, J.; Felip, E.; Juan-Vidal, O.; et al. First-line nivolumab plus ipilimumab combined with two cycles of chemotherapy in patients with non-small-cell lung cancer (CheckMate 9LA): An international, randomised, open-label, phase 3 trial. *Lancet Oncol.* **2021**, *22*, 198–211. [CrossRef]
6. Bruni, D.; Angell, H.K.; Galon, J. The immune contexture and Immunoscore in cancer prognosis and therapeutic efficacy. *Nat. Rev. Cancer* **2020**, *20*, 662–680. [CrossRef] [PubMed]
7. Vinay, D.S.; Kwon, B.S. 4-1BB signaling beyond T cells. *Cell Mol. Immunol.* **2011**, *8*, 281–284. [CrossRef]
8. Cannons, J.L.; Lau, P.; Ghumman, B.; DeBenedette, M.A.; Yagita, H.; Okumura, K.; Watts, T.H. 4-1BB ligand induces cell division, sustains survival, and enhances effector function of CD4 and CD8 T cells with similar efficacy. *J. Immunol.* **2001**, *167*, 1313–1324. [CrossRef]
9. Menk, A.V.; Scharping, N.E.; Rivadeneira, D.B.; Calderon, M.J.; Watson, M.J.; Dunstane, D.; Watkins, S.C.; Delgoffe, G.M. 4-1BB costimulation induces T cell mitochondrial function and biogenesis enabling cancer immunotherapeutic responses. *J. Exp. Med.* **2018**, *215*, 1091–1100. [CrossRef]
10. Aznar, M.A.; Labiano, S.; Diaz-Lagares, A.; Molina, C.; Garasa, S.; Azpilikueta, A.; Etxeberria, I.; Sanchez-Paulete, A.R.; Korman, A.J.; Esteller, M.; et al. CD137 (4-1BB) Costimulation Modifies DNA Methylation in CD8(+) T Cell-Relevant Genes. *Cancer Immunol. Res.* **2018**, *6*, 69–78. [CrossRef]
11. Ye, Q.; Song, D.G.; Poussin, M.; Yamamoto, T.; Best, A.; Li, C.; Coukos, G.; Powell, D.J., Jr. CD137 accurately identifies and enriches for naturally occurring tumor reactive T cells in tumor. *Clin. Cancer Res.* **2014**, *20*, 44–55. [CrossRef]
12. Gelibter, A.; Asquino, A.; Strigari, L.; Zizzari, I.G.; Tuosto, L.; Scirocchi, F.; Pace, A.; Siringo, M.; Tramontano, E.; Bianchini, S.; et al. CD137(+) and regulatory T cells as independent prognostic factors of survival in advanced non-oncogene addicted NSCLC patients treated with immunotherapy as first-line. *J. Transl. Med.* **2024**, *22*, 329. [CrossRef] [PubMed]
13. Gelibter, A.; Tuosto, L.; Asquino, A.; Siringo, M.; Sabatini, A.; Zizzari, I.G.; Pace, A.; Scirocchi, F.; Valentino, F.; Bianchini, S.; et al. Anti-PD1 therapies induce an early expansion of Ki67+CD8+ T cells in metastatic non-oncogene addicted NSCLC patients. *Front. Immunol.* **2024**, *15*, 1483182.
14. Khan, M.; Arooj, S.; Wang, H. Soluble B7-CD28 Family Inhibitory Immune Checkpoint Proteins and Anti-Cancer Immunotherapy. *Front. Immunol.* **2021**, *12*, 651634. [CrossRef] [PubMed]
15. Daassi, D.; Mahoney, K.M.; Freeman, G.J. The importance of exosomal PD-L1 in tumour immune evasion. *Nat. Rev. Immunol.* **2020**, *20*, 209–215. [CrossRef]
16. Hayashi, H.; Chamoto, K.; Hatae, R.; Kurosaki, T.; Togashi, Y.; Fukoka, K.; Goto, M.; Chiba, T.; Tomida, S.; Ota, T.; et al. Soluble immune checkpoint factors reflect exhaustion of antitumor immunity and response to PD-1 blockade. *J. Clin. Investig.* **2024**, *134*, e168318. [CrossRef]
17. Zizzari, I.G.; Di Filippo, A.; Scirocchi, F.; Di Pietro, F.R.; Rahimi, H.; Ugolini, A.; Scagnoli, S.; Vernocchi, P.; Del Chierico, F.; Putignani, L.; et al. Soluble Immune Checkpoints, Gut Metabolites and Performance Status as Parameters of Response to Nivolumab Treatment in NSCLC Patients. *J. Pers. Med.* **2020**, *10*, 208. [CrossRef]
18. Mohammadzadeh, S.; Andalib, A.; Khanahmad, H.; Esmaeil, N. Human recombinant soluble PD1 can interference in T cells and Treg cells function in response to MDA-MB-231 cancer cell line. *Am. J. Clin. Exp. Immunol.* **2023**, *12*, 11–23.
19. Kakoulidou, M.; Giscombe, R.; Zhao, X.; Lefvert, A.K.; Wang, X. Human Soluble CD80 is generated by alternative splicing, and recombinant soluble CD80 binds to CD28 and CD152 influencing T-cell activation. *Scand. J. Immunol.* **2007**, *66*, 529–537. [CrossRef]
20. Lin, Z.; Tang, Y.; Chen, Z.; Li, S.; Xu, X.; Hou, X.; Chen, Z.; Wen, J.; Zeng, W.; Meng, X.; et al. Soluble CD80 oral delivery by recombinant Lactococcus suppresses tumor growth by enhancing antitumor immunity. *Bioeng. Transl. Med.* **2023**, *8*, e10533. [CrossRef]
21. Himuro, H.; Nakahara, Y.; Igarashi, Y.; Kouro, T.; Higashijima, N.; Matsuo, N.; Murakami, S.; Wei, F.; Horaguchi, S.; Tsuji, K.; et al. Clinical roles of soluble PD-1 and PD-L1 in plasma of NSCLC patients treated with immune checkpoint inhibitors. *Cancer Immunol. Immunother.* **2023**, *72*, 2829–2840. [CrossRef] [PubMed]
22. Miao, Y.R.; Thakkar, K.N.; Qian, J.; Kariolis, M.S.; Huang, W.; Nandagopal, S.; Yang, T.T.C.; Diep, A.N.; Cherf, G.M.; Xu, Y.; et al. Neutralization of PD-L2 is Essential for Overcoming Immune Checkpoint Blockade Resistance in Ovarian Cancer. *Clin. Cancer Res.* **2021**, *27*, 4435–4448. [CrossRef] [PubMed]
23. Scirocchi, F.; Strigari, L.; Di Filippo, A.; Napoletano, C.; Pace, A.; Rahimi, H.; Botticelli, A.; Rughetti, A.; Nuti, M.; Zizzari, I.G. Soluble PD-L1 as a Prognostic Factor for Immunotherapy Treatment in Solid Tumors: Systematic Review and Meta-Analysis. *Int. J. Mol. Sci.* **2022**, *23*, 14496. [CrossRef]
24. Li, R.; Liang, H.; Shang, Y.; Yang, Z.; Wang, K.; Yang, D.; Bao, J.; Xi, W.; Zhou, D.; Ni, W.; et al. Characteristics of Soluble PD-L1 and PD-1 Expression and Their Correlations with Immune Status and Prognosis in Advanced Lung Cancer. *Asia Pac. J. Clin. Oncol.* **2025**. [CrossRef]

25. Andrzejczak, A.; Karabon, L. BTLA biology in cancer: From bench discoveries to clinical potentials. *Biomark. Res.* **2024**, *12*, 8. [CrossRef]
26. Ward, F.J.; Dahal, L.N.; Wijesekera, S.K.; Abdul-Jawad, S.K.; Kaewarpai, T.; Xu, H.; Vickers, M.A.; Barker, R.N. The soluble isoform of CTLA-4 as a regulator of T-cell responses. *Eur. J. Immunol.* **2013**, *43*, 1274–1285. [CrossRef]
27. Grohmann, U.; Orabona, C.; Fallarino, F.; Vacca, C.; Calcinaro, F.; Falorni, A.; Candeloro, P.; Belladonna, M.L.; Bianchi, R.; Fioretti, M.C.; et al. CTLA-4-ig regulates tryptophan catabolism invivo. *Nat. Immunol.* **2002**, *3*, 1097–1101. [CrossRef] [PubMed]
28. Dejean, A.S.; Beisner, D.R.; Ch'en, I.L.; Kerdiles, Y.M.; Babour, A.; Arden, K.C.; Castrillon, D.H.; DePinho, R.A.; Hedrick, S.M. Transcription factorFoxo3 controls the magnitude of T cell immune responses bymodulating the function of dendritic cells. *Nat. Immunol.* **2009**, *10*, 504–513. [CrossRef]
29. Esen, F.; Cikman, D.I.; Engin, A.; Turna, A.; Batur, S.; Oz, B.; Turna, H.Z.; Deniz, G.; Aktas Cetin, E. Functional and phenotypic changes in natural killer cells expressing immune checkpoint receptors PD-1, CTLA-4, LAG-3, and TIGIT in non-small cell lung cancer: The comparative analysis of tumor microenvironment, peripheral venous blood, and tumor-draining veins. *Immunol. Res.* **2024**, *73*, 18.
30. Castro, F.; Cardoso, A.P.; Gonçalves, R.M.; Serre, K.; Oliveira, M.J. Interferon-Gamma at the Crossroads of Tumor Immune Surveillance or Evasion. *Front. Immunol.* **2018**, *9*, 847. [CrossRef]
31. Liu, Q.; Shaibu, Z.; Xu, A.; Yang, F.; Cao, R.; Yang, F. Predictive value of serum cytokines in patients with non-small-cell lung cancer receiving anti-PD-1 blockade therapy: A meta-analysis. *Clin. Exp. Med.* **2025**, *25*, 59. [CrossRef] [PubMed]
32. Du, W.; Ouyang, Y.; Feng, X.; Yu, C.; Zhang, H.; Chen, S.; Liu, Z.; Wang, B.; Li, X.; Liu, Z.; et al. IL-10RA promotes lung cancer cell proliferation by increasing fatty acid oxidation via STAT3 signaling pathway. *Pulm. Pharmacol. Ther.* **2025**, *88*, 102344. [CrossRef]
33. Liu, H.; Zhou, C.; Jiang, H.; Chu, T.; Zhong, R.; Zhang, X.; Shen, Y.; Han, B. Prognostic role of serum cytokines level in non-small cell lung cancer patients with anti-PD-1 and chemotherapy combined treatment. *Front. Immunol.* **2024**, *15*, 1430301. [CrossRef]
34. Ritchie, M.E.; Phipson, B.; Wu, D.; Hu, Y.; Law, C.W.; Shi, W.; Smyth, G.K. Limma powers differential expression analyses for RNA-sequencing and microarray studies. *Nucleic Acids Res.* **2015**, *43*, e47. [CrossRef]
35. McKenzie, A.T.; Katsyv, I.; Song, W.M.; Wang, M.; Zhang, B. DGCA: A comprehensive R package for Differential Gene Correlation Analysis. *BMC Syst. Biol.* **2016**, *10*, 106. [CrossRef] [PubMed]
36. Bonett, D.A.; Wright, T.A. Sample size requirements for estimating Pearson, Kendall, and Spearman correlations. *Psychometrika* **2000**, *65*, 23–28. [CrossRef]
37. Haile, S.T.; Horn, L.A.; Ostrand-Rosenberg, S. A soluble form of CD80 enhances antitumor immunity by neutralizing programmed death ligand-1 and simultaneously providing costimulatio. *Cancer Immunol. Res.* **2014**, *2*, 610–615. [CrossRef]
38. Botticelli, A.; Zizzari, I.G.; Scagnoli, S.; Pomati, G.; Strigari, L.; Cirillo, A.; Cerbelli, B.; Di Filippo, A.; Napoletano, C.; Scirocchi, F.; et al. The Role of Soluble LAG3 and Soluble Immune Checkpoints Profile in Advanced Head and Neck Cancer: A Pilot Study. *J. Pers. Med.* **2021**, *11*, 651. [CrossRef] [PubMed]
39. Davoudi, F.; Moradi, A.; Sadeghirad, H.; Kulasinghe, A. Tissue biomarkers of immune checkpoint inhibitor therapy. *Immunol. Cell Biol.* **2024**, *102*, 179–193. [CrossRef]
40. Mezi, S.; Pomati, G.; Fiscon, G.; Amirhassankhani, S.; Zizzari, I.G.; Napoletano, C.; Rughetti, A.; Rossi, E.; Schinzari, G.; Tortora, G.; et al. A network approach to define the predictive role of immune profile on tumor response and toxicity of anti PD-1 single agent immunotherapy in patients with solid tumors. *Front. Immunol.* **2023**, *14*, 1199089. [CrossRef]
41. Cheng, C.C.; Chang, J.; Ho, A.S.; Sie, Z.L.; Peng, C.L.; Wang, C.L.; Dev, K.; Chang, C.C. Tumor-intrinsic IFNalpha and CXCL10 are critical for immunotherapeutic efficacy by recruiting and activating T lymphocytes in tumor microenvironment. *Cancer Immunol. Immunother.* **2024**, *73*, 175. [CrossRef] [PubMed]
42. Belardelli, F.; Ferrantini, M.; Proietti, E.; Kirkwood, J.M. Interferon-alpha in tumor immunity and immunotherapy. *Cytokine Growth Factor. Rev.* **2002**, *13*, 119–134. [CrossRef]
43. Jorgovanovic, D.; Song, M.; Wang, L.; Zhang, Y. Roles of IFN-γ in tumor progression and regression: A review. *Biomark. Res.* **2020**, *8*, 49. [CrossRef]
44. Wu, F.; Xu, J.; Huang, Q.; Han, J.; Duan, L.; Fan, J.; Lv, Z.; Guo, M.; Hu, G.; Chen, L.; et al. The Role of Interleukin-17 in Lung Cancer. *Mediators Inflamm.* **2016**, *2016*, 8494079. [CrossRef] [PubMed]
45. Wynn, T.A. IL-13 effector functions. *Annu. Rev. Immunol.* **2003**, *21*, 425–456. [CrossRef] [PubMed]
46. Zizzari, I.G.; Di Filippo, A.; Botticelli, A.; Strigari, L.; Pernazza, A.; Rullo, E.; Pignataro, M.G.; Ugolini, A.; Scirocchi, F.; Di Pietro, F.R.; et al. Circulating CD137+ T Cells Correlate with Improved Response to Anti-PD1 Immunotherapy in Patients with Cancer. *Clin. Cancer Res.* **2022**, *28*, 1027–1037. [CrossRef]
47. Cirillo, A.; Zizzari, I.G.; Botticelli, A.; Strigari, L.; Rahimi, H.; Scagnoli, S.; Scirocchi, F.; Pernazza, A.; Pace, A.; Cerbelli, B.; et al. Circulating CD137(+) T Cell Levels Are Correlated with Response to Pembrolizumab Treatment in Advanced Head and Neck Cancer Patients. *Int. J. Mol. Sci.* **2023**, *24*, 7114. [CrossRef]

48. Hao, Q.; Vadgama, J.V.; Wang, P. CCL2/CCR2 signaling in cancer pathogenesis. *Cell Commun. Signal* **2020**, *18*, 82. [CrossRef]
49. Zhang, C.; Fan, Y.; Che, X.; Zhang, M.; Li, Z.; Li, C.; Wang, S.; Wen, T.; Hou, K.; Shao, X.; et al. Anti-PD-1 Therapy Response Predicted by the Combination of Exosomal PD-L1 and CD28. *Front. Oncol.* **2020**, *10*, 760.

MDPI AG
Grosspeteranlage 5
4052 Basel
Switzerland
Tel.: +41 61 683 77 34

Cancers Editorial Office
E-mail: cancers@mdpi.com
www.mdpi.com/journal/cancers

www.ingramcontent.com/pod-product-compliance
Lightning Source LLC
Chambersburg PA
CBHW061631210926
51850CB00009B/2510
9783725866700